BF311 .L2523 2001

0 1 3 4 1

Language
 cogniti
    c2(

Language, Brain, and Cognitive Development

Language, Brain, and Cognitive Development
Essays in Honor of Jacques Mehler

edited by Emmanuel Dupoux

A Bradford Book
The MIT Press
Cambridge, Mass
London, England

© 2001 Massachusetts Institute of Technology

All rights reserved. No part of this book may be reproduced in any form by any electronic or mechanical means (including photocopying, recording, or information storage and retrieval) without permission in writing from the publisher.

This book was set in Sabon by Achorn Graphic Services, Inc., and was printed and bound in the United States of America.

Library of Congress Cataloging-in-Publication Data

Language, brain, and cognitive development ; essays in honor of Jacques Mehler / edited by Emmanuel Dupoux.
 p. cm.
 "A Bradford Book."
 ISBN 0-262-04197-9 (alk. paper)
 1. Cognition. 2. Cognitive science. I. Dupoux, Emmanuel. II. Mehler, Jacques.

BF311.L2523 2001
153—dc21

2001030606

Contents

Preface   ix
Contributors   xiii

I Introduction   1

1 Portrait of a "Classical" Cognitive Scientist: What I Have Learned from Jacques Mehler   3
Massimo Piatelli-Palmarini

2 Cognition—Some Personal Histories (with Pinker's appendix)   23
Thomas G. Bever, Susana Franck, John Morton, and Steven Pinker

II Thought   39

Representations, Psychological Reality, and Beyond   41
Luca L. Bonatti

3 In Defense of Massive Modularity   47
Dan Sperber

4 Is the Imagery Debate Over? If So, What Was It About?   59
Zenon Pylyshyn

5 Mental Models and Human Reasoning   85
Philip N. Johnson-Laird

6 Is the Content of Experience the Same as the Content of Thought?   103
Ned Block

III Language 121

Introduction 123
Christophe Pallier and Anne-Catherine Bachoud-Lévi

7 About Parameters, Prominence, and Bootstrapping 127
Marina Nespor

8 Some Sentences on Our Consciousness of Sentences 143
Thomas G. Bever and David J. Townsend

9 Four Decades of Rules and Associations, or Whatever Happened to the Past Tense Debate? 157
Steven Pinker

10 The Roll of the Silly Ball 181
Anne Cutler, James M. McQueen, Dennis Norris, and A. Somejuan

11 Phonotactic Constraints Shape Speech Perception: Implications for Sublexical and Lexical Processing 195
Juan Segui, Ulricht Frauenfelder, and Pierre Hallé

12 A Crosslinguistic Investigation of Determiner Production 209
Alfonso Caramazza, Michele Miozzo, Albert Costa, Niels Schiller, and F.-Xavier Alario

13 Now You See It, Now You Don't: Frequency Effects in Language Production 227
Merrill Garrett

14 Relations between Speech Production and Speech Perception: Some Behavioral and Neurological Observations 241
Willem J. M. Levelt

IV Development 257

How to Study Development 259
Anne Christophe

15 Why We Need Cognition: Cause and Developmental Disorder 263
John Morton and Uta Frith

16 Counting in Animals and Humans 279
Rochel Gelman and Sara Cordes

17 On the Very Possibility of Discontinuities in Conceptual Development   303
    Susan Carey

18 Continuity, Competence, and the Object Concept   325
    Elizabeth Spelke and Susan Hespos

19 Infants' Physical Knowledge: Of Acquired Expectations and Core Principles   341
    Renée Baillargeon

20 Learning Language: What Infants Know about It, and What We Don't Know about That   363
    Peter W. Jusczyk

21 On Becoming and Being Bilingual   379
    Núria Sebastián-Gallés and Laura Bosch

V   Brain and Biology   395

    On Language, Biology, and Reductionism   397
    Stanislaus Dehaene, Ghislaine Dehaene-Lambertz, and Laurent Cohen

22 Cognitive Neuroscience: The Synthesis of Mind and Brain   403
    Michael I. Posner

23 What's So Special about Speech?   417
    Marc Hauser

24 The Biological Foundations of Music   435
    Isabelle Peretz

25 Brain and Sounds: Lessons from "Dyslexic" Rodents   447
    Albert M. Galaburda

26 The Literate Mind and the Universal Human Mind   463
    José Morais and Régine Kolinsky

27 Critical Thinking about Critical Periods: Perspectives on a Critical Period for Language Acquisition   481
    Elissa L. Newport, Daphne Bavelier, and Helen J. Neville

28 Cognition and Neuroscience: Where Were We? 503
John C. Marshall

Appendix: Short Biography of Jacques Mehler   513
Afterword   529
Index   531

Preface

The history of the term "cognition" is rather short, even if the underlying intellectual issues have been with us for quite a while. When I arrived in Jacques Mehler's Paris laboratory in 1984, "cognition" was either unknown or had pretty bad press among most of my fellow graduate students or professors at the Ecole Normale Supérieure. I was advised that there were much more serious matters to be pursued, like, for instance, psychoanalysis or artificial intelligence. Fortunately enough, I was also directed to Jacques's lab where I discovered that there existed a domain, called *cognitive science*, which project was boldly to put the human mind under the scrutiny of rational inquiry, and to do so through the conjoined fire of philosophy, linguistics, computer science, psychology, and neuroscience. Further, I discovered that this field of inquiry had started more than twenty years ago in the United States, and that Jacques was one of its prominent protagonists.

Jacques's contribution to this field is uncontested. He made important discoveries both in adult and infant cognition, some of which are discussed in this book. He created and still is the editor-in-chief of an international journal, *Cognition*, one of the most innovative and prestigious in the field (see the chapter by Bever, Franck, Morton, and Pinker). He started a lab at the Ecole des Hautes Etudes en Sciences Sociales in Paris, featuring one of the very few newborn testing units in the world, and trained with enthusiasm, warmth, and rigor several generations of scientists, who now work in some of the most interesting places in Europe. All of this was achieved in the Paris of the sixties and postsixties, not a small feat considering the quite unreceptive intellectual milieu predominant then (see the chapter by Piatelli-Palmarini). Of course, cognition is now well known in France; it excites the public's attention, attracts governmental money. Everybody is doing cognition these days; however, it should be remembered that if this term is to have any substantive meaning, it is in some respectable part due to Jacques's years of uphill battle to establish it as a contentful field of inquiry.

Jacques is now leaving his Paris lab as a legacy to his former students and is starting a new intellectual venture in Italy. His sixty-fifth birthday, which coincides with the opening of his new research center in Trieste, gives us an ideal occasion for both honoring him and reflecting on cognitive science.

Why is this interesting? Where are we going with this? What does this tell us? These are some of the often embarrassing questions that Jacques typically asks his students or colleagues during conferences. In this book, these questions were posed to some of Jacques's close collaborators, friends, and former students. The outcome is a collection of chapters that forms an instantaneous snapshot, a patchwork of what is going on in the active brains of these scientists who are currently studying cognitive science. Some chapters provide critical reviews of where we have gone so far. Others offer bold and provoking hypotheses about where we ought to go. Others point to as yet unresolved paradoxes. If some chapters are in flat disagreement with others, unexpected convergence arises elsewhere in spite of apparently disconnected empirical perspectives. This should not be surprising. It corresponds to a living field which allows for divergent views and paradigms, as long as there is a principled way to settle the issue by confronting the facts.

Through the selection of contributors, however, this book reflects a certain conception of cognitive science. It is a conception that Jacques has always vehemently advocated, in spite of some resistance both from inside and outside the field. It states that it is both valid and of central importance to build a functional characterization of the mind. Such a characterization considers mentation to be essentially information processing: representations and computations over representations, as in a computer programming language. At the methodological level, information processing provides a common vocabulary that allows us to integrate the conceptual tools coming from analyical philosophy and the experimental tools of the behavioral sciences and neuroscience. At the explanatory level, representations and computations over representations enter into a causal chain and are essential to account for how a collection of neurons produce such and such behavior when immersed in a given environment (see the chapter by Morton for a similar point in the context of developmental disorders). I hope this book illustrates that such a conception has some value and can indeed help bridge the gap between the world of the neurological substrate, the world of observable behavior, and the world of phenomenal experience.

Apart from the first section of the book, which is devoted to an intellectual biography of Jacques, the book is divided into four substantive parts: Thought, Language, Development, and Brain and Biology, each one being presented by some of Jacques's close former students or colleagues. As the reader will quickly notice, many articles could appear in more than one part of the book. Such a flaw in the structure of the book is interesting for two reasons: First, it attests to the degree of cross-disciplinary integration that the field has achieved. Second, it eloquently illustrates one of Jacques's recurrent lessons, that is, that one should *not* study thought, language, development, and brain separately. Specifically, he has argued on many occasions that the study of the developing infant is intrinsically linked to that of the adult (see Christophe's introduction to Development). Second, although he has actually warned *against* too premature a link between brain and cognition, he was among the first to take the cognitive neuroscience turn and strongly advised us to take neuroimaging seriously (see the introduction to Brain and Biology by Dehaene, Dehaene-Lambertz, and Cohen). Third, although Jacques devoted most of his research efforts to psycholinguistics, he always considered that language was only one window onto more general questions regarding the architecture of thought and cognition (see Bonatti's introduction to Thought and Pallier and Bachoud-Lévi's introduction to Language).

So, where are we going with cognition? What have we learned? Where should we go? I hope this book will help the reader to gain some insights on these questions. Some chapters are optimistic, others less so. Some state enthusiastically that we are going toward a new synthesis (Posner), others foresee the end of the "classical era" (Piatelli), others view the current state of affairs as a mere *crise de croissance* (Bonatti), or even say that is there is not much new under the sun since the ancient Greeks (Marshall). I leave to the reader the pleasure of drawing his or her own conclusions.

Before closing, I would like to point out that Jacques not only provided us with a critical attitude and an urge to turn big issues into empirically testable questions. Most important, he also inoculated us with the cognitive virus, the passion and devotion attached to this quickly changing, still immature, sometimes frustrating, but always fascinating area of study. To Jacques, with all my gratitude and deeply heartfelt thanks.

Acknowledgments

I would like to thank all the contributors and the guest editors for their participation, enthusiasm, and responsiveness in making this book. I would also like to thank Evelyn Rosset for her precious help with the manuscript, as well as to Amy Brand and Tom Stone from The MIT Press for helping to work under tight timing constraints and to an unusually large number of contributors. Last but not least, I thank Susana Franck, who was, as usual, of invaluable help during the whole editorial process.

Contributors

Francois-Xavier Alario   Cognitive Neuropsychology Laboratory, Department of Psychology, Harvard University, Cambridge, Massachusetts

Anne Catherine Bachoud Lévi   Hôpital Henri Mondor, Service de Neurologie, Créteil, France

Renée Baillargeon   Department of Psychology, University of Illinois, Champaign, Illinois

Daphné Bavelier   Department of Brain and Cognitive Sciences, University of Rochester, Rochester, New York

Thomas G. Bever   Linguistics Department, University of Arizona, Tucson, Arizona

Ned Block   Department of Philosophy, New York University, New York, New York

Luca L. Bonatti   Laboratoire CNRS ESA 7021, Université de Paris, Paris, France

Laura Bosch   Department de Psicologia Bàsica, Universitat de Barcelona, Spain

Alfonso Caramazza   Cognitive Neuropsychology Laboratory, Department of Psychology, Harvard University, Cambridge, Massachusetts

Sue Carey   Psychology Department, New York University, New York, New York

Anne Christophe   Laboratoire de Sciences Cognitives et Psycholinguistique, Ecole des Hautes Etudes en Sciences Sociales, Paris, France

Laurent Cohen   Service de Neurologie, Hopital de la Salpétrière, Paris, France

Sara Cordes   Department of Psychology and Cognitive Science, Rutgers-New Brunswick, Piscataway, New Jersey

Albert Costa   Cognitive Neuropsychology Laboratory, Department of Psychology, Harvard University, Cambridge, Massachusetts

Anne Cutler   Max Planck Institute for Psycholinguistics, Nijmegen, The Netherlands

Stanislas Dehaene   Unité INSERM 334, Service Hospitalier Frédéric Joliot, Orsay, France

Ghislaine Dehaene-Lambertz Laboratoire de Sciences Cognitives et Psycholinguistique, Ecole des Hautes Etudes en Sciences Sociales, Paris, France

Emmanuel Dupoux Laboratoire de Sciences Cognitives et Psycholinguistique, Ecole des Hautes Etudes en Sciences Sociales, Paris, France

Susana Franck Laboratoire de Sciences Cognitives et Psycholinguistique, Ecole des Hautes Etudes en Sciences Sociales, Paris, France

Uli Frauenfelder Laboratoire de Psycholinguistique Expérimentale, Geneve, Suisse

Uta Frith Institute of Cognitive Neuroscience, Alexandra House, London, UK

Albert M. Galaburda, M.D. Emily Fisher Landau Professor of Neurology and Neuroscience, Harvard Medical School, Beth Israel Deaconess Medical Center, Boston, Massachusetts

Merrill Garrett Psychology Department and Cognitive Science Program, University of Arizona, Tucson, Arizona

Rochel Gelman Department of Psychology and Cognitive Science, Rutgers-New Brunswick, Piscataway, New Jersey

Pierre Hallé Laboratoire de Psychologie Expérimentale, Université René Descartes et CNRS, Paris, France

Marc Hauser Department of Psychology and Program in Neurosciences, Harvard University, Cambridge, Massachusetts

Susan Hespos Department of Brain and Cognitive Sciences, Massachusetts Institute of Technology, Cambridge, Massachusetts

Philip N. Johnson-Laird Department of Psychology, Princeton, New Jersey

Peter W. Jusczyk Departments of Psychology and Cognitive Science, Johns Hopkins University, Baltimore, Maryland

Régine Kolinsky UNESCOG, Université Libre de Bruxelles, Bruxelles, Belgium

Willem J. M. Levelt Max Planck Institute for Psycholinguistics, Nijmegen, The Netherlands

John C. Marshall Neuropsychology Unit, University Department of Clinical Neurology, Radcliffe Infirmary, Oxford, UK

James McQueen Max Planck Institute for Psycholinguistics, Nijmegen, The Netherlands

Michele Miozzo Psychology Department, Columbia University, New York, New York

José Morais UNESCOG, Université Libre de Bruxelles, Bruxelles, Belgium

John Morton Institute of Cognitive Neuroscience, Alexandra House, London, UK

Marina Nespor Facoltà di Lettere e Filosofia, Università di Ferrara, Ferrara, Italy

Helen Neville Brain Development Laboratory, University of Oregon, Eugene, Oregon

Elissa L. Newport   Department of Brain and Cognitive Sciences, University of Rochester, Rochester, New York

Dennis Norris   MRC Applied Psychology Unit, Cambridge, UK

Christophe Pallier   Laboratoire de Sciences Cognitives et Psycholinguistique, Ecole des Hautes Etudes en Sciences Sociales, Paris, France

Isabelle Peretz   Département de Psychologie, Université de Montréal, Montréal (Québec), Canada

Massimo Piatelli-Palmarini   Cognitive Science Program, University of Arizona, Tucson, Arizona

Steven Pinker   Department of Brain and Cognitive Sciences, Massachusetts Institute of Technology, Cambridge, Massachusetts

Michael I. Posner   Sackler Institute, Weill Medical College of Cornell University, New York, New York

Zenon Pylyshyn   Rutgers Center for Cognitive Science, Rutgers University, Piscataway, New Jersey

Niels Schiller   Max Planck Institute for Psycholinguistics, Nijmegen, The Netherlands

Núria Sebastián-Gallés   Departament de Psicologia Bàsica, Universitat de Barcelona, Spain

Juan Segui   Laboratoire de Psychologie Expérimentale, Université René Descartes et CNRS, Boulogne Billancourt, France

Elizabeth Spelke   Department of Brain and Cognitive Sciences, Massachusetts Institute of Technology, Cambridge, Massachusetts

Dan Sperber   Institut Jean Nicod, Ecole des Hautes Etudes en Sciences Sociales, Paris, France

David J. Townsend   Linguistics Department, University of Arizona, Tucson, Arizona

I
Introduction

# 1

Portrait of a "Classical" Cognitive Scientist: What I Have Learned from Jacques Mehler

Massimo Piattelli-Palmarini

It's my opinion that something is happening in, and to, cognitive science. We are being progressively dislodged from the neat, consistent, and exciting scheme in which we had settled for a long time. I shall call this scheme "classical" cognitive science. As I am about to tell, I mostly owe my introduction to, and settlement into, this beautiful classical universe to Jacques Mehler. The reasons why I think the scene is changing, possibly irreversibly, can be briefly summarized through a few examples, in the subdomains I happen to know best.

1. Syntactic theory is under radical reconstruction. The overall grand scheme that had reached its apex, around 1981–82, in the theory of government and binding (GB) is no longer tenable. Minimalism has supplanted it. The problem is that, in my experience at least (but I know I am not alone), minimalism is not directly teachable as such. We *have* to tell the GB story first, let it sink in, and *then* present minimalism as a reinterpretation of many of its central ideas. Does this unteachability of minimalism tell us anything at all? I do not know. But it might. Moreover, some old unsolved problems are now easily solved in the new frame, but other problems that had been solved satisfactorily by GB have turned out to be hard, or intractable, in the new model. Other styles of explanation, long forgotten, are again raising their heads, as competitors to, because suddenly awakened by, minimalism.

2. Lexical semantics is at a standstill. The deep, uniform, and explanatory links between lexical internal structures and syntactic structures, established in the pioneering works of Mark Baker, Dick Carter, Beth Levin, Malka Rappaport, Jane Grimshaw, and others, and culminating

in the remarkable paper by Hale and Keyser, is being questioned at its very roots by Jerry Fodor. It is he, not I, who has chosen as the subtitle of his 1998 book on concepts, *Where Cognitive Science Went Wrong*. In my terminology, Jerry is referring to classical cognitive science. The many brave rebuttals of his radical critique are not persuasive enough to reassert the well-foundedness of the classical approach. The solution—any solution—will have to accommodate many distinguos, and relativizations and partializations. I doubt that with some notable exceptions the field will ever be again as exciting and neat as it still was even five or six years ago.

3. Heuristics and biases, the field so firmly established by Tversky and Kahneman around 1975, is under attack. We are enjoined to go "beyond" heuristics and biases, and pay massive attention to evolutionary psychology. The classical experiments are being redesigned in myriads of variants, sometimes obtaining nonclassical results. Several authors suggest that the case for striking and systematic cognitive illusions has been overstated. I, for one, am unimpressed by these critiques, and vastly underwhelmed by the claims of evolutionary psychology. But it has become unthinkable to teach a course on judgment and decision-making these days without also covering these critiques and the new data and present cheater detectors and simple heuristics that make us smart. And let the audience decide.

4. Innatism is again under attack. The instant popularity with a lot of the high academe of Fiona Cowie's 1999 book *What's Within? Nativism Reconsidered*, and the rebuttals that are being circulated in preprint, bring us squarely back to the debate between Piaget and Chomsky. We are (as Fodor rightly remarks) pulled twenty years backward, as if nothing had happened. Some of us (surely in my case) are, alas, doubting that innatism in the domain of language and cognition will be considered acceptable by the academic world at large within our lifetime.

I'll come back to this postclassical transition of cognitive science at the end of these notes. I have to tell you first how I came onto the original scene. I have to tell you about my first encounter with Jacques.

I Enter Jacques "Mahler"

In May 1968 (yes, the famous-infamous *mai soixante-huit*), when I was a research fellow at the Institute of Physical Chemistry in the University of Rome, a colleague and friend of mine urged me to accompany her to a seminar at the Institute of Psychology. The speaker, Jacques Mahler (that's how I first misperceived his name, linking it to the composer), was a buddy of her and her husband from the old days in Buenos Aires. An interesting young fellow, she told me, who had first been a theoretical organic chemist at Oxford with the legendary Charles Coulson, then turned psychologist and linguist, having worked with Jean Piaget in Geneva, and with Noam Chomsky at MIT. These names were awesome enough to me even then, so I gladly accepted. Little could I know that this event, in the fullness of time, was bound to change my professional life, and my life as a whole.

Jacques impressed me at the time as looking more like the typical Belgian businessman than the typical scientist. He was clean-shaven and impeccably dressed in a dark three-piece suit, and spoke Italian with amazing fluency and naturalness, using an often invented, though always quite plausible, vocabulary. Over the years, many of Jacques's creative Italian expressions have stuck in our family vocabulary.[1] In hindsight, the persistence in Jacques's Italian of the setting of his phonological and syntactic parameters from Spanish is a clear instance of the resiliency of the linguistic setup of our mother tongue (this phenomenon is well analyzed by Nuria Sebastián-Gallés and Laura Bosch in chapter 21 of this book).

The topic of his talk was, somewhat strangely in hindsight, the tracking of ocular movements in reading. The central message that I retained was that a remarkable variety of underlying syntactic structures of great subtlety and richness could be revealed (according to Jacques, at the time) by tracking ocular movements. To the best of my knowledge, Jacques has not further pursued this line of inquiry, nor has it turned out to be a very fruitful method in anyone else's hands for many years (though there are interesting new results now) but that first take-home lesson from Jacques in classical cognitive science was crucial nonetheless.

Lesson 1: There are many complex underlying structures in linguistic expressions, and they can be evidenced by monitoring a subject's real-time processing.

II Getting to Know Mehler

We had a nice long chat with Jacques after his talk, outside the Institute of Psychology. I was left dumbstruck by his report of the vast disruption of ordinary Parisian life caused by the revolt of the students in that fateful May '68. The Italian press had obviously underplayed, in those initial days, the momentous impact of the revolt. Jacques told us about banks and shops being closed, about army and police patrolling the streets, about many violent clashes. The magnitude of the first spark of the events that were bound to change our existence as teachers and as citizens for many years to come suddenly sank in, thanks to Jacques.

I instantly liked him very much. He was not only obviously very bright but also fun to be with. He had a pleasant no-nonsense straight-to-the-essentials communicative style.

In April '69, having won a plush scholarship from the European Molecular Biology Organization, I moved to Paris-Orsay, to work with the Nobelist-to-be Pierre-Gilles de Gennes. I was living in Paris, and Jacques was among the acquaintances with whom I planned to try to establish better contacts. Jacques kindly invited me to dinner, on a gorgeous June evening, and introduced me to his family. He was then living near the Avenue Foche, opposite the famous Palais Rose, whose pending renovation was then fully under way. Jacques took me to visit the palace, in an after-dinner stroll. He had changed a lot physically. He wore a beard, and was dressing more like an artist than like a banker. Ten minutes into the evening I already felt as if we had always been friends.[2]

III Intimations of Cognition

We quickly became good friends, and I indulged myself in attending, now and then, under his suggestions, some lectures and seminars in his laboratory.

I was particularly impressed by a talk by Sydney Strauss, from Israel. He had clear evidence that Piaget's famous horizontal stages were not horizontal, after all. Children who clearly mastered conservation in problems dealing with volume and weight failed to apply conservation to problems dealing with temperature, concentration, or speed. The kids' exclusive focus on one dimension only of the situation, regardless of other relevant factors, mimicked exactly the results of Piaget's famous experiments on the conservation of weight and of volume, except that these kids demonstrably mastered conservation for weight and for volume. The number of sugar lumps dissolved in a water container, regardless of the size of the container, decided whether the resulting liquid was thought to be sweet, very sweet, or very, very sweet. After pouring the contents of one container judged to be "very sweet" into another, also judged to be "very sweet" the resulting liquid was judged to be "very, very sweet." The parallel progression of two toy cars along the linear projection of two trails decided how fast they were thought to travel, regardless of the fact that one trail was straight, while the other was curvilinear. Even when the second was then stretched into a longer line than the first before their eyes, these kids insisted that the two cars had traveled equally fast. This intuition persisted even when both cars moved in a straight line on trails of visibly different length, but starting together and reaching the end of their respective trails together.

Moreover, Sydney showed us an impressive instance of a cognitive illusion for grown-ups. (In the last several years I have used this demonstration in dozens of classes and lectures.)

A shoelace is tied into a loop, and the loop is then held between the index fingers and thumbs to form a square. By slightly approaching each index finger to the opposing thumb by an equal amount for each hand, the square slowly turns into a rectangle. The audience is invited to judge the ratios of the *areas* of these two figures. Having absorbed, as a physicist, a whole course on differential geometry, with plenty of cases of surface maximization, I judged that "obviously" the rectangle had a smaller area than the square. After all, since the perimeter is conserved, the area cannot be conserved too. Sydney and Jacques, much to my surprise, claimed that the majority of adult subjects, even well-educated ones, wrongly judge the two areas to be identical. And they persist in their intuition of conservation even for the areas of the more and more elongated rectangles that one slowly obtains by progressively approaching the opposing index fingers and thumbs. Until you have no area at all! Now what?

They were right. Indeed, the vast majority of subjects, of all ages, including highly educated persons, judge the area to remain constant. Some pause at the end, when no area is left at all, and are ready to retract their previous judgment. But the majority, though somewhat puzzled by the final disappearance of the area, persist in their judgment that the area is being conserved.

The cumulative lesson was clear and powerful: we have problems with judgments of conservation through all our life. There are no horizontal stages à la Piaget.

I was greatly impressed. I knew enough about psychology to be aware that Piaget was considered absolute avant-garde (in Italy, this was indeed the common opinion, then even more than now). I was being introduced to a whole post-Piagetian line of inquiry, well beyond the alleged avant-garde. Jacques patiently explained to me the notion of "cognitive strategies" (flexible strategies, not uniform horizontal stages), referring me to the works of Jerome Bruner.

Lesson 2: There are several important conceptual changes in the child's cognitive development, but none of them is "horizontal," and they do not all happen at the same age.

IV Smart, Then Less So, Then More So: Enter the U-shaped Curve

Jacques had some stunning data of his own on very young children (he systematically used, and still often uses, the colloquial Italian word *ragazzini* more aptly applied to kids above six or seven years of age). But these children were much younger (three to four), and allegedly well below Piaget's conservation stage. They were presented by Jacques with two rows of candies on a flat table. It was made clear to them that they were allowed to choose and keep the candies of one, only one, of the two rows. The number of candies was exactly the same for both rows, but one row was more elongated (the candies were more broadly spaced out than in the other). The very young chose one or the other indifferently. The older ones (four to six) indeed preferred the elongated row (as observed by Piaget, thinking it contained more candies). And then the "conserving" older children (seven and above), again chose indifferently, well aware that the number of candies was conserved by the spatial transformation. The crucial datum was the indifferent choice of the very young children, a fact never tested by Piaget, and that flew in the face of his celebrated stages. Another important lesson from Jacques.

Lesson 3: The most typical learning curve of humans in various cognitive domains is U-shaped.

Combined with the previous one, this tells us that there are many dips and drops in cognitive development, and that the troughs are not co-occurring in time. An updating of this lesson today ought to change the quantifier in the sentence: One of the most typical learning curves is U-shaped. The issue of continuity in development is still a central topic today (see Baillargeon, chapter 19; Carey, chapter 17; Gelman and Cordes, chapter 16; and Spelke and Hespos, chapter 18).

V Neonate Cognition

Another talk by Jacques impressed me greatly. With Peter Eimas, he had studied speech perception in the newborn, in the few-weeks-old, and in the few-months-old, revealing clear cases of very, very early categorial perception. A sophisticated piece of machinery (the Vocoder) could progressively vary, by a precise and programmable number of milliseconds, the time lapse between two acoustic formants of a syllable (*ba*, *da*, *ga*). In the adult, the effect was very sharp: Up to a steep threshold, the syllable was heard as being always the same (say, *ba*), then, abruptly, for any time delay above the threshold, the syllable was distinctly heard as being different (*da*), and was then perceived as the same syllable up to the next threshold (turning abruptly into *ga*), and so on. With tiny little babies, strikingly enough, the result was essentially the same. The method that Jacques and Peter used was non-nutritive sucking. The perception of sameness in a series of stimuli leads to decreasing rates of sucking (betraying increasing boredom), while the perception of novelty leads to more intense sucking (betraying increased attention and interest). The data offered very neat and very persuasive evidence of strong innate components in our linguistic capacities. After having heard so much about the ubiquitous "projection" of structures from the mind onto the external world, there I was, finally seeing a real paradigm case. Of great interest was also the fact that if these speech recordings were played backward, neither the adult nor the baby could hear any difference at all. They all just sounded like indistinct chirps.

Lesson 4: Objective physical differences in sensory stimuli may not map at all onto cognitively processable differences. And relevant cognitive differences may not map neatly onto physical differences.

The mind literally, not metaphorically, projects structure onto the external world. It cannot suck it from the world, it cannot literally learn this structure. Another capital lesson, originally due to Chomsky, but that percolated to me through Jacques, was the so-called poverty of the stimulus argument.

Lesson 5: Rich cognitive capacities that are in place very, very early, and that cannot possibly have been shaped by experience, cannot be explained through learning. They are internal to the mind.

VI Modularity (in Hindsight)

Since categorial perception is not pervasive in the realm of sounds, but is clearly present in speech, this means that our mind is naturally endowed with specialized processing units, governed by highly refined and abstract operations. In the course of a conversation Jacques also mentioned another specialized cross-modal effect, unique to speech perception (in hindsight, I think it was the McGurk effect, though the name did not stick in my memory then). It does not apply to bouncing balls, the rustling of leaves, or the screeching of tires. It is proprietary to speech perception.

Around that time, two more lessons from Jacques crystallized in my mind. I'll phrase them as closely as I can in words I might have used at the time:

Lesson 6: The mind has subdivisions into specific domains. They operate in ways that are somewhat similar to reflexes, except that they can be vastly more complex than mere reflexes.

The case of sign languages, to which Jacques introduced me, also introducing me, much later on, to Ursula Bellugi-Klima in the flesh, made the next lesson quite vivid.

Lesson 7: These cognitive domains crisscross the traditional subdivision into the five senses. Language is one such domain and much of its specialized nature is preserved regardless of whether it is presented in the acoustic or in the visual mode.

These crucial insights were not a bunch of disconnected and episodic ideas. They coalesced neatly into an articulated and rich domain of inquiry. A domain I was beginning to like more and more. Jacques had, in the space of about two years, progressively introduced me to classical cognitive science. Then something happened that steered my way into it for good.

VII   The Two Jacques

In the meantime, with de Gennes's blessing, I had moved to the Institut Pasteur and was working under Jacques Monod's supervision. In one of the precious and blessingly frequent conversations I had with Monod, I mentioned these data and these techniques. A staunch innatist himself, Monod encouraged me to invite Jacques to give a talk at the Pasteur. I objected that it was a topic lying very far from the daily concerns of the molecular geneticists. "Tant mieux, ça leur fera du bien" [So much the better, it will do them good] was Monod's instant reply. He added that he could not imagine, in perspective, a more fascinating field of inquiry *for a biologist* than the exploration of the limits of the genetic contribution to culture, the boundaries of the *enveloppe génétique* in shaping the human mind. This expression, now made famous by Jean-Pierre Changeux, who was at the time a close collaborator of Monod after having been a much beloved and esteemed pupil, was decidedly congenial to biologists of that caliber. So I invited Jacques.

He presented the data on categorial perception and reaped a great success.

Some years later, at a restricted planning meeting, Monod said emphatically (his precise words are engraved in my memory): "Ce que fait Mehler *c'est* de la science. Et, croyez-moi, j'en sais quelque-chose" [What Mehler does *is* science. And, believe me, I know something about science].

Monod eventually encouraged me to venture into this new and promising field, which we had tentatively labeled "bioanthropology." Jacques's seminar at Pasteur, in hindsight, was a milestone. It gave me the much-needed approval of Monod to enter into the biology of the mind, and, via the steady connection between Jacques and Chomsky and Luria, and the ages-deep connection between Monod and Luria, it intensified in Monod himself an active interest in this field.

A centerpiece of this connection was, of course, Changeux. I would like to hear how Jean-Pierre reconstructs his own memories, but I think that this first encounter with Mehler was quite important for both. He and Jacques were naturally destined to combine forces and interests, and ultimately share brilliant young collaborators, as they have. I, for one, am happy to think that I have had a non-negligible role in starting and, at least initially, feeding the collaboration between them. That seminar at Pasteur was the first step.

VIII The Royaumont Center

Then came the Centre Royaumont pour Une Science de l'Homme, of which I was appointed director, and Monod chairman of the board. I'll leave that complex story for another occasion, but I cannot refrain from reminiscing, with undisguised pride, that the young core of the center I had the privilege of assembling was formed by Jacques Mehler, Jean-Pierre Changeux, Dan Sperber, François Dell, and (later) Antoine Danchin. This short list speaks for itself. And it is no illusion that we were actually doing cognitive science before the label was fully consolidated.

This brings me to two curious episodes, both of which involve Jacques.

The first happened at Endicott House, a sumptuous residence outside Boston, given to MIT by a rich family of insurers. The Royaumont center, in close collaboration with Chomsky and Luria, organized a weekend-long informal workshop at Endicott House, to explore this famous biological "envelope" of the mind. If attendance by the senior component was impressive there and then (Monod, Chomsky, Luria, Zella Hurwitz, Hans Lucas Teuber, Vassily Leontieff, Daniel Bell, Michael Scriven, Daniel Lehrman, Edgar Morin), attendance by the young was no less impressive in hindsight (Jacques, Susan Carey, Ned Block, Peter and Jill De Villiers, Eric Wanner, Paula Meniuk).

At one of the sessions, Jacques had just started presenting data from psycholinguistics, including the celebrated archetypical garden-path sentence "The horse raced past the barn fell." A question was asked and before Jacques could reply, Noam sprang up, went to the blackboard, almost snatched the chalk from Jacques's hand and answered the question at length. The time allotted to the presentation by Jacques was almost entirely usurped by Noam and by the intense discussion that ensued. Jacques sank into a mood of frustrated resignation, and he later privately shared his sour grapes thoughts with me. He said: "Once a student, you are considered a student until you die." Contrary to his impression, though, the episode was quite favorably received by us all. It had shown, more than any words could, how fond of Jacques Noam was (and still is), how freely interchangeable with his own Noam then considered Jacques's data and interpretations, and how close their collaboration was.

IX Piaget's Ban of Jacques

The second episode involves Piaget during the preparation of the Royaumont debate with Chomsky. To cut a long story short, Piaget had manifested to Scott Atran his desire to meet with Chomsky at length. He wanted just the two of them, with an interpreter, possibly Guy Cellérier. I had slowly managed to persuade Piaget to turn this ultraexclusive binary exchange into a wider event, at Royaumont. One day I sent to Piaget a tentative list, for his approval, announcing my visit to him in Geneva a few days later. In his office in Geneva, Piaget was adamant in forbidding any participation by Jacques in the meeting. "Mehler n'a jamais compris ce que je dis" [Mehler has never understood what I say]. I found this claim, to say the least, unsubstantiated, but I knew well by then that their respective views on cognitive development were in stark contrast one to the other. Little did I know at that moment that Piaget was prone to assume that whoever disagreed with him could only do so because he or she could not understand what he was saying. Piaget's unshakable confidence in his being right made him assume that understanding him and believing what he said could not be two different things. This was to emerge loud and clear in the debate, and is now testified in print. Until his death, Piaget was persuaded that not even Chomsky understood what he said. But I am losing track of my story.

Piaget threatened not to come if Jacques was invited to the Royaumont debate. I tried to counter, but had to desist. I hurried to see Jacques upon my return to Paris. Jacques was equally adamant in enjoining me to stand firm. His exclusion was intolerable.

A very hot potato had landed in my lap.

I decided to play subtly and, if necessary, deviously. Some months later Scott Atran managed to assuage Piaget a bit about Mehler, transforming the veto into mere displeasure. Jacques, in the end, graciously accepted to make no presentation at Royaumont, but was officially present anyway, and then was given, as promised, ample possibility to write comments for the final proceedings.

It is not that I want to repaint, at any price, every such anecdote as positive, but I was impressed then, and maintain this impression in hindsight, by the fact that Piaget was ready to forgo a much sought-after, once-in-a-lifetime opportunity to meet with Chomsky, simply because Jacques was present. No matter how unpleasant, this was also a quite clear calibration of the scientific status that Jacques had already attained in Piaget's eyes back in the mid-seventies.[3]

X Cognition Trieste Style

I am coming now to an interesting later venture, my participation in which I also owe to Jacques: the Scuola Internazionale Superiore di Studi Avanzati (SISSA). That's where he is now, a full professor, having restarted, with unabated enthusiasm and drive, a new laboratory. The beginning of this venture dates twelve years back. Another old Argentinian-Italian buddy and relative of Jacques, the distingushed physicist Daniele Amati, director of the International School of Advanced Studies in Trieste, a graduate school in physics, mathematics, and biophysics, wanted to explore new avenues to extend and diversify the school. Cognitive science seemed a distinct possibility, so he naturally turned to Jacques.

Soon, a cognitive unit, TECS, the Trieste Encounters in Cognitive Science, was activated on the Adriatic coast. Jacques was kind enough to involve me in this enterprise, which offered me, on top of the beauty of the place, and the joy of being intermittently among distinguished colleagues and good friends, the immense pleasure of reestablishing a contact with my initial origins (as a physicist). The several periods I have spent at SISSA have been absolute bliss, and the workshops organized by Jacques there a further occasion to learn about cognitive science at its best. I'll select two of them, strictly related, for reasons that will soon be evident.

First Workshop
Languages differ in the strength, neatness, and pervasiveness with which words are parsed into syllables. Syllables have just one fixed structure in some languages (e.g., consonant-vowel, CV), a more flexible, but still stringent, small set of structures in others. And then . . . , then there is English. Jacques and Anne Cutler had set themselves to the hard task of exploring these differences systematically, tracking their possible consequences on language acquisition.

Trieste became the hub of this multilaboratory and multidisciplinary enterprise. Jacques and Anne organized an initial weeklong state-of-the-art planning seminar, which was repeated five years later, to wrap up and evaluate the results. The difference in content and climate between these two workshops, with a five-year gap, made quite palpable to me the transition to which I have alluded at the beginning of these notes, and on which I intend to close.

These are my recollections of the first workshop. Jacques had shown, among many other interesting things, that babies are sensitive to the sheer number of syllables. A repetition of bisyllabic words, or possible words, that only have in common their bisyllabicity, engenders boredom. If a three-syllable word appears in the stimulus, the baby is suddenly interested. It also works in the other direction. Being able to count syllables is crucial to being able to place accents correctly, since in many cases the tonic accent falls on the nth syllable from the beginning, or from the end, of a word. And where the accent reliably falls is a powerful clue to segmenting the stream of speech into words, especially when you are a baby, and do not know the words yet.

French, Italian, and Spanish are rather strongly syllabified languages. Adult speakers find it easy to identify a certain syllable, but are dismal at identifying the same string of phonemes, when they sometimes form a syllable, and sometimes don't. *Bal* is a syllable in "balcon," but not in "balise" and "balance." It's hard to track, and quickly detect, the sameness of the *b-a-l* sequence across these two kinds of words. The syllable constitutes a prominent, natural, and pervasive unit to the speakers of Romance languages, but not at all, or much less so, to speakers of English.

What happens when the speaker of one language parses words and sentences in the other language? Does he or she project the syllabification strategy and the accent patterns of his or her mother tongue? Initial results gave a clear positive answer. Romance speakers syllabify each and every speech sound under the sun, notably including English. English speakers, on the converse, are in trouble when having to parse a Romance language into syllables.

This has, as we just saw, interesting consequences for the identification of word boundaries in language acquisition. Anne Cutler presented indubitable evidence that the segmentation of the stream of speech into words is not physically based. The momentary gaps in the acoustic stream do not correspond *at all* to word boundaries (another fine confirmation of lesson 4 above). The child has to project these boundaries onto the stream. How does he or she do that?

Syllabification and the projection of simple, basic accent patterns are excellent devices. The hypothesis, to be later exhaustively explored by Maria-Teresa Guasti and Marina Nespor, in close collaboration with Jacques, is that the French child parses the stream tatatatatatatata as tatá/ tatá/ tatá/, while the English child parses it as tàta/ tàta/ tàta/ (one language is said to be iambic, the other trochaic, a fundamental distinction that applies to many other languages as well).

The stream prettybaby is naturally parsed in English as prétty bàby, while the stream jolibebe is equally naturally parsed in French as jolí bebé. A child that adopts this simple strategy is right most of the time. The child will get most of her "possible words" in the lexicon correctly.

In a nutshell: differential speech processing across languages fell neatly into a pattern of limited, sharp, mostly binary, differences to be set by the child once and for all, at a few specific "decision-nodes." It reminded us pleasantly of the "principles-and-parameters" scheme in syntax.

All this, and much more that I cannot summarize here, was very clear and very clean in the first workshop. That workshop was, again in my terminology, the epitome of "classical" cognitive science. Then came the second workshop, five years later.

The Second Workshop, and Beyond
Nothing was quite the same any more. An awesome quantity of data had been reaped, and many complications had arisen. Not that the neat former hypotheses were wrong. But they were not quite right either. The picture was more complex, more nuanced, with several remarkable specificities in different languages.

In a nutshell, we were already in a postclassical cognitive science.

This impression was confirmed by another workshop at SISSA, some months later, organized by Luigi Rizzi and Maria-Teresa Guasti, on the acquisition of syntax. Many interesting new data were presented, and this workshop also gave me the impression that the principles-and-parameters model, though basically still alive, needs a lot of refinements and qualifications to be viable. It hammered the point that we had entered a different phase in the development of cognitive science.

XI   On Classical, and Not-So-Classical Cognitive Science

In the spring of 1990, while organizing at MIT the annual meeting of the Cognitive Science Society, I was a chairing a small committee formed by Stephen Kosslyn, Steven Pinker, and Kenneth Wexler, all "classical" cognitive scientists to the marrow. We had before us some 300 submitted papers to be dispatched to the referees. In order to do it responsibly, we had to grasp what each paper was about. After having jointly processed the first hundred or so papers, Steve Kosslyn, rather puzzled, and somewhat exasperated, made a quite telling remark: "Gee, is *this* cognitive science?!"

The dominant topics, in fact, were flat-footed applications to a variety of very practical problems, applications of connectionism, and downright artificial intelligence, with many instances of standard problem-solving of the Pittsburgh variety. We shook our heads, in resignation, and then continued unabated.

I have been rethinking about this remark many times. As noticed by Kosslyn at that meeting, it was (and still is) another brand of cognitive science than the one I had initially learned from Jacques. A different enterprise than the one which I decided, some 25 years ago, to participate in.

I think I owe the reader a few more lines of clarification about this transition from what I have called "classical" cognitive science to a new variety, which I will call "nonclassical." A germane standard characterization of the former, in Chomsky's own term, is "rationalist cognitive science," a label that I like a lot, but that presents a delicate problem of appropriation. Cognitivists of other persuasions do not accept gladly to be dubbed, by implication, as nonrational. Perhaps calling it classical avoids this rebounding effect.

In a nutshell, and capitalizing on the lessons that I received from Jacques, I see classical cognitive science as being profoundly marked by the following overarching explanatory strategies: (1) unrestricted nativism (no capacity or content is deemed too complex or too specific to be imputed to the innate capacities of the brain/mind, if such attribution solves otherwise insoluble problems). As a consequence, we have (2), learning is essentially a triggering phenomenon (the idealization of single-trial learning *is* an idealization, but one close enough to reality). As a further consequence, it should be noted that, therefore, no historical statistics of the unfolding of the stimuli over time *can* be relevant to (classical) cognitive science. Connectionism and empiricism have reinstated the centrality of statistical analyses of the inputs, setting themselves apart (quite proudly, one can add) from classical cognitive science. The fact that Jacques, and other "classical" cognitive scientists for whom I have great respect, are presently tackling precisely such statistical analyses in the domain of phonology and syntax shows that something is changing in the picture.

The next classical core tenet, about which I can be dispensed from saying much here, is (3) massive modularity. The rich harvest of "strange" and very specific cognitive deficits witnessed by cognitive neuroscientists over many decades lends increased support to modularity. Yet, the growing resistance with which the inescapable modular conclusions of this vast literature are being met also shows, I think, a change of inclinations and explanatory standards in present-day cognitive science. From SLI (specific language impairment) to prosopagnosia, from domain-specific semantic deficits to Williams syndrome, the standard (and, in my opinion, still correct) modular interpretations are being challenged by insiders and outsiders as well. "Much more goes berserk in those cases" is the recurring critical punch line. Cognitive deficits produced by restricted brain lesions are alleged to be, after all, not so specific and circumscribed. The case is pending, and modularists now face a fight.

Finally, the signature of classical cognitive science also was (4) drastic restrictions on the number of points of possible variation, and drastic restrictions on the acceptable values at each point (the principles-and-parameters paradigm). The realm of language, phonology notably included, was the prototype case. Partly because of the raving success of optimality theory in phonology (also a parametric theory, but with unprecedented degrees of optionality built in), partly because of the difficulty in exactly spotting the points of parametric variation in syntax, the picture has been blurred somewhat. Combining some vestige of a parametric theory with statistical regularities in the input seems to be the present challenge. My expectation is that the final solution will be decidedly nonclassical.

XII Conclusion: The Demise of the Old Guard, and of Classical Cognitive Science

I cannot consider it a coincidence that Jacques, at that moment, disbanded the TECS committee and replaced it wholesale with much younger (and terrifically bright) cognitive scientists. He did the same also with the editorial board of *Cognition*. Out goes the old guard, and in come the young Turks. Adding the young and bright is splendid, and most welcome to us all. But why have them replace us wholesale? Could we not coexist with them? Are we really *that* old? Jacques's deeper agenda is somewhat inscrutable these days, to some of us but I think I know why he did this, and he may be right once again.

It's the business of scientists to determine, as best we can, what there is. It would be silly to force the data to be what we would like them to be. If our theories and schemata and styles of reasoning do not match reality, we should revise the former, not manipulate the latter. Are we, at least, entitled to feel vaguely uncomfortable? Possibly. But that should not retain us from moving ahead. But, are we *capable* of moving ahead? Jacques must have grown doubtful about this. Maybe this is why he is "retiring" us a little prematurely. And why he decided to restart from scratch, with a clean slate and with much younger collaborators.

Jacques is to me, for all the reasons I have sketched above, the prototype of the "classical" cognitive scientist. His whole career, and the first 25 years of *Cognition*, testify to this. Therefore it's a credit to his intelligence and vitality and imagination that he intends to move ahead, cost it what it may. He plans to proceed to a nonclassical phase with the young ones, unencumbered by our common past. This is the latest, and possibly the hardest, lesson I'll have to learn from Jacques.

Notes

1. We are fond of imitating his way of saying *belissimo* (one *l* only), *rrabia* (two strong *r*s, and one *b* only), and *cafoone* (with a long *o*). A lemon of a car is to us *un polmone* (an expression then current in Roman Italian to execrate a miserably underpowered and sluggish car). Except that it had an *o—polmone*, meaning "lung." But Jacques's *u* variant "pulmone" added force to the contempt, and sounded better than the original.

2. I was then riding a Triumph motorcycle, which I had parked on the pavement under the Mehlers' windows. The kids liked it very much from afar, and asked to have a closer look. So did Jacques. I think I infected him there and then with a passion for motorcycles that was destined to stick with him, and that would, years later, produce an unforgettable ride with Jacques and an Italian friend, all the way from Paris down to the Cote d'Azur, through wonderful secondary roads that Jacques was privy to (the Plateau de Mille Vaches and the *arrière pays* of Provence). In turn, he infected me with an interest (then not yet a passion) for the study of the mind that was destined to stick even more.

3. When the Royaumont center was forcibly disbanded in 1980, well after Monod's untimely death, Jacques wrote a beautiful letter to me, expressing solidarity (gladly accepted) and gratitude (a little harder for me to accept, since I felt that I was the one who owed gratitude to him).

In this letter, he said that I had played a central role in making him feel more at home in Paris, and in introducing him to eminent and interesting colleagues. I could not believe my eyes. I, the recently arrived and quite precarious visitor, had pictured *him* as firmly and stably entrenched in his splendid apartment near the Avenue Foche, and in the Parisian academic community. Moreover, it had been Jacques who had kindly introduced me, over the years, to the giants of cognitive science, and familiarized me with their work.

During all those years, most prominent in my memory, among many others, are my encounters at Jacques's house or in his laboratory with Jerry Fodor, Lila Gleitman, Tom Bever, Merrill Garrett, Richard Kayne, Ursula Bellugi, Albert Galaburda, and Peter Jucszyk). It defied my imagination how he could really think (but apparently he did) that it was *I* who had facilitated *his* entrenchment in the Parisian academe, and into the bigger world of science at large. Sweet of Jacques to think so, but it was plainly not the case. All the more reason for me to be grateful.

2

Cognition—Some Personal Histories

Thomas G. Bever, Susana Franck, John Morton, and Steven Pinker

What follows are narratives of the history of Jacques Mehler's journal *Cognition*. It is a tribute to him that no one of us can really measure how the journal has affected the field and evolved.

Present at the Creation
Thomas G. Bever

In the early 1960s, Jacques was a fresh PhD from George Miller's group, and had written one of the first dissertations that studied the "psychological reality" of the new linguistic theories. Jacques obtained a multiyear postdoc at MIT, and started sharing an office with me while I was pursuing a PhD in linguistics and an ersatz PhD in psychology. At our first meeting, I told him everything that was wrong with his thesis, which had just spawned an article in the *Journal of Verbal Learning and Verbal Behavior*. He was patiently immune to criticism on the outside. I got over it, and we became good friends forever.

Our shared office and young family lives were the basis for many conversations about nothing. In those days, "nothing" usually cashed out as "cars," since our then conventional marriages inhibited us from more interesting gossip—well, perhaps mine inhibited me more than Jacques's his. Today's young intellectuals gossip about the latest computer gadget the way we gossiped about the relative merits of Peugeot, Volvo, and Citroen.

The other frequent topic was the present and coming changes in experimental psychology. To us, it was obvious that the cognitive revolution was accomplished, and it was time to Get On With Things. We could not understand the ongoing hostility of the old behaviorist and verbal learning guard—they should just accept the truth or get out of our way. But they still kept the gates of the journals—to get published in *JVLVB*,[1] even Jacques's dissertation had to be couched primarily as an exploration of high-level memory, not a direct exploitation of linguistic theory as an explanation of behavior. Other tricks got other papers through the maze, including Chomsky's threat to resign from the board of *JVLVB* if it would not accept Fodor and Bever's really tame discovery that click location is governed by phrase structure. Conventional journal editors still recoiled at the imputation of mental representations and abstract hypothetical processes such as transformations. And there were only a few hypothetical processes such as transformations. And there were only a few journals to approach anyway. *Psychological Review, JVLVB, Journal of Experimental Psychology* (maybe), *Quarterly Journal of Experimental Psychology*, they all shared the rigid standards of turgid prose, maniacal statistics, and minimalist conclusions.

A few years later, we went to Europe in tandem to visit what we thought would be mecca—Piaget's laboratory. After all, Piaget got there first, thinking about mental life and its early emergence. We were startled and disappointed to see that he too had found his peace with empiricism—"abstractions" were still somehow frequently the result of "interiorization" of manifest external or sensorimotor experiences and behaviors. Shades of Charlie Osgood![2] The best we got out of Europe was a sedate Volvo sedan for each of us, and complete skepticism about the worth of the psychological establishment. In 1968, we returned to the United States with the proper automobiles and self-confident wisdom.

What a country we returned to! Cambridge (Massachusetts) was boiling with rage over Vietnam. We had been reading European newspapers, and we knew that a stupid and arguably genocidal war was under way. But Walter Cronkite, the *New York Times*, and precious Cambridge were just discovering this outrageous fact. There were constant teach-ins, lectures, teas; every venue was taken over by an increasing antiestablishment attitude. Chomsky was emerging as a figure in all this, suddenly doing what academic scientists are not supposed to do—applying their intelligence and proselytizing skills to political matters. He wrote a lasting article in the *New York Review of Books*, "The Responsibility of Intellectuals," which burned a place in our consciousness both by example and by force of argument. It was an eye-opener to Jacques and me among many others. We asked ourselves, what can *we* do to unify our political and intellectual lives? Noam, as usual, had taken the high ground, and was so much better at it than we could dream of becoming that it seemed silly to join him directly.

What to do? Jacques had an establishmentarian streak even then, and I was easy. One day out of the blue, he proposed that we *start a journal*. It would be devoted to a combination of scientific and political writings by the politically correct: we had three goals:

• to publish articles written in good English without the standard pseudo-scientific constraints of current journalese;

• to publish articles that blended good science with political applications related to psychology;

• to publish articles that would move the new cognitive psychology forward.

We were both in transition to permanent jobs, so we did not return to this idea until Jacques was ensconced in Paris at the CNRS, and I was at Rockefeller University in New York. What followed was a series of meetings with publishers. First, we went to Germany and met with Dr. S., head of Springer-Verlag in Heidelberg. We had a good series of meetings (though God knows what he thought about these two kids proposing a journal), and I went back to the states thinking that it was going to go. But Jacques felt he might have a difficult time dealing with S., and briefly we considered the possibility of me being the editor, since S. and I got along fine. Fortunately, that thought lasted only a few weeks, and was quickly rejected—even then, it was clear that I was not organized in the right way for such a job.

Jacques then turned to Mouton, and a series of meetings with Mr. B. ensued. B. was like a character from a Dutch mystery story. He met us in Paris several times, always in public places, always with some furtiveness. Jacques did most of the negotiating, which included a lot of concern about getting sufficient support for our offices. We agreed that having an American location for people to send manuscripts to was an important ingredient in starting the journal. Rockefeller University was lavish in supplying me with administrative support, so I was pretty sure I could handle that part. But Jacques needed specific financial support to set up the main editorial bureau in Paris. Eventually, after a lot of whimpering from B., Jacques got it, and we were in business. B. found a clever Dutch designer, who created the first logo as an impossible three-dimensional cube, and we set about putting together the first issues.

We had a lot of help from our friends. Noam contributed an article, as did others with a mixture of political skills and goals. The first few years of articles were a steady mix of the best of standard cognitive psychology, politically relevant articles such as those devoted to the IQ controversy, and even more politically wide-ranging discussions.

Gradually, as the post-sixties world calmed down, so did the journal. I moved to Columbia University, where there was far less administrative support, and it became clear to all that I was not suited to the managerial aspects of being an editor. We consolidated all the editorial activities in Paris.

The View from the Boiler Room
Susana Franck

In our preface to volume 50, which later became *Cognition on Cognition* published by MIT Press, Jacques and I described the intellectual climate in Paris in the late sixties and seventies when we were trying to get the journal off the ground. What we did not describe, however, was how the climate translated into the practicalities of actual functioning.

Jacques interviewed me in 1971, following an advertisement placed in the *International Herald Tribune* that asked for a "fully-bilingual person with a British or US University degree to work on a scientific journal." He was totally undisturbed when I told him that I knew nothing about cognitive psychology as it was then known. "Good," he said, "that way you won't have any opinions!"

At the time I thought this was a rather odd remark, but I soon realized that what Jacques really meant was that he did not want someone working with him who would take sides in the ideological battle then raging around him. This battle concerned the fact that he and Tom Bever were starting an English-language journal domiciled in France, far from the psychoanalytic ambiance reigning in Paris at the time. He wanted someone to help him and Tom get the operation off the ground and wanted no arguments about form and content.

I had arrived in France only a year before and was blissfully unaware of how difficult it was going to be to integrate a French university environment with an American degree and working experience. I was amazed when I discovered that I was non grata because I was associated with an English-language journal being published by a commercial publisher. The members of the laboratory (Centre d'Etude des Processus Cognitifs et du Langage) had received strict instructions not to talk to me and to ignore me completely. I was not to benefit from any of the services provided by the laboratory or the institution that housed us. Cognition was the Trojan horse of foreign, imperialist science.

But Jacques thought that the then incipient French cognitive science needed contact with the outside scientific community and needed to learn that our domain was as concerned with truth as chemistry, biology, or physics. In this he was helped by Massimo Piattelli-Palmarini and the Centre Royaumont who made a very concerted effort to establish a dialogue between scientists. The atmosphere generated by the center was a tremendous catalyst that people today seem almost to have forgotten. But the journal nonetheless was fated to fairly isolated entrenchment for a number of years.

Luckily, the members of our board and colleagues around the world really helped during those first difficult years. For some strange reason they seemed to believe in the enterprise, submitted papers, gave us their time and advice, and did not seem to mind overly much that we were not always as efficient as we could have been. Jacques put the accent on novelty and creativity and he and I were far less systematic than we have had to become over the years. Computers were not to be part of our operation for yet another decade. Articles arrived through the mail, went out for review through the mail, and the worldwide postal system lost its share of articles or referee reports and underwent several strikes. We accumulated mountains of paper that we could not file because we had no space, and our office, from the time operations were consolidated in Paris, counted a personnel of two: one chief and one Indian. To add to all this, there was a fairly idiosyncratic approach to content that no doubt gave the journal its particular flavor but got a number of backs up. To top all this off, Jacques rode a motorbike in those days and managed to have several manuscripts that he was supposed to read simply fly off the baggage rack.

But Jacques and I, each for our own reasons no doubt, had a common goal and that was to make the journal not only function but function well. Sadly, as we became more truly operational and discarded many aspects of our amateurish operation, the more amusing and more unorthodox parts of *Cognition* drifted out of the pages.

Antidisestablishmentarianism
John Morton

The drift into orthodoxy was inexorable. I had been on the editorial board from the beginning and was lifted up to associate editor in 1979 mainly as a hit man. When, as occasionally happens, two referees gave violently opposing opinions on a paper, the third referee remained obdurately neutral, and the authors were clamoring at the gates, I would get a little package with a polite request to solomonize. My other task was to draft letters designed to stop the persistent rejected author from continuing to protest. These were invariably people in other fields who had had the great insight about the mind but had read nothing more recent than 1939 and were indignant that we did not cherish their world-shattering breakthroughs.

I also kept on trying to recruit good British authors. Over the years my approaches led to a number of objections. These usually had to do with the amount of time it took to get articles through the system. *Cognition* was, in part, a victim of its own success. The number of articles being submitted increased from year to year, and, in response, the number of issues per year increased from four to six and then to twelve. Through all this time, Jacques kept control of the system, though it has become increasingly clear that Susana, the editorial assistant for the journal, is essential for all except the initial and final stages of the journal's process. The first stage is the assignment of referees. Jacques's knowledge of people working, first in cognition, then in cognitive science, and finally in cognitive neuroscience is awesome. There was no paper for which Jacques could not generate a list of potential reviewers from around the world based on a glance at the title. The final stage is that of decision-making and, with few exceptions, Jacques has remained in control. Unlike virtually every other journal editor, he has never used action editors. Further, he actually uses the contents of the journal. If you had a paper published in *Cognition,* then it is almost certain that Jacques would have made overheads from your figures and used them in a lecture somewhere.

The other problem with the journal, so far as new authors were concerned, was the length of the reviewing process. The intention of the journal has always been to turn papers around inside three months. However, the reviewers were very often members of the editorial board who, inevitably, were the best-known and most productive people in their field. Some, of my acquaintance, would be in tears when they were sent yet another paper to add to the pile of papers to review. There were also, and I shudder to admit this, some rogue reviewers who could take six months. The consequence of all this was that I would be approached by total strangers in the street who would offer untold riches if I could facilitate the passage of their precious offspring into the journal. Alas, I knew that the sticking point would be beyond my reach, some chaotic desk in a minor university on the East Coast, as likely as not. However, things have changed, especially as younger reviewers are increasingly used. Our average time to reply on regular submissions is now under three months and our average time on Brief Articles is four weeks. This, together with the fact that we have the shortest publication delays in the field today, means that *Cognition* is very attractive to urgent authors.

As our times became postrevolutionary, the manifest social content of the journal dried up, to Jacques's regret. One of the further preoccupations of Jacques, as with other occidental writers, has been the problem of getting scientists in the United States to cite work published in Europe. With such citations comes our position on the citation index list and a continuing flow of excellent articles from an increasingly status-bound clientele. (See the appendix to this chapter for a list of the top fifty articles from the journal.)

Some idea as to the nature of the shift in attitudes in the scientific community comes from the editorials over the last twenty-five years. In the editorial for volume 1 (1972) we have:

It is an illusion to view psychological science as an independently nurtured way of understanding the human mind that can cure the problems that are due to our misunderstanding of it.

Further,

it is our duty to discuss not only the practical value of our scientific conceptions in the light of the problems faced by people and societies but also to evaluate possible applications in the light of what we know about ourselves. This discussion is one that scientists cannot relinquish: if they claim that it falls outside their realm of competence, they side politically with views which they have not consciously examined.

Issue 1/1 opened with Chomsky's "Psychology and Ideology"[3] and the vigorous Discussion section opened in the second issue with Herrnstein's response.[4] Volume 1 closed with Tim Shallice's spirited discussion of the Ulster depth interrogation techniques with warnings against the Orwellian use of psychology.[5] The second volume opened with another editorial by Jacques and Tom, again on the theme of social responsibility in science. They ended the rather provocative piece with a plea to readers to consider the issues and respond to them. There followed an article on ideology in neurobiology by Stephen Rose and Hilary Rose[6] and further discussions of nature/nurture in language and intelligence, ethics in research, and other philosophical issues. It also rapidly became clear that *Cognition* was not just a vehicle for leading-edge science but was also an interesting read, with papers focusing on theoretical issues or with novel techniques, with language sometimes colloquial and even humorous! It was actually fun to read! My own "On Recursive Reference"[7] was thought by some to have pushed things too far, but it did result in more personal correspondence than any other paper I have ever written, though not the indefinitely large number of citations we had expected!

Jacques and Tom published obituaries and the occasional book review, and the Discussion section continued to crackle from issue to issue on scientific and social issues. Volume 4 opened with a wonderfully reflective comparison of value in art criticism and value in the evaluation of articles submitted to *Cognition*. I suspect that I appreciate this little editorial gem far more now than I did at the time and would like to see it reprinted.

Volume 7 (1979) announced the creation of a new Brief Reports and Letters section. Neither idea resulted in anything publishable. In his editorial, Jacques also bemoaned the fact that many submissions had been written for no other purpose "than that of advancing the author's career." These were themes which figured often in conversations with Jacques over the following decade. It must be recalled that Jacques and Tom had been concerned with the very definition of the field named by the journal. The idea that there was a level of scientific description between brain and behavior had been alien. One of the reasons for this is that the very notion is intellectually challenging and philosophically difficult to handle. The concepts are abstract and the nature of the enterprises of theorizing and modeling are ambiguous at best. It took a great deal of courage and imagination to oppose the authorities of the day.

Volume 10 (1981) was an anniversary volume. The editorial, written by Jacques and Susana, was both celebratory and lamenting. On the one hand they proudly claimed that "*Cognition* has . . . altered the nature of the material published in the domain." On the other hand, they noted, and tried to understand, the decline and terminal passing of "contributions on the ideological context in which science is produced." They commented further that "if the disappearance of the more ideologically oriented part of the journal has been a source of some disappointment to us, we must confess frustration with the uniformity of the submissions we tend to receive." The editorial ends in anticipating an increasingly important role both for computers and neuroscience in cognition. While "it appears probable that any causal account of development will be couched in terms that are more biological than psychological," there is the caution that "we will also have to resist the temptation of reducing phenomena to neurological terms that should only be accounted for in terms of processing."

The battle continues, of course, for it is easy to think that, in the colored pictures we get from functional magnetic resonance imaging and positron emission tomography, we see the mind rather than the results of the operation of the mind. The answer to the question *where* does not answer the question *how*, which is the question that *Cognition* has devoted itself to.

Volume 50 (1994) was another celebratory issue, and Jacques and Susana's editorial, with its focus on the history of the journal, should be consulted as a corrective to the views and remembrances of his devoted associate editors.

Forever Young
Steven Pinker

Three decades after its founding, *Cognition* is in great shape. A flood of submissions has forced it to come out monthly, and the numbers continue to grow: at the time of this writing, the number of submissions received for 2000 was twice the number received in the same period for 1999. The number of citations per article for the past six years (1994–99) shows it to be among the most-cited general interest journals in cognitive science (table 2.1).

A list of its most-cited articles (see the top fifty in the Appendix) includes many that opened up new fields. Such familiar topics as the false belief task, autism and theory of mind, the cohort model, compositionality and connectionism, the cheater-detector module, flashbulb memory, visual routines, sentence parsing heuristics, and repetition blindness are among the many that were first introduced in the journal, and its authors make up an all-star team in cognitive science. And for many of us it is the most interesting journal in our field: the package we tear open upon arrival and read over breakfast.

What made it so successful—despite relative youth, the lack of an affiliation with a scientific society, being headquartered in a backwater for scientific psychology, and being saddled with an unaffordable price? The

Table 2.1
Recent Publication and Citation Statistics for the Major Journals in Cognitive Science

Journal	Number of Citations 1994–1999	Number of Papers 1994–1999	Mean citations per paper 1994–1999
Cognition	369	262	1.4
Cognitive Psychology	172	111	1.5
Cognitive Science	123	92	1.3
Journal of Experimental Psychology: Learning, Memory, and Cognition	521	515	1.0
Journal of Memory and Language	327	282	1.2
Language and Cognitive Processes	139	153	0.9
Memory and Cognition	410	414	1.0

answer is: Jacques Mehler. *Cognition* is not a journal run by a custodian doing a tour of duty for a professional society or by a rotating committee of apparatchiks. Its distinctiveness and excellence could have come only from a doting individual. For thirty years Jacques has nurtured and improved the journal the way other scholars fuss over their laboratories and manuscripts.

In 1983 Jacques asked me to implement a new idea for *Cognition*—the special issues. Each would have an overview of a hot topic and a collection of research papers, taking up a single issue or volume and reprinted as a book. We agreed that "visual cognition," then an unknown term, would be a suitable topic. Later issues covered spoken word recognition, animal cognition, connectionism, the neurobiology of cognition (what was soon to be renamed "cognitive neuroscience"), lexical and conceptual semantics, and most recently, object recognition. While we were working on the issue Jacques asked me to join the associate editors.

The special issue was one of many ideas that Jacques came up with to keep the journal from stagnating. Some were shot down by an associate editor or two; some were good ideas in theory that never worked in reality. But the ones that worked kept the journal fresh. Another of Jacques's goals was to keep the journal young. I was untenured and not yet thirty years old when Jacques asked me aboard, and the most recent associate editor, Stanislas Dehaene, is not much older. Jacques periodically demands a list of youngsters for the editorial board, a nice contrast with the gerontocracies of most other journals.[8]

Cognition is also flexible and ecumenical. Its articles have been longer, shorter, more speculative, more polemical, and more methodologically eclectic than one sees anywhere else. Many psychology journals will reject any article with a smidgen of philosophy, linguistics, logic, anthropology, politics, or evolutionary biology. Not *Cognition,* which has always offered more than the traditional psychologists' fare of rats, sophomores, and *t*-tests (and now brain scans). When I joined I suggested changing the subtitle from *International Journal of Cognitive Psychology* to *International Journal of Cognitive Science,* but the journal embraced that multidisciplinary field well before the name existed.

Insiders know that the ancient rationalist-empiricist debate is being played out today in the continuum from the East Pole, the mythical location from which all directions are west, to the West Pole, the mythical location from which all directions are east. (The terms were coined by Dan Dennett during a seminar at MIT in which Jerry Fodor was fulminating at a "West Coast theorist" who in fact taught at Yale.) Cognition has given the East Polers their day in court, not so easy to earn with other journals. (Gary Marcus and I had an article rejected by another journal because it was "another argument from MIT that everything is innate"—though our only mention of innateness was a sentence saying that "the English past-tense rule is not, of course, innate.") But the journal has also striven to be bipolar, and over the years has published, for example, many articles by the authors of the West Pole manifesto *Rethinking Innateness*. A thousand flowers have bloomed in its pages.

Finally, *Cognition* was among the first institutions to usher in a major change of the past decade: the globalization of science. When Jacques and Tom founded the journal, prominent psychologists published in the *American Journal of Psychology*, the *British Journal of Psychology*, the *Canadian Journal of Psychology*, and worse. Even the less parochially named journals drew most of their articles from a single country or at best the English-speaking world. Of course there is no such scientific topic as "British psychology," and today everyone realizes that science cannot respect national boundaries. But in the 1970s and even 1980s Jacques was unusual in reaching out to the world for submissions and editorial board members.

Cognition, of course, is facing major new challenges. The corporations that produce scientific journals have enjoyed an envious position in the publishing world, indeed, in the business world in general: they don't pay their authors, reviewers, or editors, and they charge their customers whatever they want, figuring that a journal's prestige makes demand inelastic. But this fantasy cannot last forever. Librarians, faced with fixed budgets and absurdly escalating journal costs, are up in arms, and scientists, editors, and readers are behind them. Internet publishing, with its promise of immediate, searchable, portable, and infinite content, is spawning new outlets for scientific publishing, and could upset the prestige hierarchy quickly. Jacques and Susana have expended enormous energy in giving the journal a presence in cyberspace and in negotiating with Elsevier to maintain the journal's role in these revolutionary times.

It has been a privilege for all of us to help Jacques with his great contribution to the science of mind. Long live *Cognition!*

Notes

1. *Journal of Verbal Learning and Verbal Behavior*—the evocative name by which the *Journal of Memory and Language* used to be known.
2. A historian of psychology adds: Charles Osgood was an early renegade against traditional stimulus-response behaviorism. He developed a "mediation" model, in which internal r/s pairs connected explicit stimuli to explicit responses. He applied this model to represent "meaning" in terms of the internal r/s pairs: Fodor pointed out that the internal pairs were operationally defined in terms of explicit S/R pairs, and therefore the model was limited in the same way as traditional S/R theories of meaning.
3. Chomsky, N. (1972). Psychology and ideology. *Cognition*, 1, 11–46.
4. Herrnstein, R.J. (1972). Whatever happened to vaudeville? A reply to Professor Chomsky. *Cognition*, 1, 301–310.
5. Shallice, T. (1972). The Ulster depth interrogation techniques and their relation to sensory deprivation research. *Cognition*, 1, 385–406.
6. Rose, S.P.R., and Rose, H. (1973). "Do not adjust your mind, there is a fault in reality"—Ideology in neurobiology. *Cognition*, 2, 479–502.
7. Morton, J. (1976). On recursive reference. *Cognition*, 4, 309.
8. The other two associate editors have long regarded themselves as old farts but have also assiduously cultivated young shoots.

Appendix: The Fifty Most Cited Articles in Cognition, 1971–2000

The following list was compiled by Ami Chitwood, Librarian of the Teuber Library in the Department of Brain and Cognitive Sciences, MIT, with the assistance of Marie Lamb. The ranking is based on the number of citations from the combined Science Citation Index and Social Science Citation Index. The number of citations is listed in brackets after the article.

Wimmer, H., and Perner, J. (1983). Beliefs about beliefs: Representation and constraining function of wrong beliefs in young children's understanding of deception. *Cognition*, 13, 103–128. [437]

Baron-Cohen, S., Leslie, A. M., and Frith, U. (1985). Does the autistic child have a "theory of mind?" *Cognition*, 21, 37–46. [414]

Marslen-Wilson, W., and Komisarjevsky Tyler, L. (1980). The temporal structure of spoken language understanding. *Cognition*, 8, 1–72. [370]

Fodor, J. A., and Pylyshyn, Z. W. (1988). Connectionism and cognitive architecture: A critical analysis. *Cognition, 28,* 3–72. [342]

Liberman, A. M., and Mattingly, I. G. (1985). The motor theory of speech perception revised. *Cognition, 21,* 1–36. [328]

Morais, J., Cary, L., Alegria, J., and Bertelson, P. (1979). Does awareness of speech as a sequence of phones arise spontaneously? *Cognition, 7,* 323–332. [294]

Cosmides, L. (1989). The logic of social interchange: Has natural selection shaped how humans reason? Studies with the Wason selection task. *Cognition, 31,* 187–276. [254]

Marslen-Wilson, W. D. (1987). Functional parallelism in spoken word-recognition. *Cognition, 25,* 71–102. [235]

Brown, R., and Kulik, J. (1977). Flashbulb memories. *Cognition, 5,* 73–100. [221]

Pinker, S., and Prince, A. (1988). On language and connectionism: Analysis of a parallel distributed processing model of language acquisition. *Cognition, 28,* 73–194. [219]

Damasio, A. R. (1989). Time-locked multiregional retroactivation: A systems-level proposal for the neural substrates of recall and recognition. *Cognition, 33,* 25–62. [217]

Linebarger, M. C., Schwartz, M. F., and Saffran, E. M. (1983). Sensitivity to grammatical structure in so-called agrammatic aphasics. *Cognition, 13,* 361–392. [217]

Ullman, S. (1984). Visual routines. *Cognition, 18,* 97–160. [213]

Frazier, L., and Fodor, J. D. (1978). The sausage machine: A new two-stage parsing model. *Cognition, 6,* 291–326. [197]

Farah, M. J. (1984). The neurological basis of mental imagery: A componential analysis. *Cognition, 18,* 245–272. [189]

Armstrong, S. L., Gleitman, L. R., and Gleitman, H. (1983). What some concepts might not be. *Cognition, 13,* 263–308. [186]

Hoffman, D. D., and Richards, W. A. (1984). Parts of recognition. *Cognition, 18,* 65–96. [185]

Ullman, S. (1989). Aligning pictorial descriptions: An approach to object recognition. *Cognition, 32,* 193–254. [183]

Seidenberg, M. S. (1985). The time course of phonological code activation in two writing systems. *Cognition, 19,* 1–30. [177]

Bruner, J. S. (1974–75). From communication to language—A psychological perspective. *Cognition, 3,* 255–288. [168]

Kimball, J. (1973). Seven principles of surface structure parsing in natural language. *Cognition, 2,* 15–48. [167]

Levelt, W. J. M. (1983). Monitoring and self-repair in speech. *Cognition, 14,* 41–104. [164]

Clark, H. H., and Wilkes-Gibbs, D. (1986). Referring as a collaborative process. *Cognition, 22,* 1–40. [164]

Gelman, S. A., and Markman, E. M. (1986). Categories and induction in young children. *Cognition, 23,* 183–210. [156]

Fodor, J. A., and Pylyshyn, Z. W. (1981). How direct is visual perception?: Some reflections on Gibson's "Ecological Approach." *Cognition, 9,* 139–196. [152]

Read, C., Yun-Fei, Z., Hong-Yin, N., and Bao-Qing, D. (1986). The ability to manipulate speech sounds depends on knowing alphabetic writing. *Cognition, 24,* 31–44. [145]

Mehler, J., Jusczyk, P., Lambertz, G., Halsted, N., Bertoncini, J., and Amiel-Tison, C. (1988). A precursor of language acquisition in young infants. *Cognition, 29,* 143–178. [134]

Shatz, M., Wellman, H. M., and Silber, S. (1983). The acquisition of mental verbs: A systematic investigation of the first reference to mental state. *Cognition, 14,* 301–322. [134]

Kempen, G., and Huijbers, P. (1983). The lexicalization process in sentence production and naming: Indirect election of words. *Cognition, 14,* 185–210. [129]

Osherson, D. N., and Smith, E. E. (1981). On the adequacy of prototype theory as a theory of concepts. *Cognition, 9,* 35–58. [127]

Kean, M.-L. (1977). The linguistic interpretation of aphasic syndromes: Agrammatism in Broca's aphasia, an example. *Cognition, 5,* 9–46. [121]

Reber, A. S., and Allen, R. (1978). Analogic and abstraction strategies in synthetic grammar learning: A functionalist interpretation. *Cognition, 6,* 189–222. [121]

Baddeley, A. (1981). The concept of working memory: A view of its current state and probable future development. *Cognition, 10,* 17–24. [121]

Levelt, W. J. M. (1992). Accessing words in speech production: Stages, processes and representations. *Cognition, 42,* 1–22. [121]

Altman, G., and Steedman, M. (1988). Interaction with context during human sentence processing. *Cognition, 30,* 191–238. [120]

Fairweather, H. (1976). Sex differences in cognition. *Cognition, 4,* 231–280. [118]

Baillargeon, R., Spelke, E. S., and Wasserman, S. (1985). Object permanence in five-month-old infants. *Cognition, 20,* 191–208. [117]

Turvey, M. T., Shaw, R. E., Reed, E. S., and Mace, W. M. (1981). Ecological laws of perceiving and acting: In reply to Fodor and Pylyshyn (1981). *Cognition, 9,* 237–304. [115]

Karmiloff-Smith, A. (1986). From meta-processes to conscious access: Evidence from children's metalinguistic and repair data. *Cognition, 23,* 95–148. [112]

Clark, E. V. (1973). Non-linguistic strategies and the acquisition of word meanings. *Cognition, 2,* 161–182. [111]

Kanwisher, N. G. (1987). Repetition blindness: Type recognition without token individuation. *Cognition, 27,* 117–144. [109]

Morais, J., Bertelson, P., Cary, L., and Alegria, J. (1986). Literacy training and speech segmentation. *Cognition, 24*, 45–64. [109]

Hirsh-Pasek, K., Kemler Nelson, D. G., Jusczyk, P. W., Wright Cassidy, K., Druss, B., and Kennedy, L. (1987). Clauses are perceptual units for young infants. *Cognition, 26*, 269. [108]

Castles, A., and Coltheart, M. (1993). Varieties of developmental dyslexia. *Cognition, 47*, 149–180. [107]

Spelke, E., Hirst, W., and Neisser, U. (1976). Skills of divided attention. *Cognition, 4*, 215–230. [107]

Pinker, S. (1984). Visual cognition: An introduction. *Cognition, 18*, 1–64. [106]

Leslie, A. M., and Thaiss, L. (1992). Domain specificity in conceptual development: Neuropsychological evidence from autism. *Cognition, 43*, 225–252. [104]

Gopnik, M., and Crago, M. B. (1991). Familial aggregation of a developmental language disorder. *Cognition, 39*, 1–50. [104]

Stuart, M., and Coltheart, M. (1988). Does reading develop in a sequence of stages? *Cognition, 30*, 139–182. [104]

Cheng, K. (1986). A purely geometric module in the rat's spatial representation. *Cognition, 23*, 149–178. [104]

II
Thought

Representations, Psychological Reality, and Beyond

Luca L. Bonatti

It is a fact that for centuries philosophers, and not psychologists, have had the most interesting things to say about the mind. With few exceptions, this has been true well into our century, until the beginning of the cognitive revolution that Chomsky ignited and that a score of people contributing to this book (and one who inspired it) helped to unfold.

At least since Aristotle, for philosophers, thinking about the mind meant thinking about content and about how content is represented. But everybody does what she or he is best at, and so philosophers have carried out this millennia-long meditation about the nature of mental representations according to the accepted standards of their community. For them, a theory is good if it is coherent, sound, and possibly true, although the last requirement is certainly not the most highly priced one. So if a philosopher's theory required the postulation of a complex mental ontology, or if it demanded awkward kinds of representations almost unthinkable for the layperson, the philosopher would say, so be it. Not that much changed when mathematicians, linguists, and mathematical linguists started getting interested in the mind. They all happened to share the same opinions about the sociologically acceptable standards for a good theory.

In an important sense then, doing philosophy of mind has been pretty much like doing mathematics. But there is an extra annoying element. The mind is an empirical object, and the discovery of its structure must largely be an empirical enterprise, an unfortunate accident that philosophers have often preferred to leave aside.

To be fair, philosophers and linguists had and have plenty of reasons not to take empirical psychology that seriously. After all, the "mathematical" way has worked well. Many of the ideas that lay at the foundations of what is now cognitive science, like modularity (Sperber, chapter 3), functionalism (Block, chapter 6), models (Johnson-Laird, chapter 5), or images and language of thought (Pylyshyn, chapter 4), can be developed largely independently of empirical studies of the mind. Furthermore, when the "cognitive revolution" set in, linguistics imposed itself as the model of a promising theory about the mind. The goal of linguistics was to articulate an axiomatic theory of the speaker's competence independent of empirical considerations about performance and performance mechanisms. Because of the crucial role of linguistics in cognition, there was no reason to suppose that what was good for language was not good for the whole of the mind:

> It is our view that cognitive theory must present an axiomatic formulation of the processes which are involved in creative behavior. This definition of the goal of cognitive psychology is analogous to the goal of generative linguistic theories. (Mehler and Bever, 1968, p. 273)

But axiom systems are not enough. Mental representations are empirical objects, and sooner or later theories about the content and form of the representation have to face the tribunal of experience. How could this be done?

This is the problem of the psychological reality of representational theories of mind, a problem that has concerned Mehler since the very beginning of his scientific work (Mehler, 1968; Mehler and Bever, 1968).

The first attempt at testing the psychological reality of the representational constructs postulated by the new theories of mind has been the easiest, the simplest, and the most elegant. The idea at the core of the derivational theory of complexity (DTC) for grammar was to find a direct relation between representational structures and operations postulated by the theoretical and behavioral indices. Mehler was one of the first to try to get this approach working (Mehler, 1963, 1964; Mehler and Carey, 1967). And once again, because of the central role of linguistics among cognitive theories, the strategy at the core of DTC was soon applied to other cognitive domains (e.g., see reasoning).

This was a philosopher's paradise. It was the dream of getting to know the mind without bothering about the biological nature of the organism embedding it. It was a way to render psychology, if not dispensable, at least very peripheral in the enterprise of the discovery of the structure of the mind.

It is unfortunate that such an approach failed. When theories of performance started being developed in the details, it became clear that the relations between representations in the theories and representations in the mind are orders of magnitude more complex than the simple direct causal correlations assumed by DTC and its variations. On the one hand, representations may be constructed and computed without leaving direct detectable traces in the mind. On the other hand, behavioral data that have been taken as evidence for the existence of certain kinds of representations (U-shaped curves, S-shaped curves, etc.) may be the effect of the network activity of the brain, rather than of those representations (Elman et al., 1996).

I think that the heated, often purely ideological debates about the foundations of cognition and the sense of drifting that they have engendered (see Piattelli-Palmarini's comments in chapter 1 about the fate of "classical" cognitive science) are still a hangover from the failure of something like a DTC, of a general strategy to find the correspondence between representations in the theory and representations in the brain. Here is an example of how deep the problem is. Back in 1962 Mehler argued that when subjects remember the general sense of a sentence, they encode its "kernel" and the syntactic transformations separately. There were theories (albeit not the right ones) about what a syntactic transformation could be. But how was the "kernel" represented? Mehler offered no answer, but a range of possible options:

Exactly how the kernel is encoded, of course, is not established; it might be an image, an abstract set of symbols, or anything else capable of regenerating the kernel sentence on demand. (Mehler, 1963, p. 350)

One may have hoped that in the four decades that went by after that sentence, some sort of progress would have been made. After all, several theories of how content is represented, arguably psychologically motivated, have been proposed. But the reading of Pylyshyn's and Johnson-Laird's papers will show that the question is still up for grabs. Most people have argued that content (if we may so interpret the "kernel" of a sentence) is represented in the form of images, or of mental models (of which images are a subclass; see Johnson-Laird, 1983). Few others have objected that such theories are dangerously close to nonsense, and have taken the other option suggested by Mehler, arguing that content is represented propositionally (see, e.g., Bonatti, 1998a,b; Pylyshyn, 1981, 1986; Pylyshyn, chapter 4). What kind of evidence, if any, is appropriate to solve the issue is still under dispute.

In this case, as in many others, representations in the mind have remained ineffable.

What to do then? One way to proceed is to just turn one's back to representational theories of mind, or to try to read the structure of representation directly out of the structure of neurons and synaptic connections. It is a popular approach these days, bolstered by the undeniable progress of neuroscience. Because nobody takes such an approach in this book (and because I find it completely off the mark), I will not pursue it any further.

Another way to proceed is to keep the old philosopher's attitude, to disregard completely (or almost completely) empirical science, and to keep trying to infer the nature and structure of mental representations by way of arguments. Somebody, sooner or later, will find out the right way to test the psychological reality of representations. In favor of this point of view there is its previous success (cognitive science, or for that matter, computers, would not be here if it hadn't been for a score of bold thinkers who felt unbound by the state of empirical investigations of their time) and the perception that some of the current directions in psychological theorizing are deeply misguided. Fodor, as I understand him, has taken this route.

A third way to proceed is to put one's hand in the mud as deep as is necessary, and still try to find ways to test the psychological reality of representations by using any available empirical means, be it designing reaction time experiments, or doing statistical analyses of corpora, or running experiments in comparative cognition, or trying applications of the latest tricks of brain imagery to the mind.

Such an approach comes with its own dangers. It renders theories about the mind strongly dependent on the available tools of investigation, as it were, "here and now." Perhaps, one is left wondering, with a better imagery machine or with a more detailed description of how brain activations eventuate in reaction times, the whole perspective on the essence of the mind might change. The casual eclecticism of this position, together with its extreme dependency on empirical details of that sort, is likely to offend the philosopher's taste.

But one can see things in another way. The attempt at tackling the problem of psychological reality with all available tools comes, after all, from the recognition that in its essential lines, representationalism must have gotten it right, and that a better investigation of the empirical consequences of the postulation of a representational construct can only produce better representational theories in the long run. This position changes the balance between psychology and the philosophy of mind. Whereas psychological investigations were often undertaken as logical consequences of a representational theory, such a position sends the ball back into the field of the philosophers. It is now up to them to adjust and fine-tune representational theories to the results of a deeply enriched empirical cognitive psychology. Indeed, part II contains several contributions that show that the influence of this change in balance is already felt among philosophers. So Sperber (chapter 3) argues for a kind of mental architecture—massive modularity—by trying to extend the implications of a series of empirical findings in reasoning and perception to its limits. And Block (chapter 6) asks the important question of the respective limits of functional theories and neurological facts in the explanation of the content of experience. He forcibly suggests that there is something more than just entertaining content in a psychological experience, and that this experience of "feeling like" depends on the physical neurological embodiment of our minds. If he is right, then no psychological explanation is complete without an account of our neurological organization in principle.

This third way is the bet that Jacques has taken throughout his career. He is an excellent scientist, and he knows that a scientist must be an opportunist, act like a thief, and steal some bit of truth any time secretive Mother Nature is caught off guard. But he has used all the means he could think of to always tackle the same problem, the problem of the psychological reality of representations, the hard problem that any representationalist has to face in the absence of the paradise of something like DTC. It is a risky bet, but it is also a lot of fun. Its results may look more like Sisyphus' chores than like progress, but, as a philosopher would say, so be it.

References

Bonatti, L. (1998a). On pure and impure representations. Mental models, mental logic and the ontology of mind. Unpublished manuscript.

Bonatti, L. (1998b). Possibilities and real possibilities for a theory of reasoning. In Z. Pylyshyn (Ed.), *Essays on representations* (pp. 85–119). Norwood, NJ: Ablex.

Elman, J. L., Bates, E., Johnson, M. H., Karmiloff-Smith, A., Parisi, D., and Plunkett, D. (1996). *Rethinking innateness.* Cambridge, MA: MIT Press.

Johnson-Laird, P. N. (1983). *Mental models.* Cambridge, MA: Harvard University Press.

Mehler, J. (1963). Some effects of grammatical transformations on the recall of English sentences. *Journal of Verbal Learning and Verbal Behavior, 2,* 346–351.

Mehler, J. (1964). *How some sentences are remembered.* Cambridge, MA: Harvard University Press.

Mehler, J. (1968). La grammaire generative a-t'elle une réalité psychologique? *Psychologie Française, 13,* 137–156.

Mehler, J., and Bever, T. G. (1968). The study of competence in cognitive psychology. *International Journal of Psychology, 3,* 273–280.

Mehler, J., and Carey, P. (1967). Role of surface and base structure in the perception of sentences. *Journal of Verbal Learning and Verbal Behavior, 6,* 335–338.

Pylyshyn, Z. W. (1981). The imagery debate: Analogue media versus tacit knowledge. *Psychological Review, 88,* 16–45.

Pylyshyn, Z. W. (1986). *Computation and cognition: Toward a foundation for cognitive science.* Cambridge, MA: MIT Press.

3

In Defense of Massive Modularity

Dan Sperber

In October 1990, a psychologist, Susan Gelman, and three anthropologists whose interest in cognition had been guided and encouraged by Jacques Mehler, Scott Atran, Larry Hirschfeld, and myself, organized a conference on "Cultural Knowledge and Domain Specificity" (see Hirschfeld and Gelman, 1994). Jacques advised us in the preparation of the conference, and while we failed to convince him to write a paper, he did play a major role in the discussions.

A main issue at stake was the degree to which cognitive development, everyday cognition, and cultural knowledge are based on dedicated domain-specific mechanisms, as opposed to a domain-general intelligence and learning capacity. Thanks in particular to the work of developmental psychologists such as Susan Carey, Rochel Gelman, Susan Gelman, Frank Keil, Alan Leslie, Jacques Mehler, Elizabeth Spelke (who were all there), the issue of domain-specificity—which, of course, Noam Chomsky had been the first to raise—was becoming a central one in cognitive psychology. Evolutionary psychology, represented at the conference by Leda Cosmides and John Tooby, was putting forward new arguments for seeing human cognition as involving mostly domain- or task-specific evolved adaptations. We were a few anthropologists, far from the mainstream of our discipline, who also saw domain-specific cognitive processes as both constraining and contributing to cultural development.

Taking for granted that domain-specific dispositions are an important feature of human cognition, three questions arise:

1. To what extent are these domain-specific dispositions based on truly autonomous mental mechanisms or "modules," as opposed to being

domain-specific articulations and deployments of more domain-general abilities?

2. What is the degree of specialization of these dispositions, or equivalently, what is the size of the relevant domains? Are we just talking of very general domains such as naive psychology and naive physics, or also of much more specialized dispositions such as cheater-detection or fear-of-snakes?

3. Assuming that there are mental modules, how much of the mind, and which aspects of it, are domain-specific and modular?

As a tentative answer to these three questions, I proposed in some detail an extremist thesis, that of "massive modularity" (Sperber, 1994). The expression "massive modularity" has since served as a red cloth in heated debates on psychological evolution and architecture (e.g., see Samuels, 1998; Murphy and Stich, 2000; Fodor, 2000). I was arguing that domain-specific abilities were subserved by genuine modules, that modules came in all formats and sizes, including micromodules the size of a concept, and that the mind was modular through and through. This was so extremist that archmodularist John Tooby gently warned me against going too far. Jacques Mehler was, characteristically, quite willing to entertain and discuss my speculations. At the same time, he pointed out how speculative indeed they were, and he seemed to think that Fodor's objections against "modularity thesis gone mad" (Fodor, 1987, p. 27) remained decisive. I agreed then, and I still agree today, that our understanding of cognitive architecture is way too poor, and the best we can do is try and speculate intelligently (which is great fun anyhow). I was not swayed, on the other hand, by the Fodorian arguments. To his objections in *The Modularity of Mind* (1983), Fodor has now added new arguments in *The Mind Doesn't Work That Way* (2000). I, however, see these new considerations as weighing on the whole for, rather than against, massive modularity.

Modularity—not just of the mind but of any biological mechanism—can be envisaged at five levels:

1. At a morphological or *architectural level,* what is investigated is the structure and function of specific modules, and, more generally, the extent to which, and the manner in which the organism and its subparts,

in particular the mind/brain, are an articulation of autonomous mechanisms.

2. At the *developmental level,* modules are approached as phenotypic expressions of genes in an environment. Cognitive modules in particular are hypothesized to explain why and how children develop competencies in specific domains in ways that could not be predicted on the basis of environmental inputs and general learning mechanisms alone.

3. At the *neurological level,* modules are typically seen as dedicated brain devices that subserve domain-specific cognitive functions and that can be selectively activated, or impaired.

4. At the *genetic level,* what is at stake are the pleiotropic effects among genes such that relatively autonomous "gene nets" (Bonner, 1988) get expressed as distinct phenotypic modules. Genetic modularity is more and more seen as crucial to explaining, on the one hand, phenotypic modularity, and on the other the evolution of specific modules (Wagner, 1995, 1996; Wagner and Altenberg, 1996).

5. At the *evolutionary level,* hypotheses are being developed about the causes of the evolution of specific modules, and of genetic modularity in general. Understanding the causes of the evolution of modules helps explain the known features of known modules and also aids the search for yet-to-be discovered features and modules.

In cognitive science, discussions of mental modularity received their initial impetus and much of their agenda from Fodor's pioneering work (Fodor, 1983). Fodor's take on modularity had two peculiar features. First, while he was brilliantly defending a modularist view of input systems, Fodor was also—and this initially attracted less attention—decidedly antimodularist regarding higher cognitive processes. Incidentally, whereas his arguments in favor of input systems modularity relied heavily on empirical evidence, his arguments against modularity of central processes were mostly philosophical, and still are. Second, Fodor focused on the architectural level, paid some attention to the developmental and neurological levels, and had almost nothing to say about the genetic and evolutionary levels. Yet the very existence of modularity and of specific modules begs for an evolutionary explanation (and raises difficult and important issues at the genetic level). This is uncontroversial in the case of nonpsychological modular components of the organisms, for example, the liver, eyes, endocrine glands, muscles, or enzyme systems, which are generally best understood as adaptations (how well developed, convincing, and illuminating are available evolutionary hypotheses varies, of course, with cases). When, however, an evolutionary perspective on psychological modularity was put forward by evolutionary psychologists (Barkow, Cosmides, and Tooby, 1992; Cosmides and Tooby, 1987, 1994; Pinker, 1997; Plotkin, 1997; Sperber, 1994), this was seen by many as wild speculation.

I will refrain from discussing Fodor's arguments against the evolutionary approach. They are, anyhow, mostly aimed at grand programmatic claims and questionable illustrations that evolutionary psychologists have sometimes been guilty of (just as defenders of other ambitious and popular approaches). As I see it, looking from an evolutionary point of view at the structure of organisms, and in particular at the structure of the mind of organisms, is plain ordinary science and does not deserve to be attacked or need to be defended. Like ordinary science in general, it can be done well or badly, and it sometimes produces illuminating results and sometimes disappointing ones.

There is, moreover, a reason why the evolutionary perspective is especially relevant to psychology, and in particular to the study of cognitive architecture. Apart from input and output systems, which, being linked to sensory and motor organs, are relatively discernible, there is nothing obvious about the organization of the mind into parts and subparts. Therefore, all sources of insight and evidence are welcome. The evolutionary perspective is one such valuable source, and I cannot imagine why we should deprive ourselves of it. In particular, it puts the issue of modularity in an appropriate wider framework. To quote a theoretical biologist, "The fact that the morphological phenotype can be decomposed into basic organizational units, the homologues of comparative anatomy, has . . . been explained in terms of modularity. . . . The biological significance of these semi-autonomous units is their possible role as adaptive 'building blocks'" (Wagner, 1995). In psychology, this suggests that the two notions of a mental module and of a psychological adaptation (in the biological sense), though definitely not synonymous or coextensive, are nevertheless likely to be closely related. Autonomous mental mechanisms that are also very plausible cases of evolved adaptations—face recognition or mind reading, for instance—are prime examples of plausible modularity.

Fodor is understandably reluctant to characterize a module merely as a "functionally individuated cognitive mechanism," since "anything that would have a proprietary box in a psychologist's information flow diagram" would thereby "count as a module" (Fodor 2000, p. 56). If, together with being a distinct mechanism, being plausibly a distinct adaptation with its own evolutionary history were used as a criterion, then modularity would not be so trivial. However, Fodor shuns the evolutionary perspective and resorts, rather, to a series of criteria in his 1983 book, only one of which, "informational encapsulation," is invoked in his new book.

The basic idea is that a device is informationally encapsulated if it has access only to limited information, excluding some information that might be pertinent to its producing the right outputs and that might be available elsewhere in the organism. Paradigm examples are provided by perceptual illusions: *I* have the information that the two lines in the Müller-Lyer illusion are equal, but my visual perceptual device has no access to this information and keeps "seeing" them as unequal. Reflexes are in this respect extreme cases of encapsulation: given the proper input, they immediately deliver their characteristic output, whatever the evidence as to its appropriateness in the context. The problem with the criterion of encapsulation is that it seems too easy to satisfy. In fact, it is hard to think of any autonomous mental device that would have unrestricted access to all the information available in the wider system.

To clarify the discussion, let us sharply distinguish the informational inputs to a cognitive device (i.e., the representations it processes and to which it associates an output) from its database, that is, the information it can freely access in order to process its inputs. For instance, a word-recognition device takes as characteristic inputs phonetic representations of speech and uses as a database a dictionary. Nonencapsulated devices, if there are any, use the whole mental encyclopedia as their database. Encapsulated devices have a restricted database of greater or lesser size. A reflex typically has no database to speak of.

In his new book, Fodor attempts to properly restrict the notion of encapsulation. For this, he discusses the case of modus ponens (Fodor,

2000, pp. 60–62). Imagine an autonomous mental device that takes as input any pairs of beliefs of the form {p, [If p, then q]} and produces as output the belief that q. I would view such a modus ponens device as a cognitive reflex, and therefore a perfect example of a module. In particular, as a module would, it ignores information that a less dumb device would take into account: suppose that the larger system has otherwise information that q is certainly false. Then the smarter device, instead of adding q to the belief box, would consider erasing from the belief box one of the premises, p, or [If p, then q]. But our device has no way of using this extra information and of adjusting its output accordingly.

Still, Fodor would have this modus ponens device count as unencapsulated, and therefore as nonmodular. Why? Modus ponens, he argues, applies to pairs of premises in virtue of their logical form and is otherwise indifferent to their informational content. An organism with a modus ponens device can use it across the board. Compare it with, say, a bridled modus ponens device that would apply to reasoning about number, but not about food, people, or plants, in fact about nothing other than numbers. According to Fodor, this latter device would be encapsulated. Yet, surely, the logical form of a representation contributes to its informational content. [If p, then q] does not have the same informational content as [p and q] or [p or q], even though they differ only in terms of logical form and their nonlogical content is otherwise the same. Moreover, the difference between the wholly general and the number-specific modus ponens devices is one of inputs, not one of database. Both are cognitive reflexes and have no database at all (unless you want to consider the general or restricted modus ponens rule as a piece of data, the only one then in the database).

The logic of Fodor's argument is unclear, but its motivation is not. Fodor's main argument against massive modularity is that modules, given their processing and informational restrictions, could not possibly perform the kind of general reasoning tasks that human minds perform all the time, drawing freely, or so it seems, on all the information available. An unrestricted modus ponens device looks too much like a tool for general reasoning, and that is probably why it had better not be modular. Still, this example shows, if anything, that a sensible definition of informational encapsulation that makes it a relatively rare property of mental devices is not so easily devised.

Naively, we see our minds—or simply ourselves—as doing the thinking and as having unrestricted access to all our stored information (barring memory problems or Freudian censorship). There is in this respect a change of gestalt when passing from naive psychology to cognitive science. In the information-processing boxological pictures of the mind, there is no one box where the thinking is done and where information is freely accessible. Typically, each box does its limited job with limited access to information. So who or what does the thinking? An optimist would say, "It is the network of boxes that thinks (having, of course, access to all information in the system), and not any one box in particular." But Fodor is not exactly an optimist, and, in this case, he has arguments to the effect that a system of boxes cannot be the thinker, and that therefore the boxological picture of the mind cannot be correct.

Our best theory of the mind, Fodor argues, is the Turing-inspired computational theory according to which mental processes are operations defined on the syntax of mental representations. However such computations are irredeemably local, and cannot take into account contextual considerations. Yet, our best thinking (that of scientists) and even our more modest everyday thinking are highly sensitive to context. Fodor suggests various ways in which the context might be taken into consideration in syntactic processes, and shows that they fall short, by a wide margin, of delivering the kind of context-sensitivity that is required. He assumes that, if the computational theory of mind is correct, and if, therefore, there are only local operations, global contextual factors cannot weight on inference, and in particular cannot contribute to its rationality and creativity. Therefore the computational theory of mind is flawed.

Has Fodor really eliminated all possibilities? Here is one line that he has not explored. Adopt a strong modularist view of the mind, assume that all the modules that have access to some possible input are ready to produce the corresponding output, but assume also that each such process takes resources, and that there are not enough resources for all processes to take place. All these potentially active modules are like competitors for resources. It is easy to see that different allocations of resources will have different consequences for the cognitive and epistemic efficiency of the system as a whole.

I wake and see through the window that the street is wet. I normally infer that it has been raining. For the sake of simplicity, let us think of this as a modus ponens deduction with [if the street is wet, it has been raining] and [the street is wet] as premises. This is a process readily described in syntactic terms. Of course, the major premise of this deduction is true only by default. On special occasions, the street happens to be wet because, say, it has just been washed. If I remember that the street is washed every first day of the month and that today is such a day, I will suspend my acceptance of the major premise [if the street is wet, it has been raining] and not perform the default inference. Again, this process is local enough and easily described in syntactic terms. Whether or not I remember this relevant bit of information and make or suspend the inference clearly affects the epistemic success of my thinking.

Does there have to be a higher-order computational process that triggers my remembering the day and suspending the default inference? Of course not. The allocation of resources among mental devices can be done in a variety of non computational ways without compromising the computational character of the devices. Saliency is an obvious possible factor. For instance the premise [the street is washed on the first day of the month] may be more salient when both the information that it is the first day of the month, and that the street is wet are activated. A device that accepts this salient premise as input is thereby more likely to receive sufficient processing resources.

It is not hard to imagine how the mere use of saliency for the allocation of computing resources might improve the cognitive and epistemic efficiency of the system as a whole, not by changing local processes but by triggering the more appropriate ones in the context. The overall inferential performance of the mind would then exhibit some significant degree of context-sensitivity without any of the computational processes involved being themselves context-sensitive. Saliency is an obvious and obviously crude possibility. Deirdre Wilson and I have suggested a subtler, complex, noncomputational factor, relevance, with two subfactors, mental effort and cognitive effect (Sperber and Wilson, 1995, 1996). Relevance as we characterize it would in particular favor simplicity and conservatism in inference, two properties that Fodor argues cannot be accommodated in the classical framework. This is not the place to elaborate. The general point is that a solution to the problems raised for a computational theory of mind by context-sensitive inference may be found in terms of some "invisible hand" cumulative effect of noncomputational events in the mind/brain, and that Fodor has not even discussed, let alone ruled out, this line of investigation.

Of course, the fact that a vague line of investigation has not been ruled out is no great commendation. It would not, for instance, impress a grant-giving agency. But this is where, unexpectedly, Fodor comes to the rescue. Before advancing his objections to what he calls the "New Synthesis," Fodor presents it better than anybody else has, and waxes lyrical (by Fodorian standards) about it. I quote at some length:

> Turing's idea that mental processes are computations . . . , together with Chomsky's idea that poverty of the stimulus arguments set a lower bound to the information a mind must have innately, are half of the New Synthesis. The rest is the "massive modularity" thesis and the claim that cognitive architecture is a Darwinian adaptation. . . . [T]here are some very deep problems with viewing cognition as computational, but . . . these problems emerge primarily in respect to mental problems that *aren't* modular. The real appeal of the massive modularity thesis is that, if it is true, we can either solve these problems, or at least contrive to deny them center stage pro tem. The bad news is that, since massive modularity thesis pretty clearly *isn't* true, we're sooner or later going to have to face up to the dire inadequacies of the only remotely plausible theory of the cognitive mind that we've got so far. (Fodor, 2000, p. 23)

True, Fodor does offer other arguments against massive modularity, but rejoinders to these will be for another day (anyhow, these arguments were preempted in my 1994 paper, I believe). However, the crucial argument against computational theory and modularity is that it cannot be reconciled with the obvious abductive capabilities of the human mind, and I hope to have shown that Fodor's case here is not all that airtight. Now, given Fodor's praises for the new synthesis and his claim that this is "the only remotely plausible theory" we have, what should our reaction be? Yes, we might—we should—be worried about the problems he raises. However, rather than giving up in despair, we should decidedly explore any line of investigation still wide open.

As a defender of massive modularity, I am grateful to Fodor for giving us new food for thought and new arguments. I wish to express also my gratitude to Jacques without whose demanding encouragement, in spite of my anthropological training, I would never have ventured so far among the modules.

References

Barkow, J., Cosmides, L., and Tooby, J. (Eds.) (1992). *The adapted mind: Evolutionary psychology and the generation of culture.* New York: Oxford University Press.

Bonner, J. T. (1988). *The evolution of complexity.* Princeton, NJ: Princeton University Press.

Cosmides, L., and Tooby, J. (1987). From evolution to behavior: Evolutionary psychology as the missing link. In J. Dupré (Ed.), *The latest on the best: Essays on evolution and optimality.* Cambridge MA: MIT Press.

Cosmides, L., and Tooby, J. (1994). Origins of domain specificity: The evolution of functional organization. In L. A. Hirschfeld and S. A. Gelman (Eds.), *Mapping the mind: Domain specificity in cognition and culture* (pp. 85–116). New York: Cambridge University Press.

Fodor, J. (1983). *The modularity of mind.* Cambridge, MA: MIT Press.

Fodor, J. (1987). Modules, frames, fridgeons, sleeping dogs, and the music of the spheres. In J. Garfield (Ed.), *Modularity in knowledge representation and natural-language understanding* (pp. 26–36). Cambridge, MA: MIT Press.

Fodor, J. (2000). *The mind doesn't work that way: The scope and limits of computational psychology.* Cambridge MA: MIT Press.

Hirschfeld, L. A., and Gelman, S. A. (Eds.) (1994). *Mapping the mind: Domain specificity in cognition and culture.* New York: Cambridge University Press.

Murphy, D., and Stich, S. (2000). Darwin in the madhouse: Evolutionary psychology and the classification of mental disorders. In P. Carruthers and A. Chamberlain (Eds.), *Evolution and the human mind: Modularity, language and metacognition,* Cambridge, UK: Cambridge University Press.

Pinker, S. (1997). *How the mind works.* New York: Norton.

Plotkin, H. (1997). *Evolution in mind.* London: Alan Lane.

Samuels, R. (1998). Evolutionary psychology and the massive modularity hypothesis. *British Journal for the Philosophy of Science,* 49, 575–602.

Sperber, D. (1994). The modularity of thought and the epidemiology of representations. In L. A. Hirschfeld and S. A. Gelman (Eds.), *Mapping the mind: Domain specificity in cognition and culture* (pp. 39–67). New York: Cambridge University Press. [Revised version in Sperber, D. (1996). *Explaining culture: A naturalistic approach.* Oxford, Blackwell.]

Sperber, D., and Wilson, D. (1995). *Relevance: Communication and cognition,* 2nd ed. Oxford: Blackwell.

Sperber, D., and Wilson, D. (1996). Fodor's frame problem and relevance theory. *Behavioral and Brain Sciences, 19,* 530–532

Wagner, G.P. (1995). Adaptation and the modular design of organisms. In F. Morán, A. Morán, J. J. Merelo, and P. Chacón (Eds.), *Advances in artificial life* (pp. 317–328). Berlin: Springer Verlag.

Wagner, G. P. (1996). Homologues, natural kinds and the evolution of modularity. *American Zoologist, 36,* 36–43.

Wagner, G. P. and Altenberg, L. (1996). Complex adaptations and the evolution of evolvability. *Evolution, 50,* 967–976.

4

Is the Imagery Debate Over? If So, What Was It About?

Zenon Pylyshyn

Background

Jacques Mehler was notoriously charitable in embracing a diversity of approaches to science and to the use of many different methodologies. One place where his ecumenism brought the two of us into disagreement was when the evidence of brain imaging was cited in support of different psychological doctrines, such as the picture-theory of mental imagery. Jacques remained steadfast in his faith in the ability of neuroscience data (where the main source of evidence has been from clinical neurology and neuroimaging) to choose among different psychological positions. I personally have seen little reason for this optimism so Jacques and I frequently found ourselves disagreeing on this issue, though I should add that we rarely disagreed on substantive issues on which we both had views. This particular bone of contention, however, kept us busy at parties and during the many commutes between New York and New Jersey, where Jacques was a frequent visitor at the Rutgers Center for Cognitive Science. Now that I am in a position where he is a captive audience it seems an opportune time to raise the question again.

I don't intend to make a general point about sources of evidence. It may even be, as Jacques has frequently said, that we have sucked dry the well of reaction-time data (at least in psycholinguistics), and that it is time to look elsewhere. It may even be that the evidence from positron emission tomography (PET) and functional magnetic resonance imaging (fMRI) will tell us things we did not already know—who can foresee how it will turn out? But one thing I can say with some confidence is that if we do not have a clear idea of the question we are trying to answer, or a conceptually coherent hypothesis, neither reaction time nor PET nor fMRI nor repetitive transcranial magnetic stimulation (rTMS) will move us forward. So long as we are engaged in a debate in which the basic claims are muddled or stated in terms of metaphors that permit us to freely back out when the literal interpretation begins to look untenable, then we will not settle our disagreements by merely deploying more high-tech equipment. And this, I suggest, is exactly what is happening in the so-called imagery debate, notwithstanding claims that the debate has been "resolved" (Kosslyn, 1994).

The Historical Background of the Imagery Debate

This chapter is in part about the "debate" concerning the nature of mental imagery. Questions about the nature of conscious mental states have a very long history in our struggle to understand the mind. Intuitively, it is apparent that we think either in words or in pictures. In the last forty years this idea has been worked into theories of mental states within the information-processing or computational view of mental processes. The so-called "dual code" view has been extremely influential in psychology since about 1970, due largely to the early work of Allan Paivio (Paivio, 1971) (for an early critique of this work, see Pylyshyn, 1973). Shortly after this renaissance of research on mental imagery, the emphasis turned from the study of learning and the appeal to "imagability" as a intervening variable to an attempt to work out the nature of mental images themselves (this work is summarized in Kosslyn 1980) (for a critique of this later work, see Pylyshyn, 1981). Led by the influential work of Stephen Kosslyn, researchers investigated the structure of mental images, including their metrical properties. For example, images seemed to actually have distance, since it took longer to scan greater distance in an image; they seemed to have size inasmuch as it took longer to report small features in a small image than in a large one; the "mind's eye" that inspects images also seemed to have limited resolution which fell off with eccentricity like that of the real eye. In addition, it appears that we can manipulate images much the way we can manipulate physical objects; we can rotate mental images (in three dimensions), we can fold them and watch what happens, we can draw things on them or superimpose other images on them or on our percepts, and so on. Such abilities suggested to researchers that images must have a special form or underlying instantiation in the brain and many researchers proposed that images differ from other forms of representation (presumably "verbal" representations) in that they have spatial properties, are displayed in the brain, and represent by virtue of "depicting" or by virtue of resembling what they represent, rather than by virtue of describing their target scenes.

Throughout these developments I have maintained that we are under a collective illusion ("la grande illusion," to use Jean Renoir's phrase). The illusion is that when we experience "seeing an image with the mind's eye" we are actually inspecting a mental state, a structure that can play a role in an information-processing account of mental activity. I argued that what was going on in these studies is that subjects were being asked, in effect, what it would be like to see certain things happening (a scene being scanned by attention, looking for a small detail in a scene, or watching an object being rotated). In my critique, I suggested that what was happening in these studies was that people were using what they know about the world to simulate certain observable aspects of the sequence of events that would have unfolded in the situation being studied. In other words, the experiments were revealing what subjects believe would happen if they were looking at a certain scene, and not the inherent nature of an imagery medium or mechanism.

Such claims and counterclaims went on for two decades. Then in the past few years a new source of evidence from neuroscience was introduced which many people took to provide, in the words of the subtitle of Stephen Kosslyn's influential book (Kosslyn, 1994), "the resolution of the imagery debate." Many investigators were persuaded that here, finally, was evidence that was direct and unambiguous and proved that there were images in the brain—actual displays realized as patterns of activity in the visual cortex. What are we to make of these new results, which have persuaded a large number of researchers of the basic correctness of the picture-theory? Is it true that we now have concrete evidence about the nature of mental images when previously we had only indirect and ambiguous behavioral evidence?

Many Things to Many People

In discussing this question, I will begin by laying out what I think is the state of the "debate." Later I will come back to the question of whether neuropsychological evidence has made (or is likely to make) any difference. What is the problem? To put it as bluntly as possible, the problem over the question of the nature of mental imagery is just that some people find the experience of seeing a picture-in-the-head completely compelling, while others find it irrelevant as the basis for a theory of what is actually going on in the mind when we entertain mental images. Beyond that there is no useful general debate, only arguments about the coherence and validity of certain quite specific proposals, and about what morals can be drawn from particular experiments. This is, in the end, not about metaphysics. It is a scientific disagreement, but one in which there are very many different claims falling under the same umbrella term, "the imagery debate," and they may have little or nothing in common other than sharing the same gut reaction to the picture metaphor and to the question of what to make of the phenomenology and the empirical evidence concerning mental imagery.

Among the many things involved in this debate are a variety of substantive disagreements.

- A disagreement about whether the form of representation underlying mental imagery is in some way special, and if so, in what way it is special—whether it is special because it uses distinct mechanisms specific to imagery or whether it is special because it deals with information about how things look (i.e., because of what imaginal representations are about or their subject matter);

- a disagreement about whether images are "depictive" or "descriptive";

- a disagreement about whether or not mental imagery "involves the visual system," which itself raises the question of what exactly is the visual system and in what way it may be involved;

- a disagreement about whether certain phenomena observed during episodes of mental imagery are due to the fact that the brain evolved in a particular way resulting in a "natural harmony" between the way things unfold in one's imagination and the way they unfold in the world or in one's perception of the world (what Shepard, 1975, has called "second order isomorphism");

- a disagreement about whether some of the phenomena observed during episodes of mental imagery arise because (1) people are reasoning from what they know about the situation being imagined, and are simulating what they believe would have happened if a real event were being observed, or because (2) special image-specific mechanisms are deployed when one reasons using mental images. This is the disagreement that I wrote about in (Pylyshyn, 1981) and, I believe, remains one of the main questions about mental imagery.

One of the things that makes this debate both ironic and ill-posed is that it is hard to disagree with most of the picture-theory[1] views in this discussion, since there is something right about the claims. It is true that when we solve problems in which the geometry of a display plays a role, we usually do so by imagining the figure, and that when we do imagine the figure we are able to do things we could not do if we were to approach the problem symbolically—say, by thinking in words. This much is not in dispute. What is in dispute is what is going on in our head when we engage in the sort of activity we call "imaging," and in particular what an adequate theory of this process will need to postulate as the bearer of the imaginal information. This is not a case of believing that images do not exist or are "epiphenomenal." It is a question of whether theories of mental imagery that posit two-dimensional (2D) displays or "depictive representations" are empirically correct, or perhaps even coherent. In every case I have looked at, hypothesizing pictures or depictions does not provide any explanatory advantage over what I will call the "null hypothesis," that image content is represented as symbolic expressions (see below), even though it may feel more comfortable because it comports with one's subjective impression of what is going on. I, for one, get very nervous when I find a theory in psychology making claims that are consonant with how it looks to me from the inside—I know of too many examples where how it feels on the inside is exactly the wrong kind of theory to have. Things rarely are how they seem in any mature science, and this is especially true in a nascent science like psychology or psychobiology.

Intrinsic and Extrinsic Constraints

The basic dilemma is that while the following two claims may both seem to be true, they are incompatible—at least in their naive form:

1. Since the mental images we have are of our own making we can make them have whatever properties we wish, and we can make them behave in any way we choose. Having an image is like having a thought—it seems as though we can think any thought there is to think. Consequently which image property or thought we have depends on what we want to do, and this generally depends on what we believe.

2. Mental images appear to have certain inherent properties that allow them to parallel many aspects of the perceived world. For example, images do not seem to unfold the way thoughts do, following principles of

inference (including heuristic rules), but in a way that directly reflects how we perceive the world. An image looks a certain way to us, therefore we "see" things in it independent of our explicit knowledge about what we are imaging. If we imagine a geometrical figure, such as a parallelogram, and imagine drawing its diagonals, we can see that one of the diagonals is longer than the other and yet that the two cross at their mutual midpoints, and we can see that this is so apparently without having to infer it from our knowledge of geometry, and certainly without having to derive it from Euclidean axioms. Similarly, when we imagine a dynamic situation or event unfold in our mind (such as a baseball being hit by a batter), the imagined event behaves in ways that appear to be at least partially outside our voluntary control, and maybe even outside our intellectual capacity to calculate.

There are anecdotes to illustrate each of these two perspectives and the opposing factions in the "imagery debate" generally emphasize one or the other of the above claims. The picture-theorist argues that if you have a visual image of an object, then you have no choice but to imagine it from a particular viewpoint and having a particular shape, size, orientation, and so on. Similarly, if you imagine an object as small in size, it follows from the inherent nature of mental imagery that it will be harder to "see" and therefore to report small visual details than if you imagined it as large; or that if you focus your attention on a place in your imagination and then try to report a property that is far away, it will take longer than if you attempted to report a property that was nearer to where you were initially focused. The critics of picture-theories argue equally cogently that you can just as easily imagine an object without imagining it as having any particular properties, that there is no reason (other than the implicit requirement of the instruction to "imagine something small") why you can't imagine something as small but highly detailed and therefore not take longer to report its visible details, or to imagine switching your attention from one place on an image to another in a time independent of how far away the two places are, as long as you are not attempting to simulate what would happen if you were looking at the real object being imagined or are not attempting to simulate a situation in which you believe it would take more time to get from one place to another if the places were further apart (in fact I reported the results of several experiments showing that this is exactly what happens; see Pylyshyn, 1981). The imagery-as-general-reasoning adherents can point to many examples where the way a dynamic image unfolds is clearly under our voluntary control. For example, imagine sugar being poured into a glass full of water (does it overflow?), or imagine yellow and blue transparent colored filters being moved together so they overlap (what color do you see where they overlap?). The answer you give to these questions clearly depends on what you know about the physics of solutions and what you know (or remember) about the psychophysics of color mixing, as well as your understanding of the task.

It is easy enough to come up with examples that go either way; some empirical phenomena appear to be a consequence of inherent properties of an image representation and others appear to arise because of what you know or believe (perhaps falsely) about how the visually perceived world works. The substantive empirical question is, Which properties are inherent properties of the imagery system (or the medium or mechanisms of mental imagery) and which are properties that the person doing the imaging creates in order to simulate what he or she believes would be true of a real situation corresponding to the one being imagined.

I prefer to put this opposition slightly differently: some properties of the image or of the imagined situation are cognitively penetrable by knowledge and beliefs (many of which are held tacitly and cannot be reported on demand, as is the case with knowledge of grammar or of many social conventions).[2] Other properties may be due to intrinsic causes of various sorts, to the architecture of mind. The inherent nature of mental images might be one of the determinants of certain experimental phenomena reported in the literature, but so might the way in which you have learned certain things you know and the way in which you have organized this knowledge (which may have nothing to do with properties of imagery itself). For example, you have learned the alphabet in serial order so in order to tell whether L comes after H you may have to go through the list, and in order to tell whether certain things happen when you fold a certain paper template to make a cube, you may have to go through a sequence of asking what happens after individual folds (as in the study by Shepard and Feng, 1972, in which it was found that when observers judged whether two arrows in a paper template would touch when folded, it took them longer under just those conditions when it would have taken more folds to actually fold the template). Problems are generally solved by the application of a sequence of individual operations so this in itself says nothing special about mental imagery. It's true that in order to recall how many windows there are in your living room you may have to count them because the numerical fact is not stored as such. But this has nothing to do with the nature of imagery per se, any more than that in order to recall the second line of a poem you may need need to recall the first line, or that in order to tell how many words it has you may need to recall the line and count the words. There are also many reasons why you might observe certain reliable patterns whenever the subjective experience of "seeing with the mind's eye" occurs. The burden of proof must fall on those who wish to argue in favor of some particular special mechanism to show that it is at least unlikely that the general mechanism, which we know exists because it has to be used in nonimagery contexts, will not do.

The reason that there has been so much talk (by me and others) about the representations underlying mental imagery being propositional is that there are very good reasons for thinking that much of cognition depends on a language of thought (Fodor, 1975; Fodor and Pylyshyn, 1988; Pylyshyn, 1984). For example, propositions, or more correctly, language-like tree-structured symbolic encodings, are the only form of representation that we know that exhibit the properties of compositionality, productivity, and systematicity that are essential characteristics of at least human thought (see Fodor and Pylyshyn, 1988). Although that does not entail that mental images are propositions, the propositional proposal serves as the natural null hypothesis against which to compare any proposal for a special form of representation for mental imagery. It's not that the idea of images having the form of a set of sentences in some mental calculus is a particularly attractive or natural alternative, but it is the only one so far proposed that is not seriously flawed.

Here is the crux of the problem that picture-theorists must face if they are to provide full explanatory accounts of the phenomena. They must show that the relevant empirical phenomena, whether it is the increased time it takes to switch attention to more distant places in an image or the increased time it takes to report details from smaller images, follow from the very nature of mental images or of the mechanisms involved in their use. In other words it must be that these phenomena reveal a constraint attributable to the intrinsic nature of the image, to its form or neural implementation, or to the mechanisms that it uses, rather than to some other extrinsic constraint arising from the knowledge that the subject possesses, or from the way this knowledge is structured, or from the subject's goals or understanding of the task. If, in order to account for the regularities, one has to appeal to something other than the inherent pictorial constraints of the imagery system then the picture-theory does not play an explanatory role. That is because it is the extrinsic factors that are doing the work and they can equally be applied to any form of representation, including one that is propositional. So, for example, if a picture-theory is to explain why it takes longer to switch attention between more distant places in an image, one must show that this is required by the imagery mechanism or medium or format because of its very nature or causal structure (e.g., because of the physical laws that apply). Otherwise the appeal to the picture-theory of mental imagery carries no explanatory weight. Any form of representation can give the same result merely by adding the stipulation that switching attention between representations of more distant places requires more time (during which, for example, one might entertain thoughts of the form "now it is here," "now it is there," and so on, providing a sequence of thoughts that simulate what might happen if one were looking at a scene). So if you can show empirically that it is unlikely that the properties you observe are due to inherent properties of the image, as opposed to properties of the world envisioned, the reason for preferring the picture-theory would evaporate.

Although this is a simple point it turns out to be one that people have a great deal of difficulty in grasping, so I will try to provide an additional example. Consider the proposal that images need not be literally written on a 2D surface, but rather may be implemented in a functional space such as a matrix data structure in a computer. Note that physical laws do not apply to a functional space. There is nothing about a matrix data structure that requires that in order to get from one cell to another you have to pass through intervening cells. In the matrix a "more distant cell" is not actually farther away so no physical law requires that it take more time; in fact, in a computer one can get from any cell to any other cell in constant time. So if we do require that the process pass through certain other cells, then we are appealing to a constraint extrinsic to the nature of the matrix or "functional space." Of course, one might find it natural to assume that in order to go from one cell to another the locus of attention must go through intervening ones. But the intervening cells are not in any relevant sense located between two other cells except by virtue of the fact that we usually show matrices as 2D tables or surfaces. In a computer we can (though we don't have to—except again by extrinsic stipulation) go from one cell to another by applying a successor function to the coordinates (which are technically just ordered names). Thus we can require that in going from one cell to another we have to step through the cells that fall between the two, where the relation "between" is defined in terms of the ordering of their names. Thus we can ensure that more such cells are visited when the distance being represented is greater. But this requirement does not follow from the intrinsic nature of a matrix data structure; it is an added or extrinsic requirement, and thus could be imposed equally on any form of representation, including a nonpictorial one. All that is required is (1) that there be some way of representing potential (or empty) locations and of identifying them as being "in between," and (2) that in accessing places in the representation, those marked as "in between" have to be visited in getting from the representation of one place to the representation of another place. As regards requirement (1), it can be met by any form of representation, including a propositional or symbolic one, so long as we have names for places—which is what Cartesian coordinates (or, for that matter, any compressed form of digital encoding of pictures such as GIF or JPEG) give us.

The test of whether any particular phenomenon is attributable to the intrinsic nature of images or to tacit knowledge is to see whether the observations in question change in a rationally comprehensible way if we change the relevant knowledge, beliefs, or goals. Take, for example, the robust finding that the time it takes to switch from examining one place on an image to examining another increases linearly with the distance being imagined, a result consistently interpreted to show that images have metrical properties like distance. One can ask whether this time-distance relation arises from (a) an intrinsic property of an image or (b) the observers' understanding that they are to simulate what happens when looking at a particular display. It is clear that observers can scan an image at a particular speed, or they can scan it at a different speed, or they can simply not scan it at all when switching their attention from one place to another. In our own research, we showed that when observers are given a task that requires focusing on distinct places but that does not emphasize imagining getting from one place to another, the scanning phenomenon vanishes (Pylyshyn, 1981). As in the original scanning experiments, the setup always involved focusing on a place on a mental map and then focusing at another place on the map. But in one experiment the ostensible task in focusing on the second place was to judge the direction of the first place from it (by naming a clock direction). In this and other similar tasks[3] there is no effect of image distance on the time to switch attention between places.

I might note in passing that it is not by any means obvious that people do, in fact, represent a succession of empty spaces in scanning studies or in any dynamic visualization. We have obtained some preliminary data (Pylyshyn and Cohen, 1999) suggesting that when we imagine continuously scanning a space between two locations we do not actually traverse a succession of intermediate places unless there are visible features at those locations. When there are such features, it appears that we carry out a sequence of time-to-contact computations to selected visible features along the scan path. Thus it may well be that scanning involves computing a series of times between intermediate visible features and simulating the scanning by waiting out the appropriate amount of time for each transition.[4] Note also that while extrinsic requirements, such as those stipulating that switching attention from one location to another must occur by stepping through intermediate locations [as in (2) above], may seem unnatural and unmotivated when applied to a list of sentences, it is exactly as well motivated, no more and no less, as it is when applied to a matrix or other "functional space." In both cases the constraint functions as a free empirical parameter, filled in solely to match the data for the particular case. The same is not true, of course, when the space is a real physical space rather than a "functional" space since there is, after all, a physical law relating time, distance, and speed, which applies to real space but not to "functional space." This is no doubt why there has been so much interest in finding a real spatial representation of images, a pursuit to which I now turn.

Neuropsychological Evidence and the "New Stage" of the Debate

The arguments I have sketched, when fleshed out in detail, as I have done elsewhere (Pylyshyn, 1981; Pylyshyn, submitted) should make it clear that a picture-theory that appeals to inherent properties of a "picture-like" 2D display is no better at explaining the results of mental imagery studies than is the "null hypothesis," that the content of our image is encoded in some symbolic form which serves as the basis for inferences and for simulating various aspects of what it would be like to see some situation unfold (including the relative times taken for different tasks). The basic problem is that the phenomena that have attracted people to the picture-theory (phenomena such as mental scanning or the effect of image size on reaction times for detecting features) appear to be cognitively penetrable and thus cannot be attributed to the intrinsic geometric nature of the image itself—to how it is spatially instantiated in brain tissue—as opposed to what people know or infer or assume would happen in the real referent situation.

Any attempt to minimize this difficulty, say, by postulating that images are only "functionally" like 2D pictures, is powerless to explain the phenomena at all since functional spaces are whatever we want them to be and are thus devoid of explanatory force. But what about the literal conception of images as real 2D displays in the brain? This is the view that is now popular in certain neuropsychology circles and has led to what Kosslyn (1994) has described as the "third phase" of the imagery debate—the view that the evidence of neuroscience can reveal the "display" or "depictive" nature of mental images. So where does such evidence place the current state of understanding of mental imagery?

Neuropsychological evidence has been cited in favor of a weak and a strong thesis with little care taken to distinguish between them. The weak thesis is that mental imagery in some way involves the visual system. This claim is weak because nobody would be surprised if some parts of visual processing overlap with virtually any cognitive activity; much depends on what one takes to be "the visual system" (for more on this question, see Pylyshyn, 1999). The strong claim is that not only is the visual system involved but the input to this system is spatially laid out as a "picture-like" pattern of activity. Yet the evidence cited in favor of the weak thesis, that imagery involves the visual system, is often also taken (sometimes implicitly and sometimes explicitly) to support the stronger thesis, that images are structurally different from other forms of thought because they are laid out spatially the way pictures are, and therefore that they are not descriptive but depictive (whatever exactly that might entail beyond being topographically laid out).

An argument along the following lines has been made in the recent neuropsychology literature (Kosslyn et al., 1999; see also the accompanying News of the Week article). Primary visual cortex (area 17) is known to be organized retinotopically (at least in the monkey brain). So if the same retinotopic region of visual cortex is activated when subjects generate visual images, it would tend to suggest that (1) the early visual system is involved in visual mental imagery, and (2) during imagery the cognitive system provides topographically organized inputs such as those assumed to be normally provided by the eyes—in other words we generate a display that is laid out in a spatial or "depictive" form (i.e., like a 2D picture). This interpretation was also assumed to be supported by the finding (Kosslyn et al., 1995) that "smaller" images generated more activity in the posterior part of the medial occipital region and "larger" images generated more activity in the anterior parts of the region, a pattern that is similar to the activation produced by small and large retinal images, respectively.

There are plenty of both empirical and logical problems with this argument[5] which I will not address in this chapter (but do address in Pylyshyn, submitted). For present purposes, I will put aside these (often quite serious) concerns and assume that the conclusions reached by the authors of these recent studies are valid and that not only is the visual system involved in mental imagery but also (1) a retinotopic picture-like display is generated in visual cortex during imagery, and (2) it is by means of this spatial display that images are processed and patterns "perceived" in mental imagery. In other words I will assume that mental images literally correspond to 2D displays projected onto primary visual cortex, there to be reperceived by the visual system in the course of reasoning about imaginary situations. We can then ask whether such a conclusion would help explain the large number of empirical findings concerning mental imagery (e.g., those described in Kosslyn, 1980) and thus help to clarify the nature of mental imagery. The purpose of this exercise is mainly to make the point that neuroscience evidence has no more claim to primacy in resolving disputes concerning mental processes than does behavior evidence, and indeed neuroscience evidence is of little help in clarifying conceptually ill-posed hypotheses, such as those being considered in the research on mental imagery.

What if We Really Found Pictures in Primary Visual Cortex?

Note that what the neuropsychological evidence has been taken to support is the literal picture-in-the head story that people over the years have tried to avoid. It is no accident that the search for concrete biological evidence for the nature of mental imagery should have led us to this literal view. First of all, our search for neural evidence for the form of a representation can be no better than the psychological theory that motivates it. And the motivation all along has been the intuitive picture view. Even though many writers deny that the search is for a literal 2D display (e.g., see Denis and Kosslyn, 1999), the questions being addressed in this research show that it is the literal view of images as 2D somatotopic displays that is driving this work. Secondly, if we were looking for support for a descriptivist view, it is not clear what kind of neural evidence we would look for. We have no idea at all how codes for concepts or sentences in mentalese might be instantiated in the brain. Even in concrete, apparently well-understood systems like computers, searching the physical properties for signs of data structures would be hopeless. Similarly, if our search was for a "functional space," which some people have suggested as the basis for images, we would still have no idea what to look for in the brain to confirm such a hypothesis. It is because one is searching for a literal 2D display that the research has focused on showing imagery-related activity in cortical area 17 because this area is known to be, at least in part, topographically mapped.

The kind of story being pursued is clearly illustrated by the importance that has been attached to the finding described in Tootell, Silverman, Switkes, and de Valois (1982). In this study, macaques were trained to stare at the center of a target-like pattern consisting of flashing lights, while the monkeys were injected with radioactively tagged 2-deoxydextroglucose (2-DG), whose absorption is known to be related to metabolic activity. The animal was then sacrificed and a map of metabolic activity in its cortex was developed. This 2-DG map in primary visual cortex showed an impressive retinotopic display of the pattern that the monkey had been staring at, with only cortical magnification distorting the original pattern. In other words, it showed a picture in visual cortex of the pattern that the monkey had received on its retina, written in the ink of metabolic activity. This has led many people to believe that we now know that a picture in primary visual cortex appears during visual perception and constitutes the basis for vision. Although no such maps have been found for imagery, there can be no doubt that this is what the picture-theorists believe is there and is responsible for both the imagery experience and the empirical findings reported when mental images are being used. I have gone into these details because many people who cite the neuroscience results nevertheless continue to deny that they believe in the literal picture view. But the lines of inference that are marshaled in the course of discussing the evidence clearly belie this denial. And it is just as well, since there is no other coherent way to interpret talk about images being "depictive."

So we seem to be confronted with the notion, which is claimed to be supported by neurophysiological data, that when we entertain an image we construct a literal picture in our primary visual cortex which, in turn, is manipulated by our cognitive system and examined by our visual system. Given how widespread this view has become one ought to ask whether it makes sense on internal grounds and how well it fits the large body of data that has been accumulated over the past 30 years concerning the psychophysics of mental imagery. There are several problems with this literal picture view.

First of all, if images correspond directly to (or are isomorphic to) topographically organized pictorial patterns of activity in the visual cortex, this pattern would have to be three-dimensional (3D) to account for the imagery data. After all, the content and function (as well as the phenomenology) of images is clearly 3D; for example, the same mental scanning results are obtained in depth as in 2D (Pinker, 1980) and the phenomenon of "mental rotation"—one of the most popular demonstrations of visual imagery—is indifferent as to whether rotation occurs in the plane of the display or in depth (Shepard and Metzler, 1971). Should we then expect to find 3D displays in the visual cortex? The retinotopic organization of the visual cortex is not 3D in the way required (e.g., to explain scanning and rotation in depth). The spatial properties of the perceived world are not reflected in a volumetric topographic organization in the brain; as one penetrates deeper into the columnar structure of the cortical surface one does not find a representation of the third dimension of the scene. In fact, images are really multidimensional, insofar as they represent other spatially registered properties besides spatial patterns. For example, they represent color and luminance and motion. Are these also to be found displayed on the surface of the visual cortex?

Secondly, part of the argument for the view that a mental image consists of a topographic display in visual cortex is that the same kind of 2D cortical pattern plays a role in vision, so the visual system can serve the dual role of examining the display in vision as well as in imagery. But it is more than a little dubious that visual processing involves examining such a 2D display of information. It may well be that the visual cortex is organized retinotopically, but nothing follows from this about the form of the functional mental representations involved in vision. After all, we already knew that vision started with a 2D display of activity on the retina, yet nobody supposed that we could infer the nature of our cognitive representation of the perceptual world from this fact. The inference from the physical structure of activity in the brain to the form of its functional representations is no more justified than would the parallel inference from a computer's physical structure to the form of its data structures. From a functional perspective, the argument for the involvement of a picture-like structure in visual processing is at least as problematic as the argument that such a structure is involved in mental imagery. Moreover, the fact that our phenomenal percepts appear to be laid out in a phenomenal space is irrelevant because we do not see our internal representation; we see the world as represented and it is the world we see that appears to us to be laid out in space, and for a very good reason—because it is! The evidence is now quite clear that the assumption of an inner display being constructed in visual processing is simply untenable

(O'Regan, 1992; Pylyshyn, in preparation). Years of research on transsaccadic integration have shown that our percepts are not built up by superimposing the topographic information from individual glances onto a global image; indeed, very little information is even retained from glance to glance and what is retained appears to be much more schematic than any picture (Irwin, 1996).

Thirdly, the idea that either vision or mental imagery involves examining a topographic display also fails to account for the fact that examining and manipulating mental images is qualitatively different from manipulating pictures in many significant ways. For example, it is the conceptual rather than graphic complexity of images that matters to how difficult an image superposition task is (see Palmer, 1977) and how quickly shapes are mentally "rotated" (see Pylyshyn, 1979).[6] Although we appear to be able to reach for imagined objects, there are significant differences between our motor interaction with mental images and our motor interaction with what we see (Goodale, Jacobson, and Keillor, 1994). Also accessing information from a mental image is very different from accessing information from a scene, as many people have pointed out. To take just one simple example, we can move our gaze as well as make covert attention movements relatively freely about a scene, but not on a mental image. Try writing down a 3 × 3 matrix of letters and read them in various orders. Now memorize the matrix, form an image of it, and try reading the letters on it. Unlike in the physical 2D display, some orders (e.g., the diagonal from the bottom left to the top right cell) are extremely difficult to scan on the image. If one can scan one's image the way it is alleged one does in the map-scanning experiment (Kosslyn, Ball, and Reiser, 1978), one should be able to scan the matrix equally freely. Moreover, images do not have most of the signature properties of early vision; if we create images from geometric descriptions we do not find such phenomena as spontaneous interpretation of certain 2D shapes as representing 3D objects, spontaneous reversals of bistable figures, amodal completion or subjective contours, and visual illusions, as well as the incremental construction of visual interpretations and reinterpretations over time as different aspects are noticed, and so on.[7]

I would turn the claims about the parallels between vision and imagery around and suggest that the fact that the parallel between processing mental images and processing diagrams appears to be so close renders this entire line of argument suspect, given that a real diagram viewed by the eye is very different from a mental entity being inspected by the mind. For example, some of the psychophysical evidence cited in support of a parallel between vision and mental imagery appears to attribute to the "mind's eye" many of the properties of our own eyes. Thus, it seems that the mind's eye has a visual angle like that of a real eye (Kosslyn, 1978) and that it has an acuity profile that is also the same as our eyes; it drops off with eccentricity according to the same function and inscribes the same elliptical resolution profile as that of our (real) eyes (Finke and Kosslyn, 1980; Finke and Kurtzman, 1981), and it even exhibits the "oblique effect" wherein the discriminability of closely spaced horizontal and vertical lines is superior to that of oblique lines (Kosslyn et al., 1999). Since, in the case of the eye, such properties are due primarily to the structure of our retinas and to the distribution of early receptors, these findings seem to suggest that the "mind's eye" has a similarly structured "mind's retina." Does the mind's eye then have a blind spot as well? Of course, these close parallels could be just a coincidence, or it could be that the distribution of neurons and connections in the visual cortex comes to reflect the type of information it receives from the eyes. But it is also possible that such phenomena reflect what people have implicitly come to know about how things generally look to them, a knowledge which the experiments invite them to use in simulating what would happen in a visual situation that parallels the imagined one. Such a possibility is made all the more plausible inasmuch as the instructions in these imagery experiments invite observers to "imagine" that they are looking at a certain situation and to imagine what they would see (e.g., what they would see in their peripheral vision). The fact that subjects often profess ignorance of what would happen does not establish that they do not have tacit knowledge or simply an implicit memory of similar cases that they have encountered before (see note 2).

The picture that we are being presented, of a mind's eye gazing upon a display projected onto the visual cortex, is one that should arouse our suspicion. It comes uncomfortably close to the idea that properties of the external world, as well as of the process of vision (including the resolution pattern of the retina and the necessity of moving one's eyes around the display to foveate features of interest), are built into the imagery system. If such properties were built in, our imagery would not be as plastic and cognitively penetrable as it is. We can, after all, imagine almost any properties and dynamics we like, whether or not they are physically possible, so long as we know what the situation we are imagining would look like (we can't imagine a four-dimensional world because we lack precisely this type of knowledge about it; we don't know where the contours, occlusions, shadows, etc. would fall). The picture-theory also does not even hint at a possible neural or information-processing basis for most of the interesting phenomena of mental imagery uncovered over the past several decades, such as the efficacy of visual mnemonics, the phenomena of mental rotation, and the apparent close parallels between how things work in the world and how we imagine them to work—which makes it possible for us to plan by visualizing a process and its outcomes. The properties exhibited by our imagery do not arise by magic: if we have false beliefs about how things work, our images will exhibit false dynamics. This is exactly what happens when we imagine light of different colors being mixed, or when we imagine an object in free fall. Because most people tacitly believe in the Aristotelian mechanics of constant-velocity free fall, our imagining of free fall is inaccurate and can be shown to follow a constant-velocity trajectory (for more such examples, see Pylyshyn, 1981).

Where Do We Stand Now?

Where, then, does the "imagery debate" stand at present? As I suggested at the beginning of this chapter, it all depends on what you think the debate is about. If it is supposed to be about whether reasoning using mental imagery is somehow different from reasoning without it, who can doubt that the answer must be yes? If it is about whether in some sense imagery involves the visual system, the answer there too must be affirmative, since imagery involves experiences similar to those produced by (and, as far as we know, only by) activity in some part of the visual system (though not in V1, according to Crick and Koch, 1995). The big question, of course, is what part of the visual system is involved and in what way? Answering that will require a better psychological theory of the decomposition of the visual system itself. It is much too early and much too simplistic to claim that the way the visual system is deployed in visual imagery is by allowing it to look at a reconstructed retinotopic input of the sort that comes from the eye.

Is the debate, as Kosslyn claims, about whether images are depictive rather than descriptive? That all depends on what you mean by "depictive." Is any representation that encodes geometric, spatial, metrical, or visual properties depictive? If so, then any description of how something looks, what shape and size it is, and so on, is thereby depictive. Does being depictive require that the representation be organized spatially? That depends on what restrictions are placed on "being organized spatially." Any physically instantiated representation is organized spatially—certainly both computer memories and books are. Does being depictive require that images "preserve metrical spatial information," as has been claimed (Kosslyn et al., 1978)? Again, that depends on what it means to "preserve" metrical space. If it means that the image must represent metrical spatial information, then any form of representation will have to do that to the extent that people do encode them. But any system of numerals, as well as any analog medium, can represent magnitudes in a useful way. If the claim that images preserve metrical spatial information means that an image uses spatial magnitudes directly to represent spatial magnitudes, then this is a form of the literal picture-theory. And a literal picture requires not only a visual system but a literal mind's eye because the input is an uninterpreted layout of features.

Is there an intermediate position that we can adopt, somewhere between imagery being a symbolic representation and being a picture? This sort of representation has been the holy grail of many research programs, especially in artificial intelligence. In the case of mental imagery, the hope has been that one might develop a coherent proposal which says, in effect, that in mental imagery the visual system (or some early stage in the visual system) receives retinotopically organized information that is nonetheless more abstract (or more conceptual) than a picture, but that still preserves a measure of spatial isomorphism. There is no principled reason why such a proposal could not work, if it could be properly fleshed out. But so far as I am aware nobody has even come close to making a concrete proposal for a type of representation (or a representational language) in which geometric relations are encoded geometrically while other properties retain their symbolic force. Schemata such as the mental models many people have discussed *represent* spatial relations but do not have them. To have a geometric relation would presumably require that the representation be laid out in some spatial medium, which gets us right back to the literal display view. The geometric properties encoded in this way would then have to be cognitively impenetrable since they would be part of the fixed architecture. In any case this sort of "spatial schema" view of mental images would no longer be "depictive" in the straightforward intuitive sense. It would be more like a traditional semantic network or a schema, except that geometric relations would be encoded in terms of spatial positions in some medium. Such a representation would have to be "read" just the way that sentences are read, except perhaps that proximity in the representation would have a geometric interpretation (note that written sentences too are typically encoded spatially, yet they do not use the space except to individuate and order the words). Moreover, such a spatial schema is unlikely to provide an account of such empirical phenomena as the ones described earlier—for example, where smaller images take longer to see and distant places on an image take longer to scan to. But that is just as well since these are just the sorts of phenomena that are unlikely to be attributable to the nature of the image but to the knowledge that people have about the perceived world functions.

Acknowledgment

This research was supported in part by NIMH grant 1R01-MH60924.

Notes

1. For the sake of expository simplicity I will refer to the set of ideas motivated by the lure of the phenomenology of pictures-in-the-head as picture-theories, recognizing that this may be a very heterogeneous set of theories.
2. There has been a great deal of misunderstanding about the notion of tacit knowledge (or tacit beliefs). It is a perfectly general property of knowledge that it can be tacit, or not consciously available (Fodor, 1968). We can have knowledge about various aspects of the social and physical world that (1) qualifies as real knowledge, in the sense that it can be shown to enter into general inferences and to account for a wide range of behaviors, and (2) can't be used to answer a

direct question. Our knowledge of intuitive physics is of this sort. We have well-developed intuitions about how objects will fall, bounce, and accelerate, even though we very often cannot answer abstract questions about it (and indeed we often hold explicit beliefs that contradict our tacit beliefs as shown by the way we act toward these objects). Our knowledge of the rules of grammar and of social interaction are another example of tacit knowledge that we are generally incapable of explicitly reporting, although we can typically show that we know them through indirect means. Even the knowledge of something that is explicitly taught, such as the procedure for adding numbers, is tacit—the rules we might give are generally not the ones that play the causal role in our numerical calculations (Van Lehn, 1990).

3. Here is another experiment reported in Pylyshyn (1981). Imagine a map with lights at all the critical features, of which only one is illuminated at one time. That single illuminated light goes off at a specified time (when the trial begins). Imagine that whenever a light goes off at one place, another simultaneously goes on at another place. Now indicate when you see the next light come on in your image. In such an experiment there is no increase in the time to report seeing the light coming on as a function of the distance between lights.

4. It should be mentioned in passing that there is an important caveat to be made here concerning cases in which imagery is studied by having subjects "project" their image onto a visible scene, which includes the vast majority of mental imagery studies. In these cases there is a real space, complete with properties of rigidity and stability, in which all the Euclidean axioms hold. If subjects are able to think demonstrative thoughts such as "this" and "that," as I have elsewhere claimed they can (Pylyshyn, 2001), and to bind imagined properties to those visible features, then there is a real literal sense in which the spatial properties of the scene are inherited by the image. For example, in such a situation subjects can literally move their eyes to real places where they think of certain imagined properties as being located. Thus it is easy to see how one might get a distance-scanning effect, as well as other spatial effects (like noticing that one of the imagined objects lies between two other imagined objects). Many imagery results fall out of such a use of a real spatial backdrop (Pylyshyn, 1989; Pylyshyn, 2001).

5. For example, there is serious doubt about the main premise of the argument, namely that primary visual cortex is essential for mental imagery, since there are clear dissociations between imagery and vision—even early vision—as shown by both clinical and neuroimaging data. And even if topographically organized areas of cortex were involved in imagery, the nature of the topographic layout of the visual cortex is not what we would need to explain such results as the effect of different image sizes on time to detect visual features (e.g., larger images do not generate larger regions of activity, but only activity in different areas—areas that project from the periphery of the eyes—contrary to what would be required to explain the image-size or zoom effect, e.g., in the way it is explained in models such as that of Kosslyn, Pinker, Smith, and Shwartz, 1979).

6. It will not surprise the reader to hear that there are many ways of patching up a picture-theory to accommodate such findings. For example, one can add

assumptions about how images are tagged as having certain properties (perhaps including depth) and how they have to be incrementally refreshed from nonimage information stored in memory, and so on, thus providing a way to bring in conceptual complexity through the image generation function. With each of these accommodations one gives the actual image less and less explanatory work until eventually one reaches the point where the pictorial nature of the display becomes a mere shadow of the mechanism that does its work elsewhere, as when the behavior of an animated computer display is determined by an encoding of the principles that govern the animation.

7. A great deal of research has been devoted to such questions as whether mental images can be ambiguous and whether we can make new construals of images constructed by combining other images. In my view the preponderance of evidence shows that the only reconstruals that are possible are not visual ones but inferences based on information about shape which could be encoded in any form. These arguments are discussed in Pylyshyn (submitted).

References

Crick, F., and Koch, C. (1995). Are we aware of neural activity in primary visual cortex? *Nature, 375,* 121–123.

Denis, M., and Kosslyn, S. M. (1999). Scanning visual mental images: A window on the mind. *Cahiers de Psychologie Cognitive / Current Psychology of Cognition, 18.*

Finke, R. A., and Kosslyn, S. M. (1980). Mental imagery acuity in the peripheral visual field. *Journal of Experimental Psychology: Human Perception and Performance, 6,* 126–139.

Finke, R. A., and Kurtzman, H. S. (1981). Mapping the visual field in mental imagery. *Journal of Experimental Psychology: General, 110,* 501–517.

Fodor, J. A. (1968). The appeal to tacit knowledge in psychological explanation. *Journal of Philosophy, 65,* 627–640.

Fodor, J. A. (1975). *The language of thought.* New York: Crowell.

Fodor, J. A., and Pylyshyn, Z. W. (1988). Connectionism and cognitive architecture: A critical analysis. *Cognition, 28,* 3–71.

Goodale, M. A., Jacobson, J. S., and Keillor, J. M. (1994). Differences in the visual control of pantomimed and natural grasping movements. *Neuropsychologia, 32,* 1159–1178.

Irwin, D. E. (1996). Integrating information across saccadic eye movements. *Current Directions in Psychological Science, 5,* 94–100.

Kosslyn, S. M. (1978). Measuring the visual angle of the mind's eye. *Cognitive Psychology, 10,* 356–389.

Kosslyn, S. M. (1980). *Image and mind.* Cambridge, MA: Harvard University Press.

Kosslyn, S. M. (1994). *Image and brain: The resolution of the imagery debate.* Cambridge, MA: MIT Press.

Kosslyn, S. M., Ball, T. M., and Reiser, B. J. (1978). Visual images preserve metric spatial information: Evidence from studies of image scanning. *Journal of Experimental Psychology: Human Perception and Performance, 4,* 46–60.

Kosslyn, S. M., Pascual-Leone, A., Felican, O., Camposano, S., Keenan, J. P., Thompson, W. L., Ganis, G., Sukel, K. E., and Alpert, N. M. (1999). The role of area 17 in visual imagery: Convergent evidence from PET and rTMS. *Science, 284,* 167–170.

Thompson, W. L., Ganis, G., Sukel, K. E., and Alpert, N. M. (1999). the role of area 17 in visual imagery: Convergent evidence from PET and rTMS. *Science, 284,* 167–170.

Kosslyn, S. M., Pinker, S., Smith, G., and Shwartz, S. P. (1979). On the demystification of mental imagery. *Behavioral and Brain Science, 2,* 535–548.

Kosslyn, S. M., Thompson, W. L., Kim, I. J., and Alpert, N. M. (1995). Topographical representations of mental images in primary visual cortex. *Nature, 378,* 496–498.

O'Regan, J. K. (1992). Solving the "real" mysteries of visual perception: The world as an outside memory. *Canadian Journal of Psychology, 46,* 461–488.

Paivio, A. (1971). *Imagery and verbal processes.* New York: Holt, Reinhart, & Winston.

Palmer, S. E. (1977). Hierarchical structure in perceptual representation. *Cognitive Psychology, 9,* 441–474.

Pinker, S. (1980). Mental imagery and the third dimension. *Journal of Experimental Psychology: General, 109,* 354–371.

Pylyshyn, Z. W. (1973). What the mind's eye tells the mind's brain: A critique of mental imagery. *Psychological Bulletin, 80,* 1–24.

Pylyshyn, Z. W. (1979). The rate of "mental rotation" of images: A test of a holistic analogue hypothesis. *Memory and Cognition, 7,* 19–28.

Pylyshyn, Z. W. (1981). The imagery debate: Analogue media versus tacit knowledge. *Psychological Review, 88,* 16–45.

Pylyshyn, Z. W. (1984). *Computation and cognition: Toward a foundation for cognitive science.* Cambridge, MA: MIT Press.

Pylyshyn, Z. W. (1989). The role of location indexes in spatial perception: A sketch of the FINST spatial-index model. *Cognition, 32,* 65–97.

Pylyshyn, Z. W. (1999). Is vision continuous with cognition? The case for cognitive impenetrability of visual perception. *Behavioral and Brain Sciences, 22*(3), 341–423.

Pylyshyn, Z. W. (in preparation). *Seeing: It's not what you think.*

Pylyshyn, Z. W. (2001). Visual indexes, preconceptual objects, and situated vision. *Cognition.*

Pylyshyn, Z. W. (submitted). Mental imagery: In search of a theory.

Pylyshyn, Z. W., and Cohen, J. (1999). Imagined extrapolation of uniform motion is not continuous. Presented at the Annual Conference of the Association for Research in Vision and Ophthalmology, Fort Lauderdale, FL.

Shepard, R. N. (1975). Form, formation, and transformation of internal representations. In R. L. Solso (Ed.), Information processing in cognition: The Loyola symposium. Hillsdale, NJ: Erlbaum.

Shepard, R. N., and Feng, C. (1972). A chronometric study of mental paper folding. *Cognitive Psychology, 3,* 228–243.

Shepard, R. N., and Metzler, J. (1971). Mental rotation of three dimensional objects. *Science, 171,* 701–703.

Tootell, R. B., Silverman, M. S., Switkes, E., and de Valois, R. L. (1982). Deoxyglucose analysis of retinotopic organization in primate striate cortex. *Science, 218,* 902–904.

VanLehn, K. (1990). *Mind bugs: The origins of procedural misconceptions.* Cambridge, MA: MIT Press.

5
Mental Models and Human Reasoning

Philip N. Johnson-Laird

How do human beings reason? One answer, which has many adherents, is that they follow the "laws of thought." These laws are made explicit in formal logic, the probability calculus, and the theory of rational decision-making. Individuals can certainly learn such laws, and use them in a conscious way to solve problems. But, how do naive individuals, who have had no such training, reason? They are nonetheless following the laws of thought according to some theorists (Rips, 1994; Braine and O'Brien, 1998). Although they are not aware of following the laws, they may have acquired them in some way. Empirical observations, however, provide grounds for skepticism about this thesis.

On the one hand, intelligent individuals make mistakes in reasoning in everyday life. A lawyer in a civil action, for example, got an opposing expert witness to admit the following two facts concerning a toxic substance, tricholoroethylene (TCE):

If TCE came down the river (from other factories), then TCE would be in the river bed.

TCE was not in the river bed.

At this point, the lawyer asked the witness:

The pattern is consistent with the fact that no TCE came from the river, isn't it?

What he should have pointed out is that this conclusion follows necessarily from the premises. Neither the lawyer nor the author of the book from which this report comes appears to have noticed the mistake (see Harr, 1995, pp. 361–362). On the other hand, robust observations in experiments cast doubt on the psychological reality of the laws of thought. What is particularly embarrassing for the hypothesis is that people make systematic and predictable errors in reasoning. The laws of thought, by definition, do not allow for errors of this sort. They yield only valid inferences, and so whatever errors occur should be sporadic and haphazard. An alternative theory, however, does predict systematic errors in reasoning.

The Theory of Mental Models

The conjecture that thinking depends on internal models of the external world was made by the late Kenneth Craik (1943). Craik's proposal was prescient but programmatic. Recent accounts, however, postulate that when people understand discourse they construct mental models of the relevant situations, and that when they reason they seek to draw conclusions that are supported by these models (Johnson-Laird and Byrne, 1991). The most frequently asked question about the recent theory is, What is a mental model? And the most frequent misunderstanding is that it is a visual image. In fact, a mental model is an internal representation of a possibility. It has a structure and a content that captures what is common to the different ways in which the possibility could occur. Skeptics might argue that there could not be a mental model of abstract discourse. But you can envisage an abstract situation, and then struggle to put it into words. Similarly, the phenomena of reasoning appear much the same for inferences based on abstract content. Hence, individuals appear to represent abstract content in some way that is independent of its verbal description.

In this chapter I will not pursue the details of individual models any further (but cf. Garnham and Oakhill, 1996), because its chief concern is inferences from sets of alternative models. Each model represents a possibility, and so a conclusion is necessary if it holds in all the models of the premises, possible if it holds in at least one model of the premises, and impossible if it holds in none of the models of the premises. The theory accordingly applies to modal reasoning, that is, reasoning about what is necessary and what is possible. It also extends to probabilistic reasoning. But before we consider its account of probabilities, we need to add some substance to the theory's skeleton and to describe its application to so-called sentential reasoning.

Sentential Reasoning

Sentential reasoning, which is sometimes known as "propositional" reasoning, hinges on negation and such connectives as "if," "or," and "and." These connectives have idealized meanings in logic, so that the truth-values of sentences formed with them depend solely on the truth-values of those clauses that they interconnect. Consider, for example, an exclusive disjunction of the form

There is a table in the room or else there isn't a chair

where the two clauses cannot both be true. The disjunction is true if the first clause is true and the second clause is false, and if the first clause is false and the second clause is true. In any other case, it is false.

Mental models are less explicit than this analysis, because the model theory rests on the following fundamental principle of truth:

Mental models normally represent only what is true, that is, they represent only true possibilities and within each such possibility they represent clauses in the premises only when they are true.

The advantage of the principle of truth is that it reduces the processing load on working memory. The principle implies that the preceding disjunction has two mental models:

table
  $\neg$ chair

where "table" denotes a model of the possibility in which there is a table in the room, "$\neg$" stands for negation, and so "$\neg$ chair" denotes a model of the possibility in which there isn't a chair in the room. The principle of truth has a further, less obvious, consequence. When people think about the first possibility, they neglect the fact that it is false that there is not a chair, that is, that there *is* a chair in this possibility. Likewise, when they think about the second possibility, they neglect the fact that it is false there is a table in this case.

The principle of truth postulates that individuals normally represent what is true, not what is false. It does not imply, however, that they never represent falsity. Indeed, the theory proposes that they represent what is false in "mental footnotes," but that these footnotes are ephemeral. As long as footnotes are remembered, they can be used to construct fully explicit models, which represent the true possibilities in a fully explicit way. Hence, the mental footnotes about what is false allow reasoners to flesh out their models of the disjunction to make them fully explicit:

   table   chair
¬ table   ¬ chair

These models are also the fully explicit models for the biconditional:

There is a table in the room if and only if there is a chair.

Few people notice immediately that the disjunction is equivalent to the biconditional, because they do not automatically flesh out their models to make them fully explicit.

   The meaning of conditional assertions, such as

If there is a table then there is a chair

is controversial. Their meanings depend both on their context of use and on the semantic relations, if any, between their two clauses—the antecedent clause following "if," and the consequent clause following "then" (see Johnson-Laird and Byrne, 2000). Roughly speaking, when the meaning of a conditional is independent of its context, and its antecedent and consequent have no coreferential terms, then the conditional has two mental models. One model represents the salient possibility in which both the antecedent and the consequent are true. The other model is wholly implicit, that is, it has no explicit content, but allows for possibilities in which the antecedent of the conditional is false. The mental models for the preceding conditional are, accordingly,

   table   chair
     . . .

where the ellipsis denotes the implicit model, and a mental footnote indicates the falsity of the antecedent in the implicit possibilities. A biconditional, such as

If, and only if, there is a table then there is a chair

has exactly the same mental models, but a footnote indicates that both the antecedent and the consequent are false in the possibilities that the implicit model represents. It is the implicit model that distinguishes a conditional from a conjunction, such as

There is a table and there is a chair

which elicits only a single model:

table   chair

Table 5.1 presents the mental models and the fully explicit models of propositions formed from the main sentential connectives.

How are inferences made with mental models? An example illustrates the process:

There is a table or else a chair, but not both.
There is not a table.

What follows?

The disjunction yields the mental models:

table
      chair

The categorical premise yields the model:

¬ table

Table 5.1
Models for the Sentential Connectives

Connective	Mental Models	Fully Explicit Models
A and B	A B	A B
A or else B	A	A ¬B
	B	¬A B
A or B or both	A	A ¬B
	B	¬A B
	A B	A B
If A then B	A B	A B
	...	¬A B
		¬A ¬B
If and only if A then B	A B	A B
	...	¬A ¬B

Note: The central column shows the mental models postulated for human reasoners, and the right-hand column shows fully explicit models, which represent the false components of the true cases by using negations that are true: '¬' denotes negation and '. . .' denotes a wholly implicit model. Each line represents a model of a possibility.

Table 5.2
The Procedures for Forming a Conjunction of Two Sets of Possibilities

1. For a pair of explicit items, the result conjoins their elements, and drops any duplicates:

a b and b c yield a b c

2. For a pair of items that contain an element and its contradication, the result is null (akin to the empty set):

a b and ¬ b c yield null

3. For null combined with any item, the result is null:

null and a b yield null

4. For a pair of implicit items, the result is implicit:

. . . and . . . yield . . .

5. For an implict item combined with an explicit one, the result by default is null:

. . . and b c yield null

But, if the explicit item contains no element in common with anything in the same set from which the implicit item is drawn, then the result is the explicit item:

. . . and b c yield b c

Note: The procedures apply either to individual models (based on sentential connectives) or to possible individuals (based on quantified assertions). Each procedure is presented with an accompanying example. In principle, each procedure should take into account mental footnotes, but reasoners soon forget them. Only procedures 1, 2, and 3 are required for fully explicit models.

This model eliminates the first model of the disjunction because their truth conditions cannot be jointly satisfied. But it is consistent with the second model of the disjunction, which yields the conclusion:

There is a chair.

This conclusion is valid, that is, it is necessarily true given the truth of the premises, because it holds in all the models—in this case, the single model—consistent with the premises.

The procedures by which mental models are combined to make inferences are summarized in table 5.2. A computer program implementing these procedures operates at three additional more advanced levels of expertise taking mental footnotes into account, and, at its most advanced level working with fully explicit models. It is likely that individuals do not merely combine models of premises, but update their models of an existing premise in the light of subsequent premises. The principles in table 5.2 apply both to the combination of sets of models, as in the preceding disjunctive inference, and to the combination of possible individuals in models of quantified assertions, which are presented in the next section. The principles may suggest that the process of reasoning is a simple deterministic process that unwinds like clockwork. Not so. Thinking and reasoning are governed by constraints, such as the principles in table 5.2. But, there is seldom, if ever, just a single path that they must follow. As computer scientists say, the process is "nondeterministic," that is, the model theory does not postulate a single fixed sequence of mental operations that must be followed as soon as reasoning gets under way. Indeed, there is evidence that naive reasoners develop a variety of strategies (see Johnson-Laird, Savary, and Bucciarelli, 1999).

Reasoning with Quantifiers

The quantified assertion

All the tables are with chairs

has a single mental model representing the different sorts of entities in the situation:

table   chair
table   chair
   . . .

This diagram, unlike those for connectives, denotes one model of a single situation. The number of tokens in the model is small, but arbitrary. Its first two rows show that there are tables with chairs, and a mental footnote indicates that the ellipsis represents other possible entities, not including tables, in the situation. The footnote can be used to flesh out the model in various ways by making the other possible entities explicit, bearing in mind the footnote that tables are already exhaustively represented in the model. One such case is

   table   chair
   table   chair

¬ table   chair
¬ table   ¬ chair

The quantified assertion

Some of the tables are with chairs

has the single mental model:

table   chair
table   chair
. . .

in which the explicit pairs exist in the situation. Neither tables nor chairs are exhaustively represented in the model, and so it can be fleshed out explicitly to represent the possibility of any other sort of pair, for example,

table   chair
table   chair
table   ¬ chair
¬ table   ¬ chair

The preceding assertion is false in case

None of the tables are with chairs

which has the mental model:

table
table
    chair
    chair
. . .

in which both chairs and tables are exhaustively represented.

Quantified reasoning becomes complicated when assertions contain more than one quantifier, for example,

Some psychologists have read every issue of *Cognition*.
Therefore, every issue of *Cognition* has been read by some psychologists.

But the theory has been extended to deal with such inferences, and experimental evidence corroborates its account (see Johnson-Laird and Byrne, 1991).

Modal Reasoning

Modal reasoning is about necessary, possible, and impossible cases. Logicians have developed many modal logics, and they have proved that there are a potentially infinite number of them. None of these logics, however, appears to correspond to the modal notions of everyday life. The difficulty was identified by Karttunen (1972). He pointed out the bizarre nature of the following sort of assertion:

There isn't a table in the room, but there may be.

And he suggested that it violates the following informal principle:

What is cannot possibly be otherwise.

This principle corresponds to the following assertion in which A is a variable that can take any proposition as its value, and 'pos' is the modal operator, "possibly":

¬(A & pos¬A)

Hence, the formula asserts that it is not the case that A and possibly not A. The formula is equivalent in modal logic to the implication

If A then nec A

where "nec" is the modal operator, "necessarily." That is, whatever is true is necessarily true. But if this principle were added to an orthodox modal logic, it would have a disastrous effect: modal operators would cease to have any force whatsoever, because A would be equivalent to necessarily A. The way out of this problem is to recognize a critical difference between modal logic and the everyday use of modality. Modal logic recognizes only one modality of possibility. Ordinary language, however, distinguishes between real possibilities and counterfactual possibilities (Johnson-Laird and Byrne, 1991; Byrne, 1997). Real possibilities are events that may occur given the actual state of the world. Counterfactual possibilities were once real possibilities, but are so no longer because they did not occur:

George Bush won the Presidential election in 1992.

The way in which to combine a fact and a contrary possibility is accordingly by way of a counterfactual:

There isn't a table in the room, but there could have been.

Here is a typical modal inference in daily life:

There is a table in the room or there is a chair in the room, or both. Therefore, it is possible that there's both a table and a chair in the room.

This inference illustrates the difficulty of proving modal arguments using formal rules of inference. Osherson (1974–76) formulated some rules for modal reasoning, but his system, as he recognized, is not complete. It does not include a rule of a form required by the inference above:

A or B, or both.
Therefore, pos (A & B)

Indeed, no modal logic should include such a rule, because, as Geoffrey Keene (personal communication, 1997) has pointed out, if A implies not B, then the result of the rule would be a conclusion stating that a self-contradiction is possible:

pos (¬ B & B)

What is psychologically simple thus turns out to be surprisingly complex in a modal logic.

The mental model theory of modal reasoning provides a simple account of the preceding inference (Bell and Johnson-Laird, 1998). The premise:

There is a table or there is a chair, or both

yields the following mental models:

table
      chair
table   chair

and the third of these models supports the conclusion:

Possibly, there is both a table and a chair in the room.

In general, a conclusion is possible provided at least one model of the premises satisfies it.

Probabilistic Reasoning

Extensional reasoning about probabilities occurs when individuals infer the probability of an event from a knowledge of the different independent ways in which the event might occur. In the past, it has been studied in experiments on so-called Bayesian reasoning in which the participants have to infer a conditional probability from information given to them.

These inferences are sophisticated, and hitherto there was no account of the foundations of extensional reasoning. The model theory has recently been extended to fill this gap (Johnson-Laird et al., 1999). The theory assumes that individuals treat each mental model as equiprobable unless they have reasons to the contrary. Granted this assumption of equiprobability, the probability of a event depends on the proportion of the models in which it occurs. The theory also postulates that models can be tagged with numerals that denote probabilities or frequencies, and that simple arithmetic operations can be carried out on them.

According to the theory, a premise such as

In the box, there is a black ball or a red ball or both

elicits the mental models:

Black

      Red

Black Red

In order to answer the question:

What is the probability that in the box there is both a black ball and a red ball?

reasoners should assume that each model is equiprobable and respond on the basis of proportionality, that is, the probability is 1/3. An experiment corroborated this and other predictions based on the mental models for the various connectives shown in table 5.1 (Johnson-Laird et al., 1999).

Conditional probabilities are on the borderline of naive competence. They are difficult because to think about them properly you need to consider several fully explicit models. Consider, e.g., the following problem:

The suspect's DNA matches the crime sample. If he is not guilty, the probability of such a match is 1 in 1 million. Is he likely to be guilty?

Most people respond yes. Yet, it is an error. According to the theory, it arises because people tend to represent the conditional probability: P(match | ¬ guilty) = 1 in 1 million, by constructing one explicit model and one implicit model, which are tagged with their respective numerical frequencies:

¬ guilty   match   1
        ...   999,999

The converse conditional probability has same mental models. Hence, people tend to assume:

P(¬ guilty | match) = 1 in 1 million

They then infer that since the suspect's DNA does match the crime sample, he is likely to be guilty because the chances that he is not guilty are so tiny. In order to reach the correct conclusion, they need to envisage the complete partition, that is, the full set of mutually exclusive and exhaustive possibilities with their respective frequencies. Given the probabilities of the complete partition, naive individuals should infer the correct conditional probability by relying on the *subset* principle (Johnson-Laird et al., 1999):

Granted equiprobability, a conditional probability, P(A|B), depends on the subset of B that is A, and the proportionality of A to B yields the numerical value. If the models are tagged with their absolute frequencies or chances, then the conditional probability equals their value for the model of A and B divided by their sum for all the models containing B.

The complete partition and probabilities for the DNA problem might be as follows:

¬ guilty	match	1
¬ guilty	¬ match	999,999
guilty	match	4
guilty	¬ match	0

The first two rows show that P(match | ¬ guilty) = 1 in 1 million. If you examine the proportion within the two possibilities of ¬ guilty, the subset of matches has a probability of 1 in 1 million. But, the subset of ¬ guilty within the two possibilities of a match (rows 1 and 3 of the partition) has a probability of 1 in 1 + 4. In other words, given a DNA match, the probability that the suspect is not guilty is quite high, that is, 1/5.

The moral of this analysis is that knowledge of a conditional probability tells you almost nothing about the value of the converse conditional probability. But if you know the full partition and the respective probabilities, then you know everything necessary to infer any probability concerning the situation. This claim is important for studies of Bayesian reasoning (see, e.g., Shimojo and Ichikawa, 1989; Falk, 1992; Nickerson, 1996). A typical Bayesian problem is as follows:

The chances that Pat has a certain disease are 4 out of 10. If she has the disease, then the chances are 3 out of 4 that she tests positive. If she does not have the disease, then the chances are 2 out of 6 that she tests positive. Pat tests positive. So, what are the chances that she has the disease?

"Evolutionary" psychologists have argued that naive individuals can cope with such problems if they are stated in terms of the natural frequencies of repeated events (Gigerenzer and Hoffrage, 1995; Cosmides and Tooby, 1996). In contrast, the model theory predicts reasoning will be optimal whenever the problem is stated so that it is easy to construct the fully explicit models of the partition, to recover the subset relation for the required conditional probability, and to carry out the correct arithmetic. This prediction holds whether a problem is about the frequencies of repeated events or the chances of an individual event. In the example above about Pat, for instance, naive reasoners may well be able to construct the following partition:

disease	positive	3
disease	¬ positive	1
¬ disease	positive	2
...		4

Given that Pat tests positive (the first and third rows), the subset relation shows that Pat has 3 chances out of 5 of having the disease. Girotto and Gonzalez (in press) have corroborated the prediction. They have shown that naive participants can infer conditionals in cases that fit the model theory's predictions, regardless of whether the problems concern the frequencies of repeated events or the chances of unique events. Conversely, problems stated in terms of frequencies are difficult when it is hard to reconstruct the fully explicit partition from them. Steven Sloman (personal communication, 1999) has shown that if it is easy to construct the partition, participants can even infer the correct conditional probability when the numbers concern probabilities. In summary, there is no reason to suppose that individuals who are untrained in the probability calculus have access either to Bayes's theorem or to components of the probability calculus. They rely on the subset principle.

A Synopsis of Evidence

The model theory accounts for the reasoning of logically untrained individuals in a variety of inferential domains. The theory makes five main predictions, and each of them has been corroborated experimentally. They are in summary:

1. Inferences that depend on multiple models are harder than those that depend on single models. This prediction holds for sentential reasoning (e.g., see García-Madruga et al., in press), for modal and quantified reasoning (Bell and Johnson-Laird, 1998; Evans et al., 1999), and for reasoning about spatial and temporal relations (e.g., see Byrne and Johnson-Laird, 1989; Schaeken, Johnson-Laird, and d'Ydewalle, 1996).

2. Reasoners can establish that a conclusion is invalid by finding a counterexample, that is, a model in which its premises are true but its conclusion is false. This prediction has been corroborated for quantified reasoning (see Bucciarelli and Johnson-Laird, 1999; Neth and Johnson-Laird, 1999; Roberts, submitted).

3. Erroneous conclusions that reasoners draw for themselves tend to correspond to subsets of the models of premises, often just a single model (e.g., see Bauer and Johnson-Laird, 1993; Ormerod, Manktelow, and Jones, 1993; Bara, Bucciarelli, and Johnson-Laird, 1995; Sloutsky and Goldvarg, submitted).

4. Beliefs influence the process of reasoning: individuals who reach an unbelievable conclusion search harder for a counterexample than those who reach a believable conclusion (e.g., see Oakhill, Garnham, and Johnson-Laird, 1990; Quayle and Ball, 1997; Santamaria, García-Madruga, and Johnson-Laird, 1998; and Cherubini et al., 1998).

5. The principle of truth predicts the occurrence of illusory inferences, that is, inferences that are compelling, but invalid. For example,

One of the following assertions is true about a particular hand of cards and one of them is false:

There is a king in the hand or an ace, or both.
There is a queen and an ace in the hand.
Is it possible that there is a queen and an ace in the hand?

Nearly everyone responds yes, but the response is an illusion (Goldvarg and Johnson-Laird, 2000). If there were a queen and an ace in the hand then both of the assertions would be true, contrary to the rubric that one of them is true and one of them is false. Illusory inferences have also been demonstrated in sentential reasoning (Johnson-Laird and Savary, 1999) and quantified reasoning (Yang and Johnson-Laird, 2000).

Conclusion

Other theories of reasoning can account for some of the phenomena reviewed in the previous section, but no current theory, apart from the model theory, predicts them all. Readers should not conclude, however, that the model theory is a paragon. It has many gaps and omissions. Indeed, distinguishable variants of the model theory have been designed to make good some of these defects (see, e.g., Ormerod et al., 1993; Evans, 1993; Polk and Newell, 1995; van der Henst, 2000). If human beings followed the "laws of thought," they would make only haphazard errors, analogous to slips of the tongue. If they relied on invalid rules of inference, which would account for the illusory inferences, they would be irredeemably irrational. Yet, people understand the explanation of the illusions. They merely lack the capacity to model both truth and falsity.

Acknowledgments

Those of us who work on the psychology of thinking and reasoning owe a large debt to Jacques Mehler. He founded the journal, *Cognition*, which has been a favorite and most hospitable destination for our papers. Heartfelt thanks, Jacques, for all your long labors on the journal and best wishes for a successful and stimulating retirement.

My own work owes much to many colleagues. They include Victoria Bell, Bruno Bara, Monica Bucciarelli, Ruth Byrne, Jean-Paul Caverni, Jonathan Evans, Alan Garnham, Yevgeniya Goldvarg, Vittorio Girotto, Uri Hasson, Markus Knauff, Jim Kroger, Paolo Legrenzi, Maria Sonino Legrenzi, Juan García Madruga, Hansjoerg Neth, Mary Newsome, Jane Oakhill, David Over, Cristina Quelhas, Carlos Santamaria, Walter Schaeken, Vladimir Sloutsky, Patrizia Tabossi, Jean-Baptiste van der Henst, and Yingrui Yang. Thanks to you all, and an extra thanks to Ruth Byrne and her colleagues for: www.tcd.ie/Psychology/People/Ruth_Byrne/mental_models/.

References

Bara, B., Bucciarelli, M., and Johnson-Laird, P. N. (1995). The development of syllogistic reasoning. *American Journal of Psychology, 108,* 157–193.

Bauer, M. I., and Johnson-Laird, P. N. (1993). How diagrams can improve reasoning. *Psychological Science, 4,* 372–378.

Bell, V., and Johnson-Laird, P. N. (1998). A model theory of modal reasoning. *Cognitive Science, 22,* 25–51.

Braine, M. D. S., and O'Brien, D. P. (Eds.) (1998). *Mental logic.* Mahwah, NJ: Erlbaum.

Bucciarelli, M., and Johnson-Laird, P. N. (1999). Strategies in syllogistic reasoning. *Cognitive Science, 23,* 247–303.

Byrne, R. M. J. (1997). Cognitive processes in counterfactual thinking about what might have been. In D. K. Medin (Ed.), *The psychology of learning and motivation. Advances in research and theory.* Vol. 37 (pp. 105–54). San Diego: Academic Press.

Byrne, R. M. J., and Johnson-Laird, P. N. (1989). Spatial reasoning. *Journal of Memory and Language, 28,* 564–575.

Cherubini, P., Garnham, A., Oakhill, J., and Morley, E. (1998). Can any ostrich fly? Some new data on belief bias in syllogistic reasoning. *Cognition, 69,* 179–218.

Cosmides, L., and Tooby, J. (1996). Are humans good intuitive statisticians after all? Rethinking some conclusions from the literature on judgment under uncertainty. *Cognition, 58,* 1–73.

Craik, K. (1943). *The nature of explanation.* Cambridge, UK: Cambridge University Press.

Evans, J. St. B. T. (1993). The mental model theory of conditional reasoning: Critical appraisal and revision. *Cognition, 48,* 1–20.

Evans, J. St. B. T., Handley, S. J., Harper, C. N. J., and Johnson-Laird, P. N. (1999). Reasoning about necessity and possibility: A test of the mental model theory of deduction. *Journal of Experimental Psychology. Learning, Memory, and Cognition, 25,* 1495–1513.

Falk, R. (1992). A closer look at the probabilities of the notorious three prisoners. *Cognition, 43,* 197–223.

García-Madruga, J. A., Moreno, S., Carriedo, N., Gutiérrez, F., and Johnson-Laird, P. N. (in press). Are conjunctive inferences easier than disjunctive inferences? A comparison of rules and models. *Quarterly Journal of Experimental Psychology.*

Garnham, A., and Oakhill, J. V. (1996). The mental models theory of language comprehension. In B. K. Britton and A. C. Graesser (Eds.), *Models of understanding text* (pp. 313–339). Hillsdale, NJ: Erlbaum.

Gigerenzer, G., and Hoffrage, U. (1995). How to improve Bayesian reasoning without instruction: Frequency formats. *Psychological Review, 102,* 684–704.

Girotto, V., and Gonzalez, M. (in press). Solving probabilistic and statistical problems: A matter of information structure and question form. *Cognition.*

Goldvarg, Y., and Johnson-Laird, P. N. (2000). Illusions in modal reasoning. *Memory & Cognition, 28,* 282–294.

Harr, J. (1995). *A civil action.* New York: Random House.

Johnson-Laird, P. N., and Byrne, R. M. J. (1991). *Deduction.* Hillsdale, NJ: Erlbaum.

Johnson-Laird, P. N., and Byrne, R. M. J. (2000). Conditionals: A theory of meaning, pragmatics, and inference. Under submission.

Johnson-Laird, P. N., and Savary, F. (1999). Illusory inferences: A novel class of erroneous deductions. *Cognition, 71,* 191–229.

Johnson-Laird, P. N., Legrenzi, P., Girotto, P., Legrenzi, M. S., and Caverni, J-P. (1999). Naive probability: A mental model theory of extensional reasoning. *Psychological Review, 106,* 62–88.

Johnson-Laird, P. N., Savary, F., and Bucciarelli, M. (1999). Strategies and tactics in reasoning. In W. S. Schaeken, A. Vandierendonck, G. De Vooght, and G. d'Ydewalle (Eds.), *Deductive reasoning and strategies.* Mahwah, NJ: Erlbaum.

Karttunen, L. (1972). Possible and must. In J. P. Kimball (Ed.), *Syntax and Semantics.* Vol. 1 (pp. 1–20). New York: Seminar Press.

Neth, H., and Johnson-Laird, P. N. (1999). The search for counterexamples in human reasoning. In *Proceedings of the Twenty-First Annual Conference of the Cognitive Science Society,* p. 806.

Nickerson, R. S. (1996). Ambiguities and unstated assumptions in probabilistic reasoning. *Psychological Bulletin, 120,* 410–433.

Oakhill, J., Garnham, A., and Johnson-Laird, P. N. (1990). Belief bias effects in syllogistic reasoning. In K. Gilhooly, M. T. G. Keane, R. H. Logie, and G. Erdos, (Eds.), Lines of thinking, Vol .1, (pp. 125–38). London: Wiley.

Ormerod, T. C., Manktelow, K. I., and Jones, G. V. (1993). Reasoning with three types of conditional: Biases and mental models. *Quarterly Journal of Experimental Psychology, 46A,* 653–678.

Osherson, D. (1974–76). *Logical abilities in children.* Vols. 1–4. Hillsdale, NJ: Erlbaum.

Polk, T. A., and Newell, A. (1995). Deduction as verbal reasoning. *Psychological Review, 102,* 533–566.

Quayle, J. D., and Ball, L. J. (1997). Subjective confidence and the belief bias effect in syllogistic reasoning. In *Proceedings of the Nineteenth Annual Conference of the Cognitive Science Society,* pp. 626–631.

Rips, L. (1994). *The psychology of proof.* Cambridge, MA: MIT Press.

Roberts, M. J. (submitted). Strategies in relational inference.

Santamaría, C., García-Madruga, J. A., and Johnson-Laird, P. N. (1998). Reasoning from double conditionals: The effects of logical structure and believability. *Thinking and Reasoning, 4,* 97–122.

Schaeken, W. S., Johnson-Laird, P. N., and d'Ydewalle, G. (1996). Mental models and temporal reasoning. *Cognition, 60,* 205–234.

Shimojo, S., and Ichikawa, S. (1989). Intuitive reasoning about probability: Theoretical and experimental analyses of the "problem of three prisoners." *Cognition, 32,* 1–24.

Sloutsky, V. M., and Goldvarg, Y. (submitted). Representation and recall of determinate and indeterminate problems.

van der Henst, J-B. (2000) Mental model theory and pragmatics. *Behavioral and Brain Sciences, 23,* 283–286.

Yang, Y., and Johnson-Laird, P. N. (2000) Illusions in quantified reasoning: How to make the impossible seem possible, and vice versa. *Memory and Cognition, 28,* 452–465.

6

Is the Content of Experience the Same as the Content of Thought?

Ned Block

There are two different perspectives on consciousness that differ on whether there is anything in the phenomenal character of conscious experience that goes beyond the intentional, the cognitive, and the functional. A convenient terminological handle on the dispute is whether there are qualia, or qualitative properties of conscious experience. Those who think that the phenomenal character of conscious experience goes beyond the intentional, the cognitive, and the functional believe in qualia.[1]

The debates about qualia have recently focused on the notion of representation, with issues about functionalism always in the background. All can agree that there are representational contents of thoughts, for example, the representational content *that virtue is its own reward*. And friends of qualia can agree that experiences at least sometimes have representational content too, for example, *that something red occludes something blue*. The recent focus of disagreement is on whether the phenomenal character of experience is *exhausted* by such representational contents. I say no. Don't get me wrong. I think that sensations—almost always—perhaps even always—*have* representational content in addition to their phenomenal character. Moreover, I think that it is often—maybe always—*the phenomenal character itself* that has the representational content. What I deny is that representational content is all there is to phenomenal character. I insist that phenomenal character *outruns* representational content. I call this view "phenomenism." ("Phenomenalism" is the quite different view that the external world is in some sense composed of sensations, actual or possible; this view was held, famously, by Berkeley, and more recently by Russell.) Phenomenists believe that phenomenal character outruns not only representational content but also the functional and the cognitive; hence they believe in qualia.

This chapter is a defense of phenomenism against representationism; hence issues of reduction of the phenomenal to the functional or the cognitive will be in the background. First I discuss an internalist form of representationism; I then go on to externalist forms of the view.

Internalism

Putnam (1975) and Burge (1979) famously argued for "externalism." Externalism is perhaps most easily understood by contrast with "internalism," the view that there can be no difference in the meaning of what we say and the content of what we think without a physical difference. More exactly, if X and Y are persons (or person stages) who differ in the meaning or content of some representation, then X and Y must differ physically.[2] To put it in another way, there could not be two beings that are molecule-for-molecule duplicates that differ at all in the meaning of their terms or the contents of their thoughts. Using the term "supervenes," internalism says that meaning and content supervene on physical properties of the body.[3]

That was internalism. Externalism is the negation of internalism. According to Putnam and Burge, a person's meanings and contents depend on the person's linguistic and nonlinguistic environment. Putting it in a flamboyant way, meaning and content are not wholly in the head but are partly spread out in the environment and the language community. Perhaps you have heard of Twin Earth a faraway place in which the water-like stuff in the oceans and that comes out of taps is a substance that is chemically unrelated to H_2O. Putnam argues that the water-like substance would not *be* water. If we visit Twin Earth and we say on arrival "Look, there is water here," we speak falsely (assuming that they do not also have H_2O). Their word "water" does not mean the same as ours. Further, as Burge argued, the contents of the thought that the natives of Twin Earth express when they say "Water is good for you" is not the same content that we would express by those words.

One reaction that philosophers have had to externalism is to say "OK, so there is one kind of content and meaning that is spread out in the language community and the environment; but there is another kind, *narrow* content, that is in the head." And this brings us to the topic of this chapter. This chapter is about the issue of whether phenomenal character is representational content. I will discuss "narrow content" first, and then move to wide content, the kind of content that is spread out in the world.

So the kind of representationism that is under discussion now holds that the phenomenal character of experience is its "narrow intentional content,", intentional content that supervenes on nonrelational physical aspects of the body, intentional content that is "in the head" in Putnam's phrase. That is, heads that are the same in ways that don't involve relations to the environment share all narrow intentional contents. A full-dress discussion of this view would cover various ideas of what narrow intentional content is supposed to be. But this isn't a full-dress discussion. I will simply say that all versions of this view that I can think of that have even the slightest plausibility (and that aren't committed to qualia) are functionalist. They are functionalist in that they involve the idea that narrow intentional content supervenes on internal functional organization. That is, there can be no differences in narrow intentional contents without corresponding differences at the level of causal interactions of mental states within the head. The view comes in different flavors, but one thing that all versions of functionalism have in common is that the realizations of the functional organization don't matter to the identity of the functional organization itself: there is a level of "grain" below which brain differences make no difference. Functional organization can be understood in terms of the interactions of commonsense mental states or in terms of the causal network of computational states. In both cases, neurological differences that don't affect the causal network make no difference. One functional organization is multiply realizable physicochemically in ways that make no difference in narrow intentional content.

In other words, there is a level of organization above the level of physiology ("mental" or "computational") that determines narrow intentional content. (Tye, 1994, takes this view and I understand Rey, 1992a,b, and White, 1995, as endorsing it.) Perhaps a computer analogy will help: Two computers that work on different principles can nonetheless compute the same function, even the same algorithm. Differences in hardware below a certain level don't matter to the question of what algorithm is being computed. No one would think the computation of an alogrithm could be identified with a hardware process, for the same algorithm could be computed on different hardware. Similarly, narrow intentional content finds its place above the level of hardware. (Note that this is not to say that hardware is causally inefficacious. Even if the operation of a hydraulic computer can be duplicated at the algorithmic level by an electronic computer, that wouldn't show that the liquid in the hydraulic computer has no causal efficacy.)

Of course phenomenists can (and should) be internalists about phenomenal character too. But phenomenists can allow that phenomenal character depends on the details of the physiology or physicochemical realization of the computational structure of the brain. (This is my view, a form of neural reductionism.) Of course, there are also dualist forms of phenomenism, but both the physicalist and dualist forms of phenomenism agree that there is no need to suppose that qualia supervene on functional organization.

You may ask, 'What is the difference between a view that says that phenomenality is such and such a neural state and a view that says that phenomenality is the functional role of such and such a neural state? It is the difference between a role (or the property of having a certain role) and the neural *occupant* of that role. An analogy: dormitivity is a role—the possession of some chemical property or other that causes sleep. But there are different chemicals that can play that role, the benzodiazepines, the barbiturates, and so on. The application to phenomenality could be put like this: phenomenality may require actual diffusion of ions across a boundary or actual phase-linked oscillations, not just the role that those things play, for that role could perhaps be played by something else. Those in artificial intelligence who favor computational theories of phenomenality (a special case of a functional theory) have often felt that the implementation of the computations could be safely ignored. After all, a program that is run on one hardware is the same as that program run on another—at least if one ignores hardware variables such as speed, failure characteristics, and such. But the neurological view says that the fundamental nature of phenomenality *may be found in the implementational details.*

There is a very simple thought experiment that raises a serious (maybe fatal) difficulty for any such (functionalist) internalist form of representationism. Suppose that we raise a child (call her Erisa) in a room in which all colored surfaces change color every few minutes. Further, Erisa is allowed no information about grass being green or the sky being blue, and so on. The result is that Erisa ends up with no standing beliefs that distinguish one color from another. Suppose further that she is not taught color words, nor does she make them up for herself. (There are many languages which have only two words for colors, for example, the language of the Dani famously studied by Rosch, 1972). Now we may suppose that the result is that there is little in the way of abiding functional differences among her color experiences. Most important, Erisa has no associations or behavioral inclinations or dispositions toward red that are any different from her associations or inclinations or dispositions toward blue. Of course, she responds to color similarities and differences—for example, she groups blue things together as having the same color, and she groups red things together as having the same color. But her ability to group under the *same color* relation does not distinguish her reaction to red from her reaction to blue. The experience as of red is vividly different from the experience as of blue. But what difference in function in this case could plausibly constitute that difference?

The challenge to the internalist representationist, then, is to say what the difference is in internal intentional content between the experience of red and the experience of blue. The only resources available to the internalist representationist are functional. There is a difference in phenomenal character, so the internalist representationist is committed to finding a difference in function. But the example is designed to remove all abiding differences in function.

The functionalist can appeal to *temporary* differences. Erisa will say "The wall is now the same color that adorned the table a second ago," and "For one second, the floor matched the sofa." But these beliefs are fleeting, so how can they constitute the *abiding* differences between the phenomenal character of her experience of red and green? The differences between these phenomenal characters stay the same (for us) from moment to moment, from day to day, and there is no reason to suppose that the same cannot be true for Erisa. The point of the thought experiment is to make it plausible that color experiences can remain just as vivid and the differences between them just as permanent as they are for us even if the functional differences between them attenuate to nothing that could plausibly constitute those differences.

Of course, there is one abiding difference in functional role between the experience of red and the experience of green—the properties of the stimuli. Since we are talking about *internalist* representationism, the stimuli will have to be, for example, light hitting the retina rather than colored surfaces. But these differences in the stimuli are what *cause* the differences in the phenomenal character of experience, not what *constitute* those phenomenal differences. I don't expect die-hard functionalists to recant in response to this point, but I really don't see how anyone with an open mind could take *being caused by certain stimuli* as constituting phenomenal characters of color experiences.

Perhaps there are innate behavioral differences between the experience of red and the experience of blue. Perhaps we are genetically programmed so that red makes us nervous and blue makes us calm. In my view it is a bit silly to suppose that the phenomenal character of the experience as of blue is completely constituted by such factors as causing calmness. But since some philosophers think otherwise, I will also note that despite claims of this sort (Dennett, 1991), assertions of innate color differences are empirical speculations. (See Dennett's only cited source, Humphrey, 1992, which emphasizes the poor quality of the empirical evidence.) And, of course, anyone who holds that representationism is a conceptual truth will be frustrated by the fact that we don't know without empirical investigation whether it is a truth at all. The upshot is that we don't know whether normal humans in different cultures need share such properties as being made nervous by red and calm by blue.

Of course, there may be culturally conditioned differences in color experience among such cultures. There are certainly culturally conditioned differences in phonological experience that are produced by environmental differences. For example, an American's experience of the sound '/y/' as in the French *rue* is very different from a French speaker's experience of that sound. And French speakers are "deaf" to suprasegmental dimensions like pitch and duration which are not exploited in their language.

Further, language differences can result in the perception of sounds that are considered illusory from the point of view of a different phonology. For example, Japanese subjects hear 'sphinx' as 'sufuNKusu.' (See chapter 11 by Segui, Frauenfelder and Hallé in this book. My thanks to Emmanuel Dupoux for drawing my attention to these cases.) Dennett (1991) gives a good example of how cultural factors can influence experience. He imagines that a long-lost Bach cantata is discovered which is ruined by the fact that the initial sequence of notes is the same as that of "Rudolf the Red-Nosed Reindeer." We can't hear the cantata the way Bach's contemporaries heard it.

Moving back to the case of color, I am happy to allow that culture can *influence* color experience. The issue is whether there is a sensory core of color experience, a phenomenal similarity that is the same across cultures and that resists functional or representational analysis. The sensory core *has* a functional description—for example, in terms of the color solid—but that functional description does not determine the nature of the experience. (There could be more than one phenomenology compatible with the color solid.) Whatever the culture, a normal perceiver sees colors in a way that is categorizable in terms of hue, saturation, and brightness. And there is every reason to think that Erisa would share that core.

Perhaps the internalist will say that there would be no differences among her experiences. Red would of necessity look just the same to her as yellow. But this is surely an extraordinary thing for the internalist to insist on. It *could* be right, of course, but again it is surely an unsupported empirical speculation.

I claim that an Erisa of the sort I described is conceptually possible. The replies I have just considered do not dispute this, but instead appeal to unsupported empirical claims.

But physicalism is an empirical thesis, so why should the representationist be embarrassed about making an empirical claim? Answer: physicalism is a very general empirical thesis having to do with the long history of successes of what can be regarded as the physicalist research program. Internalist representationism, by contrast, depends on highly specific experimental claims. For example, perhaps it will be discovered that newborn babies hate some colors and love others before having had any differential experience with these colors. I doubt that very many opponents of qualia would wish their point of view to rest on speculations as to the results of such experiments.

To avoid misunderstanding, let me be clear that even if newborn babies hate some colors and love others, the representationists should not suppose there are no qualia. I have been focusing on the empirical issue because it ought to embarrass functionalists (of the Dennettian stripe) that they rest an argument against qualia on the existence of asymmetries that are so subtle that we are not now sure whether they exist at all. Still, the empirical issue is just a side issue. For an experience that has the sensory core of the human experience of red could possibly calm one person yet make another anxious. Thus the defender of qualia does not depend on a prediction about the results of such experiments. Our view is that even if such experiments do show some asymmetries, there are *possible* creatures—maybe genetically engineered versions of humans—in whom the asymmetries are ironed out (see Shoemaker, 1981). And those genetically engineered humans could nonetheless have color experience much like ours in its sensory core.

In commenting on an earlier version of this chapter, Emmanuel Dupoux suggested that the representationist might say that what color experience represents is triples of hue, brightness, and saturation. I think there are a number of ways one could pursue that suggestion.

1. One version would be externalist, close to the view that what color experience represents is the *colors* themselves. This is an "externalist" view because it holds that it is the nature of something in the external world that determines phenomenal character. Why couldn't two different phenomenal characters represent the same triple of hue, brightness, and saturation? Indeed, why couldn't something represent such triples without any phenomenal character at all? Externalism will be discussed in detail in the next section.

2. Another version would cash out hue, saturation, and brightness phenomenally, that is, in terms of what it is like to have the experiences themselves. There are a number of ways that this could be done, the most interesting I think pursued in recent papers by Sydney Shoemaker

(1994a,b; see Block, 1999, for a discussion of Shoemaker's proposals). For my purposes, it suffices to say that these proposals all acknowledge qualia and are thus compatible with the main thesis of this chapter.

3. Another version would cash out the representation of hue, saturation, and brightness physiologically, in terms of the neural bases of the phenomenal experiences of hue, saturation, and brightness. The idea would be that the experiences of hue, saturation, and brightness are neural states (or at least supervene on neural states). This is the view I favor. The lover of qualia needn't be disturbed if qualia turn out to *be* something neurological. It is important to distinguish between *reduction* and *replacement* as we can see from a bit of attention to the history of science. Thermodynamics can be reduced to statistical mechanics, allowing the reduction of heat to molecular kinetic energy, but physics *replaced* references to caloric, phlogiston, and ether with references to other things. Heat is real and turns out to be something molecular, but caloric, phlogiston, and ether turned out not to exist. If qualia turn out to be neural states, the lover of qualia might reasonably rejoice, for it is a great confirmation of the reality of something to discover what it is in more basic terms.

In sum, I have canvassed three ways of interpreting the idea that representationism can be saved by taking the phenomenal character of experience as representing triples of hue, saturation, and brightness. Two of them are internalist and compatible with the pro-qualia view that I am defending. The third is externalist, and is treated below.

Externalism

That is all I will have to say about internalist representationism. Now I will move to externalist representationism. I will take the view to be that the phenomenal character of a particular color experience consists in its representing a certain color (hue, brightness, and saturation). Let me reiterate that I agree that all or most of our phenomenal experience does have such representational contents. In particular, I agree that our color experience does represent such triples of hue, saturation, and brightness. The issue is whether that is all there is to the phenomenal character of color experience.

Supervenience

Recall that what it means to say that phenomenal character supervenes on the neurological is that there can be no phenomenal difference without a neurological difference. Or, in other words, two beings who are neurologically the same are necessarily phenomenally the same. *If* phenomenal character supervenes on the brain, there is a straightforward argument against externalist representationism. For arguably, there are brains that are the same as yours in internal respects (molecular constitution) but whose states represent *nothing* at all. Consider the swampman, the molecular duplicate of you who comes together by chance from particles in the swamp. He is an example of total lack of representational content. He doesn't have innate representational contents because he didn't evolve. (His formation was a chance event.) And how can he refer to Conan O'Brien given that he has never had any causal connection with him; he hasn't ever seen Conan or anyone else on TV, never seen a newspaper. (This is an extreme version of a Twin Earth example.)

Although he has no past, he does have a future. If, at the instant of creation, his brain is in the same configuration that yours is in when you see a ripe tomato, it might be said that the swampman's state has the same representational content as yours in virtue of that state's role in allowing him to "track" things in the future that are the same in various respects (e.g., color) as the tomato. It might be said that he will track appearances of Conan O'Brien in the future just as well as you will. But this invocation of the swampman's future is not very convincing. Sure, the swampman will track Conan O'Brien if he (the swampman) materializes in the "right" environment (an environment in which Conan O'Brien exists and has the superficial appearance of the actual Conan O'Brien), but in the wrong environment he will track someone else or no one at all. And the same point applies to his ability to track water and even color. If put on Twin Earth, he will track twin-water, not water. Of course, you will have the same tracking problems if suddenly put in "wrong" environments, but your references are grounded by the very past that you have and the swampman lacks.

If we can assume supervenience of phenomenal character on the brain, we can refute the representationist. The phenomenal character of the swampman's experience is the same as yours but its experiences have no representational content at all. So phenomenal character cannot be representational content. If phenomenal character supervenes on the brain, and if phenomenal character is representational content, then representational content supervenes on the brain. Thus there could not be two brains that are neurologically the same but representationally different. However, you and the swampman have a pair of brains that are neurologically the same but representationally different. So representationism of the supervenience variety is refuted.

If this point is right, there will be a great temptation for the representationist to deny that phenomenal character supervenes on the brain. And in fact, that's what representationists often do (Dretske, 1995; Lycan, 1996a,b; McDowell, 1994). These respected figures actually claim that there can be neurologically—even molecularly—identical brains that differ in phenomenal character. This seems to me to be a desperate maneuver with no independent plausibility. It is a good example of how a desperate maneuver with no independent plausibility can, as a matter of sociological fact, succeed in changing the focus of debate. Externalist representationism doesn't get to first base, but we have to refute it just the same. And once we are deprived of arguments that depend on supervenience—like the argument based on the swampman example—it isn't easy. Supervenience is supported by a wide variety of evidence that phenomenal character is dependent on what happens in the brain, but we will have to put that evidence to one side.

Bodily Sensations

Is the experience of orgasm completely captured by a representational content *that there is an orgasm?* Orgasm is phenomenally *impressive* and there is nothing very impressive about the representational content *that there is an orgasm.* I just expressed it and you just understood it, and nothing phenomenally impressive happened (at least not on my end). I can have an experience whose content is that my partner is having an orgasm without *my* experience being phenomenally impressive. In response to my raising this issue (Block, 1995a,b), Tye (1995a) says that the representational content of orgasm "in part, is that something very pleasing is happening down *there*. One also experiences the pleasingness alternately increasing and diminishing in its intensity." But once again,

I can have an experience whose representational content is that my *partner* is having a very pleasing experience down *there* that changes in intensity, and although that may be pleasurable for me, it is not pleasurable in the phenomenally impressive way that that graces my own orgasms. I vastly prefer my own orgasms to those of others, and this preference is based on a major-league phenomenal difference. The location of "down there" differs slightly between my perception of your orgasms and my own orgasms, but how can the representationist explain why a small difference in represented location should matter so much? Of course, which subject the orgasm is ascribed to is itself a representational matter. But is that the difference between my having the experience and my perceiving yours? Is the difference just that my experience *ascribes* the pleasure to you rather than to me (or to part of me)? Representational content can go awry in the heat of the moment. What if in a heated state in which cognitive function is greatly reduced, I *mistakenly* ascribe your orgasm to me or mine to you? Would this difference in ascription really *constitute* the difference between the presence or absence of the phenomenally impressive quality? Perhaps your answer is that there is a *way* in which my orgasm-experience ascribes the orgasm to me that is immune to the intrusion of thought, so there is no possibility of a confused attribution to you in *that way*. But now I begin to wonder whether this talk of "way" is closet phenomenism.

No doubt there are functional differences between my having an orgasm-experience and merely ascribing it to you. Whether this fact will help to defend representationism depends on whether and how representationism goes beyond functionalism, a matter which I cannot go into here.

Lycan (1996c) appeals to the following representational properties (of male orgasm): it is "ascribed to a region of one's own body," and the represented properties include "at least warmth, squeezing, throbbing, pumping and voiding. (On some psychosemantics, I suppose, impregnating is represented as well.)" (p. 136) Lycan says that it is "impracticable to try to capture detailed perceptual content in ordinary English words, at least in anything like real time," but he thinks he has said enough to "remove reasonable suspicion that there are nonintentional qualitative features left over in addition to the functional properties that are already considered characteristic of the sensation." But Lycan's list of properties represented seems to me to *increase* the suspicion rather than remove it. *Everything* that matters (phenomenally speaking) is left over.

Of course, we should not demand that a representationist be able to capture his contents in words. But if we are to try to believe that the experience of orgasm is nothing over and above its representational content, we need to be told something fairly concrete about what that representational content is. Suppose the representational content is specified in terms of recognitional dispositions or capacities. One problem with this suggestion is that the *experience* of orgasm seems on the face of it to have little to do with *recognizing* orgasms. Perhaps when I say to myself, 'There's that orgasm-experience again,' I have a different experience from the cases where no recognition goes on. But as before, there is a sensory core that stays the same. And there is no plausibility in the insistence that the sensory core of the experience must involve some sort of categorization. If you are inclined to be very intellectual about human experience, think of animals. Perhaps animals have experience with that sensory core without any recognition.

The representationists should put up or shut up. The burden of proof is on them to say what the representational content of experiences such as an orgasm are. (Alex Byrne in an unpublished paper says the content is *an orgasm*. But that answer is ambiguous between [1] a phenomenist reading, in which qualia are constitutively required to have that content and [2] a nonphenomenist reading which has yet to be filled in.)

Phosphene-Experiences
Harman (1990) says "Look at a tree and try to turn your attention to intrinsic features of your visual experience. I predict you will find that the only features there to turn your attention to will be features of the represented tree" (p. 478). But the diaphanousness of perception is much less pronounced in a number of visual phenomena, notably phosphene-experiences. (I use the cumbersome "phosphene-experience" instead of the simpler "phosphene" by way of emphasizing that the phenomenist need not have any commitment to phenomenal individuals such as sense data.) If all of our visual experiences were like these, representationism would have been less attractive. Phosphene-experiences are visual

sensations "of" color and light stimulated by pressure on the eye or by electrical or magnetic fields. (I once saw an ad for goggles that you could put on your eyes that generated phosphenes via a magnetic field.) Phosphene-experiences have been extensively studied, originally in the nineteenth century by Purkinje and Helmholtz. Close your eyes and place the heels of your hands over your eyes. Push your eyeballs lightly for about a minute. You will have color sensations.

Can you attend to those sensations? I believe I can but we won't get anywhere arguing about the first-person data. Putting this issue to one side, can one be aware of them? Perhaps there can be awareness without the kind or amount of attention that some feel we cannot have toward our sensations. For example, I have had the experience of talking to someone when there is a jackhammer drilling outside. Our voices rise in volume to counter the noise, but I am so intent on the conversation that I do not at first notice the noise. But when I do notice it, I am aware that I have been experiencing the noise for some time. Experiences like this make it plausible that one can have a kind of awareness without the kind of attention required for noticing. The question of whether we are aware of the phenomenal characters of our own sensations depends in part on what concepts one applies to them. If one applies concepts of experience rather than concepts of external world properties, then one can be aware of them. Harman's point depends for its plausibility on an appeal to a very nonphilosophical kind of circumstance in which one is not applying introspective concepts. According to the representationist, all awareness of those sensations could consist in is awareness of the colored moving expanses that are represented by them. My view is that one can be aware of something more. Harman's point depends on an introspective claim that can reasonably be challenged.

Lycan (1987) says that "given any visual experience, it seems to me, there is *some* technological means of producing a veridical qualitative equivalent—e.g., a psychedelic movie shown to the subject in a small theater" (p. 90). But there is no guarantee that phosphene experiences produced by pressure or electromagnetic stimulation could be produced by light. (Note: I don't say there is a guarantee that phosphene-experiences could *not* be produced by light, but only that there is no guarantee that they could; I have no idea whether they could or not.)

I do wonder if Lycan's unwarranted assumption plays a role in leading philosophers to suppose that the phenomenal characters of phosphene-experiences, afterimage-experiences, and the like are exhausted by their representational content.

Bach-y-Rita

According to me, in normal perception one can be aware of the mental paint—the sensory quality that does the representing. This idea can be illustrated (this is more of an illustration than it is an argument) by Bach-y-Rita's famous experiment in which he gave blind people a kind of vision by hooking up a TV camera to their backs which produced tactual sensations on their backs. Bach-y-Rita says that the subjects would normally attend to what they were "seeing." He says that "unless specifically asked, experienced subjects are not attending to the sensation of stimulation on the skin of their back, although this can be recalled and experienced in retrospect" (quoted in Humphrey, 1992, p. 80). The retrospective attention of which Bach-y-Rita speaks is a matter of attending in retrospect to a feature of one's experience that one was aware of but not attending to when the perception originally happened, as with the jackhammer example mentioned earlier. Of course, the analogy is not perfect. In attending to visual sensations, we are not normally attending to sensations of the eye (Harman, 1996).

Let us return to the challenge I raised earlier for the representationist. Just what is the representational content of bodily sensations like pain and orgasm? In vision, it often is plausible to appeal to recognitional dispositions in cases where we lack the relevant words. What's the difference between the representational contents of the experience of color A and color B, neither of which have names? As I mentioned earlier, one representationist answer is this: the recognitional dispositions themselves provide or are the basis of these contents. My experience represents A as *that* color, and I can *mis*represent some other color as *that* color. Let's accept this idea for color experience for the sake of argument.

But note that this model can't straightforwardly be applied to pain. Suppose I have two pains that are the same in intensity, location, and anything else that language can get a handle on—but they still feel different. Say they are both twinges that I have had before, but they aren't burning or sharp or throbbing. "There's *that* one again; and there's that other one" is the best I can do. If we rely on my ability to pick out *that* pain (arguably) we are demonstrating a phenomenal character, not specifying a representational content. The appeal to recognitional dispositions to fill in representational contents that can't be specified in words has some plausibility, so long as the recognitional dispositions are directed outward. But once we direct them inward, one begins to wonder whether the resulting view is an articulation of representationism or a capitulation to phenomenism.[4]

Notes

1. Actually, this is oversimple. In my view, the scientific nature of qualia could be discovered to be functional or representational or cognitive, and this is not an anti-qualia view. Thus I prefer a definition of *qualia* as features of experience that cannot be *conceptually* reduced to the nonphenomenal, in particular the cognitive, the representational, or the functional. Or, for those who reject conceptual reduction, qualia could be said to be features of experience that cannot be reduced to the nonphenomenal, except scientifically.

2. An important qualification: there is no difference in content or meaning without a difference in a *subset* of the physical properties, namely those that are not individuated with respect to things outside the body. (Conditions of individuation of X are considerations of what constitutes the boundaries between X and that which is not X, and whether Y=X.)

3. Assuming the same qualification mentioned in the foregoing note.

4. This paper overlaps perhaps 30 percent of a paper that may come out one day in a festschrift for Tyler Burge, edited by Martin Hahn and Bjorn Ramberg. I am grateful to Emmanuel Dupoux for helpful comments on an earlier draft.

References

Block, N. (1995a). 'On a confusion about a function of consciousness.' *Behavioral and Brain Sciences*, 18, 227–247. [Reprinted in Block, N., Flanagan, O., and Guzeldere, G. (1997) *The nature of consciousness: Philosophical debates.* Cambridge, MA: MIT Press.]

Block, N., (1995b). How many concepts of consciousness? *Behavioral and Brain Sciences*, 18, 272–287.

Block, N. (1999). Sexism, racism, ageism and the nature of consciousness. In R. Moran, J. Whiting, and A. Sidelle (Eds.), *The philosophy of Sydney Shoemaker. Philosophical Topics*, 26 (1,2).

Burge, T. (1979). Individualism and the mental. *Midwest Studies in Philosophy*, 4, 73–121.

Dennett, D. (1991). *Consciousness explained.* Boston: Little, Brown.

Dretske, F. (1995). *Naturalizing the mind.* Cambridge, MA: MIT Press.

Harman, G. (1990). The intrinsic quality of experience. In J. Tomberlin (Ed.), *Philosophical perspectives* 4. *Action theory and philosophy of mind* (pp 31–52). Atascadero, CA: Ridgeview. [Page numbers refer to version reprinted in W. Lycan (ed.) *Mind and Cognition* Blackwell: Oxford, 1999:474–484.]

Harman, G. (1996). Explaining objective color in terms of subjective reactions. In E. Villanueva (Ed.), *Philosophical issues 7: Perception.* Atascadero, CA: Ridgeview.

Humphrey, N. (1992). *A history of the mind* New York: Simon & Schuster.

Lycan, W. G. (1987). *Consciousness.* Cambridge, MA: MIT Press.

Lycan, W. G. (1996b). *Consciousness and experience.* Cambridge, MA: MIT Press.

Lycan, W. G. (1996c). Replies to Tomberlin, Tye, Stalnaker and Block. In E. Villanueva (Ed.), *Philosophical Issues 7: Perception.* Atascadero, CA: Ridgeview.

McDowell, J. (1994). The content of perceptual experience. *Philosophical Quarterly,* 190–205.

Putnam, H. (1975). The meaning of "meaning." in *Mind, language and reality: Philosophical papers.* Vol. 2. London: Cambridge University Press.

Rey, G. (1992a). Sensational Sentences. In M. Davies and G. Humphreys (Eds.), *Consciousness: Psychological and philosophical essays.* Oxford: Blackwell.

Rey, G. (1992b). Sensational sentences switched. *Philosophical Studies, 68,* 289–331.

Rosch, E. (1972) [E.R. Heider]. Probabilities, sampling and ethnographic method: The case of Dani colour names. *Man (n.s.), 7,* 448–466.

Shoemaker, S. (1982). The inverted spectrum. *Journal of Philosophy, 79,* 357–381.

Shoemaker, S. (1994a). Self-knowledge and "inner sense. Lecture III: The phenomenal character of experience. *Philosophy and Phenomenological Research, 54,* 291–314.

Shoemaker, S. (1994b). Phenomenal character. *Nous, 28,* 21–38.

Tye, M. (1994). Qualia, content and the inverted spectrum. *Nous, 28,* 159–183.

Tye, M. (1995a). Blindsight, orgasm and representational overlap. *Behavioral and Brain Sciences, 18,* 268–269.

White, S. (1995). Color and the narrow contents of experience. *Philosophical Topics,* 471–505.

Recommended Reading

Block, N. (1978). Troubles with functionalism. *Minnesota Studies in the Philosophy of Science, 11,* 261–325. [Reprinted (in shortened versions) in N. Block (Ed.) (1980). *Readings in philosophy of psychology.* Vol. 1. Cambridge, MA: Harvard University Press; in W. Lycan (Ed.), (1990), *Mind and cognition* (pp. 444–469).

Oxford: Blackwell; in D. M. Rosenthal (Ed.) (1991), *The nature of mind* (pp. 211–229). Oxford: Oxford University Press; in B. Beakley and P. Ludlow (Eds.) (1992), *Philosophy of mind*, Cambridge, MA: MIT Press; and in A. Goldman (Ed.) (1993), *Readings in philosophy and cognitive science.* Cambridge, MA: MIT Press.]

Block, N. (1981). Psychologism and behaviorism. *Philosophical Review, 90,* 5–43.

Block, N. (1990) Inverted earth. *Philosophical Perspectives, 4,* 51–79. [Reprinted in W. Lycan (ed.) (1999), *Mind and Cognition* (2d ed). Oxford: Blackwell, 1999; and in Block, N., Flanagan, O., and Güzeldere, G. (1997). *The nature of consciousness: Philosophical debates.* Cambridge, MA: MIT Press.]

Block, N. (1994a). Qualia. In S. Guttenplan (Ed.), *A companion to philosophy of mind.* (pp. 514–520). Oxford: Blackwell, 514–520.

Block, N. (1994b). Consciousness. In S. Guttenplan (Ed.), *A companion to philosophy of mind* (pp. 514–520). Oxford: Blackwell.

Block, N. (1995c). Mental paint and mental latex. In E. Villanueva (Ed.), *Philosophical Issues.* Atascadero, CA: Ridgeview.

Block, N. (1998). Is experiencing just representing? *Philosophy and Phenomenological Research,* 663–671.

Block, N., and Fodor, J. (1972). What psychological states are not *Philosophical Review, 91,* 159–181. [Reprinted in Block (Ed.) (1980), *Readings in Philosophy of Psychology.* Vol. 1. Cambridge. MA: Harvard University Press.]

Block, N., Flanagan, O., and Guzeldere, G. (1997). *The nature of consciousness: Philosophical debates.* Cambridge MA: MIT Press.

Dretske, F. (1996). Phenomenal externalism or if meanings ain't in the head, where are qualia? In E. Villanueva (Ed.), *Philosophical issues 7: Perception.* Atascadero, CA: Ridgeview.

Harman, G. (1982). Conceptual role semantics. *Notre Dame Journal of Formal Logic, 23,* 242–256.

Loar, B. (1990). Phenomenal states. In J. Tomberlin (Ed.), *Philosophical perspectives 4. Action theory and philosophy of mind,* (pp. 81–198). Atascadero, CA: Ridgeview. [A much revised version of this paper is to be found in N. Block, O. Flanagan, and G. Guzeldere (1997). *The nature of consciousness: Philosophical debates* (pp 587–6160. Cambridge, MA: MIT Press.

Lycan, W. G. (1990). *Mind and cognition. A reader.* Oxford: Blackwell.

Lycan, W. G. (1995). We've only just begun. *Behavioral and Brain Sciences, 18,* 262–263.

Lycan, W. G. (1996a). Layered perceptual representation. In E. Villanueva (Ed.), *Philosophical issues 7: Perception.* Atascadero, CA: Ridgeview.

Rosenthal, D. (1991). *The nature of mind.* Oxford: Oxford University Press.

Stalnaker, R., (1996). On a defense of the hegemony of representation. In E. Villanueva (ed.) *Philosophical issues 7: Perception.* Atascadero, A: Ridgeview.

Tye, M., (1995b). *Ten problems of consciousness.* Cambridge, MA: MIT Press.

III
Language

Introduction

Christophe Pallier and Anne-Catherine Bachoud-Lévi

"Looky here, Jim; ain't it natural and right for a cat and a cow to talk different from us?"
"Why, mos'sholy it is."
"Well, then, why ain't it natural and right for a Frenchman to talk different from us? You answer me that?"
"Is a cat a man, Huck?"
"No."
"Well, dey ain't no sense in a cat talkin' like a man. Is a cow a man?—er is a cow a cat?"
"No she ain't either of them."
"Well, den she ain't got no business to talk either one er the yuther of 'em. Is a Frenchman a man?"
"Yes."
"Well, den! Dad blame it, why doan' he talk like a man? You answer me"
—Mark Twain (cited in G. Miller, 1951)

In a sense, a large part of Jacques Mehler's research has been devoted to address the question raised in the above quotation: "Why doesn't a Frenchman talk like an (English) man?" (or more accurately "Why doesn't a Frenchman listen like a English man?" see Cutler, Mehler, Norris, and Segui, 1986). This question is deeper that it seems. Why aren't we born equipped with a completely innate, universal, system of communication? If one can believe that all concepts are innate, why not also the word forms, or the grammar? Of course, "why"-questions are notoriously slippery. The fact is that there are many different languages and that human beings are born equipped to acquire whichever human language is present in their environment. Tangible advances are really made when researchers ask "how" and "what" questions, such as "How different are the languages?", "What processes subserves word recognition, or production?", or "What is the impact of learning a language on the human brain?" Real advances have been made on these subjects in the last forty years, and the chapters to come will illustrate it.

One of the central question that is facing the study of language processing is how linguistic knowledge is put to use in various situations like language perception, comprehension or acquisition. Three papers adress directly this question but give quite different answers. Pinker (chapter 9) argues, from the well-discussed example of the past tense of verbs, for the psychological reality of linguistic rules operating on symbolic representations. He nevertheless acknowledges the existence of an associative-like memory system that stores irregular patterns and is responsible for some non-regular patterns of generalizations. Bever (chapter 8) proposes that listeners compute the (approximate) semantics of a sentence before the syntax. In his idea, such early semantics is in fact used by the syntactic parser to derive the structure of the sentence. Nespor (chapter 7) reflects on the possibility that infants may use the prosody of speech to discover some syntactic properties of their mother language and set syntactic parameters through prosody. Overall, the picture that emerges here is that syntactic knowledge interacts in rather complex ways with other knowledge types and performance systems.

Three papers, by Cutler et al. (chapter 10), Segui et al. (chapter 11), and Caramazza et al. (chapter 12), present experimental data that demonstrate how characteristics of various languages influence information processing in speech perception and production. Segui et al., in a series of experiments, demonstrate the phenomenon of phonotactic assimilation, and argues in favor of syllabic-like units in perception. In contrast, Cutler et al. reject the role of syllable as the unit of speech perception but rather argue for its role in lexical segmentation. Despite this disagreement, both papers outline the fact that speech perception is influenced by phonological properties of the mother tongue. The paper by Caramaza et al. shows that the stage in word production where the determiner is selected depends on whether or not the language displays interactions between the form of the determiner and the phonological shape of the noun (e.g., in French, the possessive "my" can take two forms "mon" or "ma" according to whether the following word starts with a vowel or a consonant. No such thing happens in English). These chapters illustrate how crosslinguistic studies can enrich the database, and help build a more comprehensive theory of the human's language processor. In the past, psycholinguists often proposed "universal" models of the speaker-hearer, based on data collected with speakers from one language community (e.g., Mehler et al., 1981). Gradually, as money for travel increased, more languages were included in the studies, a trend in which Jacques played an important role (see, e.g., Cutler et al., 1986, or Otake et al., 1993). Nowadays, a joke says that "no universal claim can be accepted in psycholinguistics unless it is based on at least three languages."

For many years, perception and production were studied separately. Levelt (chapter 14) discusses the possible relationships between the processes underlying speech perception and production. He proposes that production and perception share the conceptual and syntactic levels but that the input and output phonological lexicons are separate. He does so from the point of view of a theory of speech production that has been buttressed essentially by behavioral data, but in his paper, he takes "the perspective to cognitive neuroscience, not only because the recent neuroimaging literature often provides additional support for the existing theoretical notions, but also because it provides new, additional challenges for how we conceive of the production/perception relations." Yet, one may note that Levelt proposes that the output phonological lexicon involves Wernicke's area, an area considered in neuropsychology to be dedicated to language perception.

For a long time, speech production was approached by studying spontaneous speech error, tip of the tongue phenomena, tongue twisters, and pause location. Garrett (chapter 13) shows that considerable advances were made by taking into account converging evidence using more advanced paradigms like reaction time or picture-word interference studies. Yet, he shows that in the case of the frequency effect, converging evidence is not always converging, and that more research is needed to understand what the various experimental paradigms are exactly telling us. The importance of converging evidence is also discussed in the paper by Levelt in the context of the neuroimaging data.

The comprehensive approach to the language faculty exemplified in those chapters should please Jacques. Language is still often studied in different academic departments: linguists study languages, psycholinguists study language users, and neurolinguists study the brain of the language users. It is, indeed, a natural bend, in front of matters as complex as language, to try and restrict one's field of inquiry. Many researchers have adopted the strategy of splitting the subject matter until a seemingly tractable problem is reached. This is not, we think, Jacques's strategy. He has always remained close to the global issues. His taste for simple and obviously important questions may be the most important lesson his students learned from him.

References

Cutler, A., Mehler, J., Norris, D., and Segui, J. (1986). The syllable's differing role in the segmentation of French and English. *Journal of Memory and Language, 25,* 385–400.

Mehler, J., Dommergues, J. Y., Frauenfelder, U., and Segui, J. (1981). The syllable's role in speech segmentation. *Journal of Verbal Learning and Verbal Behavior, 20,* 298–305.

Otake, T., Hatano, G., Cutler, A., and Mehler, J. (1993). Mora or syllable? Speech segmentation in Japanese. *Journal of Memory and Language, 32,* 258–278.

7
About Parameters, Prominence, and Bootstrapping

Marina Nespor

To Jacques, who sees that standing on each other's shoulders is better than stepping on each other's toes.

Linguistics, psycholinguistics, and neurolinguistics aim at understanding the structure and functioning of the cognitive system responsible for language. Though these three fields investigate the mechanisms underlying the language faculty on the basis of different data, it is the same object they are seeking to understand.

The traditional separation of these disciplines, the mutual ignorance of each other's results, and often the belief in the superiority of some types of data over others, are largely responsible for the fact that little integration of these fields has occurred.

Jacques saw, before most other people, the advantage of putting together researchers from different fields to discuss each other's questions, data, and problems. Jacques started his career in psycholinguistics with the aim of testing the psychological reality of formal aspects of transformational grammar and although he abandoned this particular theory (derivational theory of complexity; Mehler, 1963), he never gave up completely trying to link linguistic theory and psycholinguistics. The two SISSA Encounters in Cognitive Science he organized in Trieste in 1991 and 1995 on Psycholinguistic Consequences of Contrastive Language Phonologies, and that on bilingualism held in Girona in 1997, are examples of how exciting and useful it can be to share knowledge coming from different approaches to the study of language. The exchange of ideas that took place on those occasions was precious, mainly because those ideas originated from different types of data. Convergence of empirical controls of theoretical hypotheses as varied as possible can only make the conclusions more reliable. It is, in addition, my belief that thanks to the integration (however partial it may be) of the three fields of linguistics, psycholinguistics, and neurolinguistics, a new set of questions has arisen.

I will not discuss the advantages of integrating research in neurolinguistics and more specifically research using brain-imaging techniques to the study of language, although this field of research has occupied Jacques a lot in the last years. I would rather discuss, with one example, what I consider some steps that have been made and others that will possibly be made in the future, thanks to the integration of research in the theory of grammar and research based on behavioral experiments, that is, thanks to the integration of research in linguistics and in psycholinguistics.

Different Linguistic Data

Linguists working in the framework of generative grammar have often discussed the logical problem of language acquisition. They have also been aware of the fact that a child cannot develop language without being exposed to one or more languages and without being able either to perceive language (auditorily, in the case of spoken language, or visually, in the case of signed language) or to form a representation of it. Although the linguistic literature of the last decades shows that much attention has been paid to different levels of representation, one finds almost no mention of perception. The psycholinguistic literature, instead, shows great attention to those aspects of language that a person can perceive, as well as to the problem of whether a person can use what she or he perceives.

Both groups of scholars aim at understanding the psychological reality of different aspects of the linguistic system, but they differ both because they use different data to develop a theory and because they use different criteria of adequacy to choose between competing, empirically adequate, theories.

As to the data, within generative grammar, the description of competence regarding the computational system is largely based on grammaticality judgments. That is, it is ideally based on intuitions about the grammaticality of the analyzer investigating the grammar of his or her native language, often compared with grammaticality judgments of other speakers. As to the description of competence regarding the phonological interpretive system, intuitions and judgments about grammaticality are often less reliable and it is thus often necessary to analyze also the acoustic signal.

Psycholinguists, by contrast, aim at understanding the psychological reality of the linguistic system through data acquired by using one of the experimental techniques typical of psychology, specifically behavioral experiments aimed at establishing either the reaction times or the error rates in a given task. Among these are perception experiments, which establish psychological representations on the basis of the subjects' perception of specific auditory stimuli.

As to criteria for adequacy in generative grammar, the choice between two empirically adequate theories about the psychological reality of a linguistic competence corresponds to the simplest description of that competence. In the case of allophonic variation, for example, the simplest description is one in which the phoneme is the segment that exhibits the most varied distribution, while the segment that exhibits a more restricted distribution is derived from it. The phoneme is also the segment which the speakers are aware of, while they are in general not aware of the derived allophones, but identify them with their phonemic representation.

In psycholinguistics, the choice between two alternative theories is made on the basis of their efficacy of computation in real time, as well as of their robustness in the face of noise. To come back to the phoneme, for example, the fact that, in an experimental situation, subjects do not discriminate two syllables like *la* and *ra* (or that they are slow in learning the distinction) is evidence that there is no phonemic contrast between the two consonants in the language of exposure.

Research based on one or the other of these different data has brought us some knowledge about grammatical competence at the steady state. I believe, however, that the one research field where the interests of linguists and psycholinguists most clearly come together is language acquisition: both must account for learnability. That is, both linguistic theories and psycholinguistic models must account for the fact that human infants, except in cases of specific neurological conditions or complete deprivation of language input, develop the language spoken around them.

It is precisely in the field of language development that the integration of the two approaches has proved useful. We know by now, in fact, that the construction of grammar starts just after birth. We also know that many steps are made in the first year of life. Since children do not make mistakes in word order when they start combining words, they must have built up a knowledge of syntax before this stage (Brown, 1973; Meisel, 1992; Clahsen and Eisenbeiss, 1993). Linguists thus need the results of psycholinguistic research because, in the case of infants, it is clearly impossible to use the data typical of generative grammar, that is, grammaticality judgments. It is also impossible to analyze the relevant signal, because infants perceive fine distinctions long before they are able to produce them.

Thanks to different types of discrimination experiments carried out in recent years, we now have some knowledge about the initial state, as well as about different stages of language development before infants start producing language. That is, what the infant perceives allows us to formulate hypotheses about the development of a representation. Thus, by testing whether infants stop perceiving the difference between two elements that are not contrastive in the language of exposure, it can be seen whether they are aware of the phonemic system of their language. That is, in the affirmative case, their phonological representation will not have two phonemes, one for each of the phones (Werker and Tees, 1984).

Psycholinguists also need the results of linguistic research because in order to understand how the human brain is attuned to a specific language, knowledge of different linguistic systems is needed. The psychological model must in fact account for the acquisition of any linguistic system. A fine description of the grammar of typologically different languages is thus necessary.

In the remainder of this chapter I briefly discuss how the two approaches combined—that of generative grammar and that of experimental psychology—allow us to formulate new hypotheses about important aspects of linguistic competence. As an example of how it is possible—and hopefully fruitful—to look at a problem from different angles, I consider the study of grammatical interfaces and what it can tell us about the development of language.

Interfaces

Within generative grammar, the study of the interfaces between grammatical components has brought us significant advances in the comprehension of the organization of grammar. If we take the syntax-phonology interface, investigations of the principles governing the phonological shape of utterances has yielded a better understanding of the interpretive role of phonology. The question is: How does the sound of utterances convey information about the syntactic structure of the sentences they contain? And also: What syntactic information does phonology convey and what is it incapable of conveying?

Prosodic phonology above the word level defines the influence of syntax on the phonological shape of a string (Selkirk, 1984; Nespor and Vogel, 1986; Inkelas and Zec, 1990; Truckenbrodt, 1999). It is thus a theory of the phonological interpretation of syntactic structure, that is, of the alignment of syntactic and phonological constituents, as well of the assignment of prominence relations. Although phonological constituents are projected from syntactic constituents, the constituents of the two hierarchies are not necessarily isomorphic. They thus signal certain aspects of syntactic structure and not others, as we shall see below.

The two constituents relevant to the present discussion are the phonological phrase (f) and the intonational phrase (I). The domain of the phonological phrase expands from the left edge of a syntactic phrase to the right edge of its head in head-complement languages, and from the left edge of a syntactic head to the right edge of its syntactic phrase in complement-head languages. In English, French, or Greek, it thus begins with the left edge of a phrase and it ends with the right edge of its head. In Turkish, Japanese, or Basque, it begins with the left edge of a head and it ends with the right edge of its maximal projection.

The intonational phrase is less well defined: specific constituents, for example, parentheticals, form intonational phrases on their own, but for the rest its domain is quite variable. Since the intonational phrase is the domain of intonational contours, and since a new contour must begin when a new breath group begins, its domain is largely dependent on the length of a string as well as on speech rate.

In (1) the analysis of a string into phonological and intonational phrases is exemplified.

(1) [(The students)f (of her class)f]I, [(as you may know)f]I, ((have read)f (the complete work)f (of Kostantinos Kavafis)f]I

Phonological constituents above the word level are signaled through various types of phonological phenomena that apply within but not across them, through the relative prominence relations established among the elements they contain, and through the anchoring of intonational contours to their edges.

Phonological phenomena may signal, in different ways, the right or left edge of a constituent. The main prominence established within a constituent is either leftmost or rightmost and thus also is a signal of the edges, and so are the limit tones that constitute a melody: these are anchored at the edges of phonological phrases and intonational phrases (Pierrehumbert, 1980; Hayes and Lahiri, 1991).

Two potentially ambiguous sentences, in the sense that they are composed of the same sequence of words, may thus be disambiguated when interpreted phonologically, as in the following Italian example, where the two readings are differently analyzed into phonological phrases.

(2) La vecchia spranga la porta
   a. [(la vecchia)f (spranga)f (la porta)f]I
      "the old lady blocks the door"
   b. [(la vecchia spranga)f (la porta)f]I
      "the old bar carries it"

That phonological interpretation is not always capable of disambiguating two sentences is shown by the example in (3): the two meanings of this sentence (high or low attachment of the prepositional phrase) have identical phonological interpretations.

(3) [(The girl)f (saw)f (the old mailman)f (with the binocular)f]I

In the phonological phrase, not only the domain but also the location of main prominence varies according to the value of the syntactic parameter, which establishes the relative order of head and complements. The main prominence established among its daughter constituents is rightmost in head-complement languages and leftmost in complement-head languages

(Nespor and Vogel, 1986). The main prominence of a phonological phrase thus gives a cue to some aspects of the internal syntactic structure of a string. The two sentences in (2), for example, are disambiguated because the prominences of the phonological phrases do not always fall on the same word.

Constituency at the level of the intonational phrase also makes disambiguation possible between two sentences with identical sequences of words, but different syntactic structures. One such case is the different prosody assigned to restrictive and nonrestrictive relative clauses, as exemplified in (4), where the analysis into intonational phrases is indicated.

(4) I like owls which come along unexpectedly
   a. (I like owls which come along unexpectedly)
   b. (I like owls) (which come along unexpectedly)

Two sentences cannot be disambiguated phonologically when their analysis into both phonological phrases and intonational phrases is identical, as is the case for the example in (3).

The relative prominence established among the phonological phrases that constitute an intonational phrase, always associated with a tone, gives a cue to the focus structure of a sentence: main prominence and focus are always aligned (Jackendoff, 1982). The syntactic constituent that bears the main prominence of the intonational phrase, as well as all the constituents that dominate it in the syntactic tree, may be interpreted as focused. When the focused constituent is as large as the entire clause, the main prominence of the intonational phrase is rightmost, that is, it falls on its last phonological phrase, independently of the syntax of the language (Hayes and Lahiri, 1991). In this case, a sentence is thus ambiguous as to its focus structure: different constituents can be interpreted as focused, as defined above. A sentence such as (5), with main prominence on the last word, can thus be interpreted as conveying new information on any of the constituents in a–c (Guasti and Nespor, 1999).

(5) John sent a letter to *Paul*
   a. Paul
   b. sent a letter to Paul
   c. John sent a letter to Paul

The different interpretations of the sentence in (5) depend on what is considered new information in a given context: (5) a–c are adequate interpretations in contexts in which the new information is as required by the shared (old) information indicated in (6) a–c, respectively.

(6) a. John sent a letter to someone
    b. John did something
    c. No shared information, i.e., "out of the blue"

If a sentence is meant to be semantically interpreted as having focus on a constituent different from the last one (the one bearing the main prominence), languages differ as to which remedies are used to eliminate the misalignment between stress and focus. If a language has the possibility of choosing a word order such that focus is final, this is the preferred option. By contrast, if such a word order is ungrammatical for independent syntactic reasons, then stress is placed on the focused constituent which remains in situ (Nespor and Guasti, submitted). English and Italian will be used to exemplify languages that remedy in different ways the misalignment of stress and focus.

Suppose we have a sentence composed of subject and verb. Focus can be on either of the two constituents.[1] Two famous examples taken from Bolinger (1965, 1982) are given in (7), where italics indicate main prominence within the intonational phrase:

(7) a. Truman *died*
    b. *Johnson* died

The first utterance is adequate in a context in which the old shared information is that Truman is ill and the new information is that he died. The second utterance is adequate in a context in which Johnson has not shown any sign of being ill and he unexpectedly dies.[2] In these two different pragmatic situations, the adequate Italian utterances are those in (8).

(8) a. Truman *è morto*
    b. E' morto *Johnson*

The reason for this difference between English and Italian lies in the fact that postverbal subjects are allowed in Italian but not in English.[3] Italian thus places the constituent with new information where the main prominence of I is; English places the main prominence where the new information is.

Imagine now a situation in which in a sentence composed of subject-verb-direct object-indirect object, as that in (9), the new information is on the direct object.

(9) I introduced Robert to Sarah

Italian places the direct object in final position, while English places the main prominence on the direct object which remains in situ, as seen in (10).[4]

(10) a. Ho presentato a Sara *Roberto*
      b. I introduced *Robert* to Sarah[5]

In this type of configurations, in English, the direct object cannot be in final position, since it must be adjacent to the verb from which it receives case. It is thus there that the main prominence is placed. In Italian, the adjacency of the direct object to the verb is not a requirement (Rizzi, 1986; Belletti and Schlonsky, 1995). The object can thus occupy the final position and be aligned with the main prominence of I (Guasti and Nespor, 1999).

In the next two sections I advance some hypotheses as to how infants may possibly use the prosodic contours of phonological phrases and intonational phrases described above to enter the syntax of their language of exposure.

Initialization of Grammar

So far, we have seen that utterances contain audible cues that signal specific syntactic properties of sentences. Besides being used to interpret sentences during speech processing (cf., e.g., the disambiguating role of prosody), these cues might be used in the bootstrapping of syntax in early language acquisition, as stated in the so-called phonological bootstrapping hypothesis (Gleitman and Wanner, 1982; Morgan, 1986; Morgan and Demuth, 1996).

According to the principles and parameters theory of grammar (Chomsky, 1980, 1981), grammars are defined by universal principles, and by parameters whose values may vary from language to language. The principles are thus part of the common endowment of the species, while the parameters must be set to the correct value for the language of exposure during language development.

Part of Jacques's experimental work in the past twenty years has concentrated on how rhythm can help the child in the development of the language organ. In Mehler et al. (1996) it is proposed that vowels, the most salient elements of speech, may be used to determine the rhythmic class of the language of exposure (cf. also Ramus, 1999; Ramus and Mehler, 1999). The identification of the language class, in turn, will allow the infant to keep languages apart in a multilingual environment, and possibly to determine the segmentation strategy to be used in language processing (Mehler et al., 1981; Cutler et al., 1986), as well as to fix the values of some parameters relating to syllable structure (Ramus, Nespor, and Mehler, 2000).

The question we may now ask is, Can infants use rhythmic regularities at higher levels to understand some other aspects of the grammar of their language? It is conceivable that vowels with different types of prominence may help bootstrapping other parameters of grammar. For example, one of the basic syntactic parameters determines whether the surface word order of a language is such that the head precedes or follows its complements. In a head-complement language, the object typically follows the verb and the language has a system of prepositions; moreover, main clauses typically precede subordinate clauses. In a complement-head language, the object typically precedes the verb and the language has a system of postpositions; moreover, main clauses typically follow subordinate clauses. Thus, by setting the head-complement parameter to the correct value, the child makes an important step into the discovery of the syntax of his or her language.

It has been proposed that the specific value of this parameter may be set at the prelexical stage, solely on the basis of prosody. In Mazuka's (1996) proposal, the setting of this parameter could be accomplished by setting the order of main and subordinate clauses through cues that signal the edges of intonational phrases. Then, deductively, the order of heads and complements can also be set. According to Nespor, Guasti, and Christophe (1996), instead, the relative prominence within phonological phrases would allow setting of the relative order of heads and complements. Deductively from this, the relative order of main and subordinate clauses may also be set. Whether infants are indeed able to discriminate two languages on the basis of one of these phonological cues awaits conclusive experimental data. Still, it has been shown that infants do discriminate sentences from two languages whose rhythm varies only in the location of main prominence within the phonological phrase (Christophe et al., submitted). In whichever way this parameter may be set, if it is set through prosody the big advantage is that a lot of information about word order can be gained at the prelexical stage, and independently of the segmentation of a string into words.

The setting of the head-complement parameter illustrates a case of a hypothesis based on data from theoretical linguistics that needs data from psycholinguistics to give plausibility to specific linguistic mechanisms that may guide language development.

If it turns out to be indeed the case that infants are sensitive to the relative prominence relations present in phonological phrases—that is, to whether the strongest element is leftmost or rightmost—and thus might use this information to set the head-complement parameter, we will have to start thinking about whether they can also use another type of prominence, specifically that within intonational phrases, to set other syntactic parameters. As Jacques once noticed, in fact, it would be surprising if just one syntactic parameter would be set through prosody, while the other ones would be set in a different way.

Some Speculations

Here are some thoughts concerning the possibility, still speculative at the moment, that other syntactic parameters may be set through prosody. This possibility must be further investigated in future research. We have seen above that the main prominence within the intonational phrase is aligned with the element in focus crosslinguistically. The location of this element, however, is not the same in all languages. Let us try to see what can be deduced from the location of the main I prominence. Before doing so, it should be noted that it is important that the nature of this prominence be different from that of the phonological phrase. Otherwise the two could never possibly be used to fix different parameters. In this respect, it can be observed that the main prominence within I is always associated with a tone and that, in general, it exhibits more pitch movement than lower-level prominences (Pierrehumbert, 1980; Hayes and Lahiri, 1991).

In languages in which the subject precedes the verb (SV), I prominence may or may not fall in initial position, as we have seen above. If prominence in initial position is allowed, it may signal that postverbal subjects are disallowed. We have seen, in fact, that if this possibility is available, it constitutes the unmarked option for stress-to-focus alignment. The possibility of having postverbal subjects is connected with the possibility of having null subjects (Rizzi 1982, 1994). By hearing main prominence at the beginning of an intonational phrase, it may thus also be deduced that the language has obligatory subjects.

In languages in which the verb precedes the object (VO), I main prominence in constituents other than the final (unmarked) or the initial, just mentioned, may signal that the direct object cannot be separated from the verb. Recall, in fact, that if the object can be separated from the verb, the preferred option to put focus on the object is to place it in I final position.

More generally, the more the prosody within an I is unmarked in a language, that is, with main prominence in final position, the more it can be deduced that the order of syntactic constituents is quite free. The more the prosody is variable within an I, the more we can deduce that word order is quite rigid. In the latter case, the specific location of nonfinal prominence may indicate the value of specific syntactic parameters. Since the main prominence of I is generally not just stronger than main prominence in f, but involves more pitch movement as well, it is not unfeasible that infants might use the two types of prominence to set different parameters.

The picture of initialization of language envisaged here is thus one in which specific characteristics of vowels allow determination of the rhythmic class of the language of exposure, establishment of the appropriate level of phonological representation, and a start to building syllabic structure. The location of vowels bearing the main prominence (stress) of the phonological phrase would allow the setting of the head-complement parameter, and vowels associated with the main prominence of the intonational phrase (pitch) would allow the setting of two additional parameters regulating word order.

Conclusions

In the beginning of this chapter, it was observed that one of the advantages of looking at one phenomenon, in this case language, from different points of view, in this case from that of generative grammar and that of psycholinguistics, is that new questions arise. The specific point exemplified here concerns the contribution that different levels of linguistic analyses (syntactic, phonological, and phonetic), on the one hand, and behavioral experiments with infants, on the other hand, could give to the understanding of the development of the language organ. That is, the behavior of infants in perception experiments designed to test their ability to discriminate entities hypothesized to be different in linguistic theories can inform us about the initialization of specific aspects or parts of grammar, in this case syntax.

Syntactic theory determines the parameters along which one language can differ from another. Phonological theory at the phrasal level determines whether and how specific syntactic variations are phonologically interpreted. Phonetics determines precisely what is there in the signal. Behavioral experiments determine whether infants can perceive the different signals to syntax and, in the affirmative case, render conceivable the hypothesis that they might use them to bootstrap different aspects of the grammar of exposure.

This way of looking at the integration of different fields of research to discover the mechanisms underlying language development is largely due to Jacques, who has always been eager to initiate and encourage interdisciplinary collaboration.

Acknowledgments

For comments on this chapter, I am grateful to Caterina Donati, Maria Teresa Guasti, Sharon Peperkamp (who gave up hours of sun and sea in Leipsoi for this purpose), and Luigi Rizzi. Most of all I thank Emmanuel

Dupoux for his insightful comments and for the care he gave to the editing of this chapter.

Notes

1. Note that focus should be distinguished from contrastive focus: the latter is used in a pragmatic context of correction (Nespor and Guasti, submitted).
2. This is not to say that the two utterances are appropriate exclusively in these pragmatic contexts.
3. There are a few exceptions to this generalization, represented by sentences starting with "here" or "there" such as
Here comes the train.
4. There are also other ways to convey the same information. For example, cleft sentences or sentences containing dislocated constituents.
5. There are cases in which double object constructions exemplified by sentences such as *I gave John a book* can also be used with the focus structure of (10)b.

References

Belletti, A., and Schlonsky, U. (1995). Inversion as focalization. Unpublished manuscript, University of Siena, Italy.

Bolinger, D. (1965). *Forms of English: Accent, morpheme, order.* Cambridge, MA: Harvard University Press.

Bolinger, D. (1982). Intonation and its parts. *Language, 58,* 505–533.

Brown, R. (1973). *A first language.* Cambridge, MA: Harvard University Press.

Chomsky, N. (1980). *Rules and representations.* New York: Columbia University Press.

Chomsky, N. (1981). Principles and parameters in syntactic structure. In N. Hornstein and D. Lightfoot (Eds.), *Explanations in linguistics* (pp. 32–75). London: Longman.

Christophe, A., Nespor, M., Guasti, T., and van Ooyen, B. (submitted). Prosodic structure and syntactic acquisition: The case of the head-direction parameter.

Clahsen, H., and Eisenbeiss, S. (1993). The development of DP in German child language. Presented at the 1993 SISSA Encounter in Cognitive Science, Trieste.

Cutler, A., Mehler, J., Norris, D., and Segui, J. (1986). The syllable's differing role in the segmentation of French and English. *Journal of Memory and Language, 25,* 385–400.

Gleitman, L. R., and Wanner, D. (1982). Language acquisition: The state of the state of the art. In D. Wanner and L. R. Gleitman (Eds.), *Language acquisition: The state of the art* (pp. 3–48). Cambridge, UK: Cambridge University Press.

Guasti, M.-T., and Nespor, M. (1999). Is syntax phonology-free? In R. Kager and W. Zonneveld (Eds.), *Phrasal phonology* (pp. 73–98). Nijmegen, Netherlands: Nijmegen University Press.

Hayes, B., and Lahiri, A. (1991). Bengali intonational phonology. *Natural Language and Linguistic Theory, 9,* 47–96.

Inkelas, S., and Zec, D. (Eds.) (1990). *The phonology-syntax connection.* Chicago: Chicago University Press.

Jackendoff, R. (1972). *Semantic interpretation in generative grammar.* Cambridge, MA: MIT Press.

Mazuka, R. (1996). Can a grammatical parameter be set before the first word? Prosodic contribution to early setting of a grammatical parameter. In J.L. Morgan and K. Demuth (Eds.), *Signal to syntax* (pp. 313–330). Mahwah, NJ: Erlbaum.

Mehler, J. (1963). Some effects of grammatical transformations on the recall of English sentences. *Journal of Verbal Learning and Verbal Behavior, 7,* 345–351.

Mehler, J., Dommergues, J. Y., Frauenfelder, U., and Segui, J. (1981). The syllable's role in speech segmentation. *Journal of Verbal Learning and Verbal Behavior, 20,* 298–305.

Mehler, J., Dupoux, E., Nazzi, T., and Dehaene-Lambertz, G. (1996). Coping with linguistic diversity. In J. L. Morgan and K. Demuth (Eds.), *Signal to syntax* (pp. 101–116). Mahwah, NJ: Erlbaum.

Meisel, J. M. (Ed.) (1992). *The acquisition of verb placement. Functional categories and V2 phenomena in language acquisition.* Dordrecht, Netherlands: Kluwer Academic Press.

Morgan, J. L. (1986). *From simple input to complex grammar.* Cambridge, MA: MIT Press.

Morgan, J. L., and Demuth, K. (Eds.) (1996). *Signal to syntax.* Mahwah, NJ: Erlbaum.

Nespor, M., and Guasti, M.-T. (submitted). Stress to focus alignment.

Nespor, M., and Vogel, I. (1986). *Prosodic phonology.* Dordrecht, Netherlands: Foris.

Nespor, M., Guasti, M.-T., and Christophe, A. (1996). Selecting word order: The rhythmic activation principle. In U. Kleinhenz (Ed.), *Interfaces in phonology* (pp. 1–26). Berlin. Academik Verlag.

Pierrehumbert, J. (1980). *The phonetics and phonology of English intonation.* PhD dissertation, MIT, Cambridge, MA. Distributed by Indiana University Linguistics Club.

Ramus, F. (1999). *Rhythme des langues et acquisition du langage.* PhD dissertation. Paris. Laboratoire de Sciences Cognitives et Psycholinguistiques. École des Hautes Etuder en Sciences Sociales.

Ramus, F., and Mehler, J. (1999). Language identification with suprasegmental cues: A study based on speech resynthesis. *Journal of the Acoustical Society of America, 105,* 512–521.

Ramus, F., Nespor, M., and Mehler, J. (2000). Correlates of linguistic rhythm in the speech signal. *Cognition, 72,* 1–28.

Rizzi, L. (1982). *Issues in Italian syntax.* Dordrecht, Netherlands: Foris.

Rizzi, L. (1986). Null objects in Italian and the theory of pro. *Linguistic Inquiry, 17,* 501–557.

Rizzi, L. (1994). Early null subjects and root null subjects. In T. Hoekstra and B. Schwartz (Eds.), *Language acquisition studies in generative grammar.* Amsterdam: Benjamin, 151–176.

Selkirk, E. O. (1984). *Phonology and syntax: The relation between sound and structure.* Cambridge, MA: MIT Press.

Truckenbrodt, H. (1999). On the relation between syntactic phrases and phonological phrases. *Linguistic Inquiry, 30,* 219–256.

Werker, J., and Tees, R. (1984). Cross language speech perception: Evidence for perception reorganization during the first year of life. *Infant Behavior and Development, 7,* 49–63.

8

Some Sentences on Our Consciousness of Sentences

Thomas G. Bever and David J. Townsend

"My end is my beginning."
—Luciano Berio, 1975

"All the . . . furniture of earth . . . have not any subsistence without a mind."
—George Berkeley, 1710

"Consciousness . . . [is] . . . a river. . . . Like a bird's flight it seems to be made of an alternation of flights and perchings. . . . Consciousnesses melt into each other like dissolving views."
—William James, 1890

"Instead of . . . a single stream [of consciousness], multiple channels . . . do their various things, creating Multiple Drafts as they go . . . some [drafts] get promoted to further functional roles . . . by the activity of a virtual machine in the brain."
—Daniel Dennett, 1991

"I see everything twice! I see everything twice."
—Joseph Heller, 1961

Overture: How Many Times Do We Perceive a Thing?

We have a problem common to those who depend on eyeglasses: each morning, there is a ritual hunt to find where we left them the night before. The Catch-22 in this, of course, is that without our glasses on, it is hard to see them, since they are a composite of spidery metal and maximally invisible lenses. We ritually wander around the house checking my habitual domestic locations, like a dog on optimistic visits to its most likely backyard bone mounds. Inevitably, we find the glasses (the first day that we do not do so may be our last).

The fact that this is an everyday perceptual event has allowed us to analyze it into those parts of which we can be conscious. We are sure that when we finally find them, we "see" the glasses twice. The first time is fleeting, but with a clear memory of an event that appears "preperceptual." We believe we saw them the first time only after we see them for the second time, a split second later. In ordinary life, the second percept is what we think of as the real percept. The first percept is almost immediately suffused by the second, and is primarily conscious as a memory.

The phenomenological difference between the two representations is striking. The first is a foggy mixture of a skeletal configuration and one or two clear parts—an earpiece, or nosepiece connected to a chimerical outline of the rest—the combination of global and particular is "the glasses" at a conceptual level rather than in visual detail. The second representation is explosive, sharp and complete overall, no part stands out more than the others, it has suddenly become the coherent visual entity, "glasses." The formation of this representation is the perceptual event that we normally would describe as, "Then, we found our glasses."

Fortunately, we are not aware of seeing everything two times. The daily eclectic search for my glasses is ingrained, and perhaps has allowed us to peer into its stages more thoroughly than into the random events of quotidien life. The cautious reader (or a scholar of the downfall of Titchnerian introspectionism) may object that our memory is tricking us, that the first "prepercept" is merely a shadow of the in a retrospective backwash. Perhaps so, but the fundamental point would remain potentially true of all perceptual acts: we perceive everything twice, once as a metonymously schematic object, once as clear image of the object.

Let's turn to sentence comprehension and see what it might tell us about the possible reason for multiple representations in perception. What follows is a brief précis of our just-published book (Townsend and Bever, 2001). We show that the long-neglected analysis-by-synthesis model of comprehension can offer an explanation for the role of successive representations in sentence comprehension. At the end of this summary, we suggest some hypothesis about the implications of this for consciousness of sentences and other objects.

Theme 1: Sentences Sound Clearer

Cl#ss#c#l s#m# th#n n r#nd#m t th# s#nt#nc#s a th#t s#mpl#rr #s wh#n w#rds f#nd#ng #n w#rds ndrst#nd #r# t# #rd#r. That is, #t #s a cl#ss#c#l f#nd#ng th#t w#rds #n s#nt#nc#s #r# s#mpl#r t# r#c#gn#z# th#n th# s#m# w#rds #n r#nd#m #rd#r. When the words are presented in a noisy acoustic background to make the perception difficult (or with visual noise added as in the preceding sentences), people get more words right when they are in sentence order. But most important for this discussion, the words also sound (or look) physically clearer as one takes them in and recognizes them.

The apparent sensory punch of words-in-sentences may not be surprising, for a number of reasons. First, sentences carry meaning, which we can relate to separately existing knowledge. It is intuitively reasonable that accessing sentence meaning provides a redundant source of information about the words. An empirical problem with this explanation is that it is also relatively easy to recognize words in sentences with little meaning ("The slithey toves did gyre and gymbal") or even with counterintuitive meaning ("The electrolytic mailman bit the religious dog"). Furthermore, even if it is intuitive, it is a scientific mystery how accessing the meaning at the sentence level could enhance the physical clarity of the individual words.

It is also a fact that sentences have formalizable syntactic structure that might provide a different kind of redundant information enhancing the acoustic clarity of the words. In "the big city-i   was rich", there is a surface hierarchical segmentation (indicated by the relative size of spaces between words). In many theories, there is also an abstract level of representation that links distant phrases and canonical locations together. This can underlie important differences in sentences that appear the same superficially: in "the big city-i   was attacked [i]", "city-i" is linked to the usual patient position immediately after the verb by the abstract coindexed trace element, "[i]." This dual representation for "city" captures the intuitive fact that it is both the topic of the sentence, and the patient of the verb. There are numerous formal models of how this linking is computed. Many models presuppose a form of "derivation"—in the most transparent case, the sentence formation involves a stage in which the words are initially placed in the marked location and subsequently moved to their apparent location (e.g., "city" would be originally in postverbal patient position of "attack" at a point when "attack" has no subject, and is then moved to the beginning of the sentence). This movement creates the notion of "derivation," in which the structure is built and deformed in a formal sequence of operations. Some models overlay the distant relations between phrases with a different kind of structure, but they are arguably the equivalent of the more transparent kind of derivational sequence of operations.

In any case, all sentence structure models involve considerable abstract symbolic manipulation. This complexity is a mixed theoretical blessing if our goal is to use syntactic structure as the explanation of why sentences sound clear. On the one hand, syntactic structure might afford another kind of representation that relates the words to each other. But it also involves a lot of computational work, which one might expect would actually take mental resources *away* from recognizing the words. And, as in the case of meaning, the syntactic model itself does not explain how syntactic relations between adjacent words and distant phrases enhance the clarity of the words themselves.

Theme 2: Syntax Is Real

Syntax accounts for an essential mental fact: with little or no training, people can distinguish grammatical from ungrammatical sentences in their language, often regardless of meaning. English speakers know that there is something wrong with "The electrolytic mailman bite the religious dog" which has nothing to do with its unusual meaning. On occasion, syntacticians rely on extremely subtle or unconvincing grammaticality discriminations; this often tempts psychologists to dismiss the mental relevance of the entire linguistic enterprise. But the clear cases of grammaticality distinctions remain, and support the relevance of syntactic structures and at least the broad outline of how they are computed.

Psychology aspires to be an experimental science, so despite the many mental facts syntax accounts for, psychologists have doggedly attempted to prove that syntax is "really" real. The goal is to show that syntax plays a role in all language behaviors, not just rendering grammaticality intuitions. Jacques Mehler's dissertation (Mehler, 1963) is a classic example, demonstrating that the abstract level of syntactic structure plays an important role in how sentences are remembered. Other studies (many by Mehler as well) revealed that both surface and abstract structures play a role in ongoing sentence comprehension.

For this discussion, the most important aspect of linguistic structure is the postulation of an abstract level at which structurally related phrases are adjacent, before they are moved to distant surface positions. This movement leaves behind a coindexed trace as a relic of the original location and relation (or a silent copy of it in some syntactic theories), and thus the complete syntactic structure provides two representations of such an element. An empirical prediction of this is that elements that have an extra trace should be relatively salient during comprehension: For example, "big" in "the big city-i was attacked [i]" should be more salient than in "the big city-i was rich." This is the case: a variety of experiments have shown it takes less time to recognize that "big" was in trace-sentences like the first than in trace-free sentences like the second. Most important for this discussion is the fact that the relative saliency of elements with trace becomes clear only after the proposition is complete: probes presented right at the trace position itself generally do not show the relative saliency immediately.

The struggle to prove the "psychological reality" of syntax is not over. Grammatical formalisms change and compete like weeds. The hapless experimental psychologist is often left behind in the formal dust. We have chosen the experiments on "trace" because that way of representing phrase movement from one level to another has corresponding formalisms in all versions of transformational grammar. Indeed, while the comprehension trace experiments have appeared only in the last decade, the corresponding predictions could have been equally well motivated by the 1957 model of syntax in *Syntactic Structures* (Chomsky, 1957). (E.g., the critical word would be represented both in the "kernel" and passive "surface" sentence structure, and hence more salient psychologically. Such studies were not done then, perhaps because of the fact that "cognition" in the sixties was preoccupied with memory and not perception.)

The upshot of much of the last forty years of psycholinguistics since Mehler's dissertation is that comprehension involves assignment of syntactic representations. Any model of language understanding must take this into account.

Theme 2, Variation: Associations Are Real

The cognitive revolution of the 1960s was a triumph of rationalist structuralism over behaviorist associationism. Abstract structures, such as the initial level of representation of sentences, were recognized by all as a dangerous challenge to the "terminal metapostulate" of behaviorism—the claim that all theoretical terms and all corresponding inner psychological entities are "grounded" in more superficial entities, ultimately in explicit behaviors and stimuli. There is no obvious way in which inner syntactic forms are extracted from outer ones. Indeed, syntactic theory suggests that the relation is in the opposite direction. This sharpened the focus of research on the behavioral role of such inner forms, and as the evidence mounted, behaviorist structures lapsed into the obscurity of irrelevance.

But arguments against behaviorism are not direct arguments against associationism. Associationism is merely the doctrine that all mental relations are associative rather than categorical, usually built up out of repeated experiences or co-occurrences. And the apparent arguments for associationism are compelling. No one can deny that a vast proportion of everyday life is based on habits. We rely on them all the time, we see how we can train them in animals, and we use the notion as a powerful explanation of many forms of natural behavior.

One of the arguments against the idea that absolutely everything is based on associative habits is the complexity of normal behaviors, such as the comprehension of language. Furthermore, the existence of inner levels of representation would seem to defy associationistic explanation. The traditional answer by associationists has been that we underestimate the subtlety of behavior that a large complex associative network can compute. Even inner representations might be connected by myriads of associations so complex as to mimic the appearance of hierarchies, part-whole relations, relations at a distance, and other kinds of categorical facts. This is the thrust of a major school of modern connectionism—to apply relatively complex networks of associatively connected units, and thereby explain away apparent categorical and rulelike structures as the result of "pattern completion" and the exigencies of associative information compression. The idea is that with enough experience, associative networks produce behavior that looks structural in nature, but we allegedly "know" it really is not because we manufactured the machine ourselves.

That, of course, is just one of many human conceits. We manufacture electric generators, but we do not make the laws or the electrons that the laws govern—rather, we construct machines that manipulate those phenomena to our own ends. Similarly, we might someday wire up an associative network as massively interconnected and complex as a human brain, and watch it learn language. The language-learning process in the machine would be as much a mystery as in a human child. We would be back to the 1960s' cognitive psychology joke: "someday the artificial intelligencers will create a plausibly human robot, and then we will have to practice psychophysics, theoretical psychology, and experimental "neuro"science on it to figure out how it works." We will have a new division of international psychological associations: "robopsychology."

Similar modesty is indicated about the mimicking achievements of today's modest connectionist models. The ability of a model to converge on 90 percent accuracy in computing a specific syntactic structure after thousands of training trials, or to differentiate lexical classes based on repeated exposure to local contexts, is an important achievement. It stands as an existence proof, that statistically reliable information is available in the language stimulus world that can support the induction of recognition patterns. But to claim that such achievements show we can account for actual categorical syntactic structure is not warranted. It would be like claiming that the amazing height of a medieval church spire shows that humanity can reach the heavens.

Yet, whatever the limits of associationism, habits exist and dominate most of life. Any adequate model of comprehension must find the appropriate computational locus for their operation and influence.

Development: Analysis by Synthesis

We are left with two truths. most of the time we do what we usually do, using experientially based superficial habits. But sometimes we create novelty, using categorical computations that reflect sequential symbolic operations. How can we reconcile these politically opposed, but valid approaches to language behavior in a model of perception?

It is fashionable today to assert that the best models of behavior are "hybrids" between connectionist and symbolic systems. Such models combine the two kinds of approaches by definition, but the real test is the particular architecture that does this. The architecture is constrained by several features:

1. Associative information operates on relatively superficial representations and is immediately available.
2. Readily available surface information includes the lexical items in sequence, a rough phrasing structure, and a likely meaning.
3. Syntax is derivational—it involves a series of steps in building up sentence structure, which can obscure the initial computational stages in the surface form.

The first two points suggest that comprehension can proceed directly from surface cues to a primitive syntax, and associative relations between words and phrases that converge on particular meanings. This is exactly what many connectionist modelers assume, as they build their toy models and extrapolate them to the entire language. For example, the sentence "the city was rich" might be interpreted directly, based on separate analyses of the words, and a rough phrase structure, as "NP be PREDicate." "Rich" is a lexical item which carries the semantic information that its syntactic subject is semantically a stative experiencer. Similarly, "the city was ruined" can be analyzed in the same way. Though it looks like a passive form, "ruined" is actually a lexical item, which acts like a normal adjective; for example, it can modify nouns, as in "the ruined/rich city lay before us." Accordingly, the word itself carries the semantic information that the syntactic subject is a stative patient.

"The city was attacked" can be initially understood on the same superficial template, "NP be PREDicate." But the syntactic analysis is actually wrong, even if the associated meaning is roughly correct. "Attacked" is not a lexical item, as reflected in the oddity of "*the attacked city lay before us" (contrast that with the acceptable, "the ruined/rich city lay before us"). Thus, the correct syntactic analysis should reflect the movement of "city" from patient to subject position, as represented in the surface form, "the city-i was attacked [i]." What is a possible model for assigning this derivationally based syntactic structure?

The derivational nature of syntax makes it difficult to go back from the surface form to the original compositional stages. Like an airplane, the grammar runs in "forward" only. And, as with an airplane, the only way to retrace where it started and how it arrived at a current position is to go back to a likely starting point and recapitulate the potential journey, marking the route and checking to make sure it arrives at the correct destination. In the case of syntax, this means re-creating possible derivations from likely starting points, and checking to be sure they arrive at the right surface structure. The initial analysis provides a number of pointers as to the likely starting configuration. Individual lexical items are recognized, as well as major phrases, along with a potential meaning that relates the phrases in functional categories. This sets rich constraints on the possible starting configuration and derivation to a surface form. In fact, it corresponds well to the initial derivational stage in current "minimalist" theories of syntax, which start with a "numeration" of lexical items and functional categories, and build up the surface form from there (see Chomsky, 1995).

So, we have two phases in comprehension: an initial analysis of a likely meaning, and a recapitulation of the complete derivational syntactic structure consistent with the form and meaning. This invites a comprehension architecture which proceeds in several major stages:

a. Analyze the string into lexical sequences broken into major phrases.

b. Analyze the structure in (a) for a likely meaning using canonical syntactic patterns and associative semantic information.

c. Take the output of (a) and (b) as input to a syntactic derivation.

d. Compare the output of (c) and (a): if they are identical, then the meaning in (b) and the structure in (c) are correct. If they are not identical, reanalyze (a) and reinitiate the process.

In words, this is equivalent to a traditional "analysis-by-synthesis" model, which enjoyed much currency for a while. On this model, candidate derivations are computed and matched to a temporarily stored representation of the surface input. When there is a match, the derivation is assigned as part of the overall representation. Such models account for the active nature of perception, but most have had a major flaw that contributed to the loss of interest in them. Since the starting form of a derivation is "abstract," it is unclear how a surface analysis can constrain potential derivations to be anywhere near correct. Our current version lacks this weakness: the analysis into lexical items and functional categories gives exactly what is needed to start a derivation in minimalist syntax that is likely to converge on the correct surface analysis.

In principle, the model works. The question is, does it correspond to real facts about comprehension? Our book presents the model and an eclectic compilation of existing and new data in its favor. Some salient features of the model and supporting arguments and predictions that we note are the following:

1. A complete syntactic structure is assigned fairly late in the process. Past studies of the formation of a complete phrase structure have suggested that it is assigned only after a fraction of a second. For example, mislocation of interrupting "clicks" is controlled by fine details of phrase structure only after a brief interval. Similarly, as mentioned above, evidence for the saliency of movement traces occurs only a few hundred milliseconds after their actual position. Finally, ungrammaticality based on movement constraints is detected relatively slowly (both in explicit grammaticality judgment tasks, and reflected in evoked brain response patterns). All three facts follow from the view that the complete syntactic details, including derivationally based traces, are assigned as part of the syntactic recapitulation of structure.

2. Initially, the meaning is computed from an incomplete, and sometimes incorrect syntax. A number of researchers have suggested that meaning can be derived from syntactically deficient analysis. Not only is this an assumption of much of connectionist modeling and also an assumption of a number of current symbolic comprehension models, it also appears to be characteristic of much comprehension by children. A typical and frequently cited example is the idea that passives are initially comprehended as "NP BE Predicate," in which the PREDicate is itself a complex adjective. This coordinates exactly with the proposal we have made. In general, it is striking to note that constructions with Nounphrase trace are just those that also have parallel nontrace constructions, as below:

The city-i was attacked [i]: the city was ruined
The city-i was likely [i] to attack: the city was eager to attack
The city-i appeared [i] to attack; the city rose to attack

This may be a chance matter, but it is also predicted by our model: Nptrace can occur only in constructions which can be initially understood by the (mis)application of an independently supported perceptual template.

3. We "understand" everything twice: once on the initial pass, and once when assigning a correct syntactic derivation. First, one might ask, If the initial meaning is successfully derived from surface cues, why does the system recapitulate a correct syntax at all? If associative and preliminary structural information is sufficient for discovery of a meaning, why gild the lily? The answer is reflected in the fact that Dehli copra pricing techniques are quite ununderstandable . . . well, no, it is actually reflected in the fact that you were able to understand that proposition about Delhi and copra prices, despite the fact that the syntax is odd and the meaning is even odder and totally out of context. The functional role of recapitulating the syntax and checking the meaning is to make sure that the initial comprehension was correct. Otherwise, we would be consigned only to understand frequent constructions that convey common and predictable meanings that make sense in the local context.

The formation of dual meaning representations is the most theoretical claim at the moment, and is largely untested. It certainly is intuitive that we understand sentences immediately as they are apprehended. Thus, the claim that we initially understand sentences online substantiates the first pass in the model. The second phase of forming a complete syntax and meaning is more controversial, but subject to empirical test. A simple demonstration is the fact that it can take some time to realize that a superficially well-formed sentence is defective. Consider your intuitions as you peruse a run-on sentence like "More people have gone to Russia than I have." At first, it seems plausible and you think you understand it. This follows from the fact that it meets a superficial template and appears to have a meaning. But then, as you reconstitute it, it does not compute, and you realize that it is not actually a sentence, and you are actually quite confused about what it really means.

There are also predictions of the two-phase model that we can test experimentally. For example, it predicts that we go through two phases of understanding syntactic passive sentences like "the city-i was attacked." In the first phase, we actually misconstrue the city as experiencer of a stative predicate; in the second phase, we understand "city" correctly as the patient of an action. We are currently testing this prediction in a variety of ways.

Recapitulation: Sentences Sound Clear because Comprehension Involves the Formation of Two Surface Representations

We now return to the classic and enduring fact about sentences mentioned in the first section: their constituents actually sound extraclear. We note that the analysis-by-synthesis model affords a direct explanation of this. In comprehension we form and compare two surface representations of the sentence: one as part of the initial analysis, one as a result of the syntactic reconstitution. Thus, comprehension contributes an extra surface representation to the perception of the word sequence, as well as adding a check for the identity of the two representations. While the details await more specification, the model involves multiple manipulations of surface representations, which affords an explanation of why the sentences seem so clear physically.

Coda: Is Recapitulation of Object Representation an Important Component of "Consciousness?"

Our finale goes back to our overture. We have sketched a comprehension model which chases its tail, at least once—it starts with a meaning, and then uses that to begin a syntactic derivation which rearrives at the same surface form and enriched meaning. This presents the comprehension process as a set of emerging representations which converge. We have sketched the representations as occurring in series, but, of course, they could be partially computed in parallel in many actual cases. In this respect, we can infer that our model of comprehension is an instance of Dennett's "multiple stage" model of consciousness (Dennett, 1991). We think we understand a sentence and are conscious of it as a unified experience. Yet, analysis and experiment suggest that this apparently unified process is actually composed of converging operations of quite different kinds.

We are left with the puzzle of why we phenomenologically give priority to the final stage of processing. Why does the "real" perception of our eyeglasses of which we are conscious seem to be the final one? Why does the "real" comprehension of the sentence seem to be the one that includes a syntactically complete representation? There are several possible answers.

- The "latest" draft seems to be the most real by co-opting the earlier ones.
- The most complete draft seems to be the most real.
- The draft that depends most on internal computations seems to be the most real.

In a real machine, these explanations tend to co-occur. But I maintain the top-down computational bias Mehler and I acquired in the heady days of the sixties "cognitive revolution." So we favor the last alternative. The appearance of reality depends most strongly on internally generated representations of it.

This is not mere rhetoric. It can be tested. And probably will be. As in other cases, the precision of linguistic theory allows us to quantify the computational complexity of sentences. In this way, we can test the relative effect of complexity on conscious clarity of the words in sentences. The prediction is bold: the more complex a sentence is, the more acoustically clear the words in it (so long as it can be understood). This is somewhat counterintuitive, which is all to the better if it turns out to be true.

Thus, we end at our beginning. Jacques, salut, and many more!

References

Chomsky, N. (1957). *Syntactic structures.* The Hague: Mouton.

Chomsky, N. (1995). *The minimalist program.* Cambridge, MA: MIT Press.

Dennett, D. (1991). *Consciousness explained.* Boston: Little, Brown.

Mehler, J. R. (1963). Some effects of grammatical transformations on the recall of English sentences. *Journal of Verbal Learning and Verbal Behavior, 2,* 346–351.

Miller, G. A., Heise, G. A., and Lichten, W. (1951). The intelligibility of speech as a function of the context of the test materials. *Journal of Experimental Psychology, 41,* 329–335.

Miller, G. A., and Isard, S. (1963). Some perceptual consequences of linguistic rules. *Journal of Verbal Learning and Verbal Behavior, 2,* 217–228.

Townsend, D. J. and Bever, T. G. (2001). *Sentence comprehension.* Cambridge, MA: MIT Press.

9

Four Decades of Rules and Associations, or Whatever Happened to the Past Tense Debate?

Steven Pinker

For fifteen years I have explored an issue that was presciently outlined in 1967 by Jacques Mehler's mentor and coauthor, George Miller. In his essay "The Psycholinguists," Miller wrote:

For several days I carried in my pocket a small white card on which was typed UNDERSTANDER. On suitable occasions I would hand it to someone. "How do you pronounce this?," I asked.
 He pronounced it.
 "Is it an English word?"
 He hesitated. "I haven't seen it used very much. I'm not sure."
 "Do you know what it means?"
 "I suppose it means one who understands."
 I thanked him and changed the subject.
 Of course, UNDERSTANDER *is* an English word, but to find it you must look in a large dictionary where you will probably read that it is "now rare." Rare enough, I think, for none of my respondents to have seen it before. Nevertheless, they all answered in the same way. Nobody seemed surprised. Nobody wondered how he could understand and pronounce a word without knowing whether it was a word. Everybody put the main stress on the third syllable and constructed a meaning from the verb "to understand" and the agentive suffix -*er*. Familiar morphological rules of English were applied as a matter of course, even though the combination was completely novel.
 Probably no one but a psycholinguist captured by the ingenuous behavioristic theory that words are vocal responses conditioned to occur in the presence of appropriate stimuli would find anything exceptional in this. Since none of my friends had seen the word before, and so could not have been "conditioned" to give the responses they did, how would this theory account for their "verbal behavior"? Advocates of a conditioning theory of meaning—and there are several distinguished scientists among them—would probably explain linguistic productivity in terms of "conditioned generalizations." They could argue that my respondents had been conditioned to the word *understand* and to the suffix -*er*; responses to their union could conceivably be counted as instances of stimulus generalization. In this way, novel responses could occur without special training.

Although a surprising amount of psychological ingenuity has been invested in this kind of argument, it is difficult to estimate its value. No one has carried the theory through for all the related combinations that must be explained simultaneously. One can speculate, however, that there would have to be many different kinds of generalization, each with a carefully defined range of applicability. For example, it would be necessary to explain why "understander" is acceptable, whereas "erunderstand" is not. Worked out in detail, such a theory would become a sort of Pavlovian paraphrase of a linguistic description. Of course, if one believes there is some essential difference between behavior governed by conditioned habits and behavior governed by rules, the paraphrase could never be more than a vast intellectual pun. (Miller, 1967, pp. 80–82)

Twenty years later, David Rumelhart and James McClelland (Rumelhart and McClelland, 1986) took up the challenge to work out such a theory in detail. They chose Miller's example of people's ability to suffix words productively, focusing on the past tense suffix *-ed* that turns *walk* into *walked,* similar to the *-er* that turns *understand* into *understander.* The past tense suffix was an apt choice because productivity had been demonstrated in children as young as four years old. When children are told, "Here is a man who knows how to *rick;* he did the same thing yesterday; he____," they supply the appropriate novel form *ricked* (Berko, 1958). Rumelhart and McClelland wanted to account for the phenomenon using the theory of stimulus generalization that Miller was so skeptical of. They recognized that the main hurdle was to carry the theory through for "all the related combinations that must be explained simultaneously," and to implement "many different kinds of generalization, each with a carefully defined range of applicability." The English past tense system makes this requirement acute. Alongside the thousands of regular verbs that add *-ed,* there are about 180 irregular verbs of varying degrees of systematicity, such as *come-came, feel-felt,* and *teach-taught.*

Rumelhart and McClelland's classic associationist model of the past tense used the then revolutionary, now familiar approach of parallel distributed processing (PDP), or connectionism, whose models are ideal for computing "many different kinds of related generalizations simultaneously," Miller's main challenge. Their model acquired the past tense forms of hundreds of verbs, generalized properly to dozens of new verbs. More strikingly, it displayed a number of phenomena known to characterize children's behavior, most notably their overregularization of irregular verbs in errors such as *breaked* and *comed.* But the model had no explicit representation of words or rules; it simply mapped from units standing for the sounds of the verb stem to units standing for the sounds of the past tense form. Its apparent success led Rumelhart and McClelland to conclude that they had answered Miller's challenge that no alternative to a rule system could account for the productivity seen when people produce and understand novel affixed words.

The Rumelhart-McClelland (RM) model irrevocably changed the study of human language. Miller's challenge has been met, and it is no longer possible to treat the mere existence of linguistic productivity as evidence for rules in the head. To determine whether modern, sophisticated associative networks are no more than "a vast intellectual pun," or whether, as Rumelhart and McClelland suggested, it is the linguist's rule system that is the pun, finer-grained analyses are needed.

In 1988 Jacques and I edited a special issue of *Cognition* (Pinker and Mehler, 1988) containing an analysis of the RM model by Alan Prince and myself (Pinker and Prince, 1988), together with papers by Jerry Fodor and Zenon Pylyshyn (1988) and by Joel Lachter and Tom Bever (1988), which also discussed the model. Prince and I heavily criticized the model on a number of grounds. Our criticisms did not go unanswered, nor did we let the answers to our criticisms go unanswered. *Connections and Symbols,* the special issue that Jacques and I edited, contained the first of about a hundred papers on the debate, sixteen of them published in *Cognition* alone (see Marcus, 2001; and Pinker, 1999, for reviews). Every empirical claim in Pinker and Prince (1988) has been further examined, and twenty-five connectionist models purporting to fix the flaws of the RM model have been reported.

Now that we are in the fourth decade of the debate that Miller began in the early days of psycholinguistics, where do we stand? Are there rules in the head, as Miller suggested, or do people only generalize by similarity to trained forms? Or are we stalemated, as many bystanders seem to think? It would be a depressing reflection of the field if all these data, verbiage, and computer code left the issues as unresolved as they were when the issues were first joined. But I believe that a coherent picture has taken shape. In this chapter I summarize my view of the current resolution of the debate, growing out of my collaboration with Iris Berent, Harald Clahsen, Gary Marcus, Alan Prince, Michael Ullman, and others.

Not surprisingly, considering the longevity of the debate, the emerging picture embraces elements of both sides. I think the evidence supports a modified version of the traditional words-plus-rules theory in which irregular forms, being unpredictable, are stored in memory as individual words, and regular forms are generated by rule, just like other productive complex constructions such as phrases and sentences. Memory, however, is not just a list of slots, but is partly associative: features are linked to features—as in the connectionist pattern associators—as well as words being linked to words. This means that irregular verbs are predicted to show the kinds of associative effects that are well modeled by pattern associators: families of similar irregular verbs (e.g., *fling-flung, cling-clung, sling-slung*) are easier to store and recall, because similar verbs repeatedly strengthen a single set of connections for their overlapping material, and people are occasionally prone to generalizing irregular patterns to new verbs similar to known ones displaying that pattern (e.g., as in *spling-splung*) because the new verbs contain features that have been associated with existing irregular families.

On the other hand, I believe the evidence shows that *regular* verbs are generated by a linguistic rule of the kind that Miller alluded to. Whereas irregular inflection is inherently linked to memorized words or forms similar to them, people can apply regular inflection to *any* word, regardless of its memory status. Many phenomena of linguistic structure and productivity can be parsimoniously explained by the simple prediction that whenever memorized forms are not accessed, for any reason, irregular inflection is suppressed and regular inflection is applied.

These diverse phenomena and their explanations have recently been summarized in Pinker (1999), but that book was aimed at exploring the linguistic and psychological phenomena more than at contrasting the connectionist and words-and-rules approaches directly. Here I lay out the major phenomena and compare the explanations of the words-and-rules theory to those proposed by connectionist modelers in the dozen years since the special issue appeared (see also Marcus, 2001). In these models, both regular and irregular forms are generated by a single pattern-associator memory; symbol concatenation operations and hierarchical linguistic structures are eschewed, as they were in the original RM model.

Before beginning it is important to realize how far backward I am bending even in comparing pattern-associator models of the past tense to the traditional words-and-rules theory as a general test case of the relative merits of connectionist and traditional linguistic theories of language. Pattern-associator models ignore so many key features of language that even if they did succeed in capturing the facts of the past tense, no one would be justified in concluding that they are viable models of language or have made rules obsolete, as many connectionists claim. Here are the rhetorical concessions we have made in comparing models of the generation of past tense forms:

• First, the models have never seriously dealt with the problem of phonological representation, which in standard linguistic theories requires a hierarchical tree. Instead, they tend to use the problematic "Wickelfeature" representation (unordered sets of feature trigrams), or to artificially restrict the vocabulary to a subset of English, such as consonant-vowel-consonant (CVC) monosyllables.[1]

• Second, the models account only for the *production* of past tense forms; they do not *recognize* such forms (e.g., for the purposes of speech production or grammaticality judgments), and therefore require a second, redundant network to do so.[2] Clearly, we need a representation of information for inflection that can be accessed in either direction, because people do not separately learn to produce and to comprehend the past tense form of every word.

• Third, the models are trained by a teacher who feeds them pairs consisting of a verb stem and its correct past tense form. This is based on the assumption that children, when hearing a past tense form in their parents' speech, recognize that it *is* the past tense form of a familiar verb, dredge the verb stem out of memory, feed it into their past tense network, and silently compare their network's output with what they just heard. How a child is supposed to do all this without the benefit of the lexical and grammatical machinery that the connectionists claim to have made obsolete has never been explained.

• Fourth, the models are studied in isolation of the rest of the language system. The modeler spoon-feeds verb stems and then peers at the model's output; the myriad problems of deciding whether to inflect a verb to start with, and if so with what inflection, are finessed. As is the process of feeding the output into the right slot in a phrase or a larger word such as a compound.

162 S. Pinker

- Fifth, the models are restricted to the relatively simple task of inflecting a single word. Complex, multiaffix morphology (as seen in polysynthetic languages) and all of syntax and compositional semantics are almost entirely ignored.

In contrast, the words-and-rules theory treats the past tense as a mere example of the kind of symbol manipulation and modular design that characterizes the language system in general. It has already been scaled down from more articulated theories, and so does not face the severe problems of scaling up that would plague the pattern-associator approach even if that approach succeeded at the past tense. That having been said, let's see whether they do succeed in five areas originally raised in *Connections and Symbols*.

1. Reliance of generalization on similarity. Prince and I noted that the RM model showed puzzling failures in generalizing the regular pattern to many novel verbs. For example, it turned *mail* into *membled,* and failed to generate any form at all for *jump* and *pump*. We conjectured that the problem came from the fact that the model generalizes by similarity to trained exemplars. It generalizes inflection to new words because they overlap the phonological input units for previously trained similar words and can co-opt their connections to phonological output units to past tense sounds. It does not process *symbols* such as "verb," which can embrace an entire class of words regardless of their phonological content. Therefore the model could not generate past tense forms for simple verbs that were not sufficiently similar to those it had been trained on. Whereas irregular forms may indeed be generalized by similarity, as in pattern-associator models, the essence of regular generalizations is the ability to concatenate symbols.

Sandeep Prasada and I (1993) (see also Egedi and Sproat, 1991; Sproat, 1992) confirmed the conjecture by showing that the trained RM model did a reasonably good impersonation of the human being when it comes to generalizing irregular patterns: both model and human being converted *spling* to *splung,* generalizing the pattern from similar *cling-clung*. But with the regular words, people and the model diverged: both people and the model could convert *plip* (similar to existing verbs such as *flip* and *clip*) to *plipped,* but only people, not the model, could convert *ploamph* (not similar to any existing verb) to *ploamphed*. The model instead produced gibberish such as *ploamph-bro, smeej-leafloag,* and *frilg-freezled.* Lacking a symbol, and confined to associating bits of sound with bits of sound, the model has nothing to fall back on if a new item doesn't overlap similar, previously trained items, and can only cough up a hairball of the bits and pieces that are closest to the ones that it *has* been trained on. People, in contrast, reason that a verb is a verb, and, no matter how strange the verb sounds, they can hang an *-ed* on the end of it.

The problem of computing coherent past tense forms for novel-sounding verbs still has no satisfactory solution in the framework of standard connectionist pattern-associator memories (see Marcus, 2001, for extensive analysis; as well as Marcus et al., 1995; Prasada and Pinker, 1993). Several modelers, stymied by the models' habit of outputting gibberish, have hardwired various patches into their model that are tailor-made for regular verbs. One team of modelers included a second pathway of connections that linked every input unit to its twin in the output, implementing by brute force the copying operation of a rule (MacWhinney and Leinbach, 1991). Another team added an innate clean-up network in which the units for *-ed* strengthen the units for an unchanged stem vowel and inhibit the units for a changed vowel, shamelessly wiring in the English past tense rule (Hare, Elman, and Daugherty, 1995). And many connectionist modelers have given up trying to generate past tense forms altogether. Their output layer contains exactly one unit for every past tense suffix or vowel change, turning inflection into a multiple-choice test among a few innate possibilities (e.g., see Hare and Elman, 1992; Nakisa and Hahn, 1996). To turn the choice into an actual past tense form, some *other* mechanism, hidden in the wings, would have to copy over the stem, find the pattern corresponding to the chosen unit, and apply the pattern to the stem. That mechanism, of course, is called a rule, just what connectionists claim to be doing without.

2. Systematic regularization. Prince and I pointed out that some irregular verbs mysteriously show up in regular garb in certain contexts. For example, you might say *All my daughter's friends are lowlifes,* not *lowlives,* even though the ordinary irregular plural of *life* is *lives.* People say Powell *ringed the city with artillery,* not *rang,* and that a politician *grandstanded,* not *grandstood.* This immediately shows that sound alone cannot be the input to the inflection system, because a given input, say, *life,* can come out the other end of the device either as *lifes* or as *lives*, depending on something else.

What is that something else? Connectionists have repeatedly suggested that it is meaning: a semantic stretching of a word dilutes the associations to its irregular past tense form, causing people to switch to the regular (e.g., see Harris, 1992; Lakoff, 1987; MacWhinney and Leinbach, 1991). But that is just false. In the vast majority of cases in which an irregular word's meaning changes, the irregular form is unchanged. For example, if you use a noun metaphorically, the irregular plural is untouched: *straw men, snowmen, sawteeth, God's children* (not *mans, tooths,* or *childs*). And English has hundreds of idioms in which a verb takes on a wildly different meaning, but in all cases it keeps its irregular past tense form: *cut a deal* (not *cutted*), *took a leak, caught a cold, hit the fan, blew them off, put them down, came off well, went nuts,* and countless others (Kim et al., 1991, 1994; Marcus et al., 1995; Pinker and Prince, 1988). So it is not enough simply to add a few units for meaning to an associative memory and hope that any stretch of meaning will cut loose an irregular form and thereby explain why people say *low-lifes* and *grandstanded*.

Equally unsatisfactory is the suggestion that people regularize words to avoid ambiguity and make themselves clear (Daugherty et al., 1993; Harris, 1992; Shirai, 1997). Many idioms are ambiguous between literal and idiomatic senses, such as *bought the farm* and *threw it up,* and some are ambiguous with other idioms as well: *blew away,* for example, could mean "wafted," "impressed," or "assassinated"; *put him down* could mean "lower," "insult," or "euthanize." But that doesn't tempt anyone to single out one of the meanings in each set by saying *buyed the farm, throwed up, blowed him away,* or *putted him down.* Conversely, the past tense of to *grandstand* is *grandstanded,* not *grandstood,* but *grandstood* would be perfectly unambiguous if anyone said it. The same is true of *Mickey Mice, high-stuck,* and *lowlives,* which would be perfectly clear, especially in context. But with these unambiguous words people are tempted, even compelled, to use a regular past tense form.

A better theory, from linguistics (Kiparsky, 1982; Selkirk, 1982; Williams, 1981), says that *headless* words become regular (Kim et al., 1991, 1994; Marcus et al., 1995; Pinker, 1999; Pinker and Prince, 1988). The point of rules of grammar is to assemble words in such a way that one can predict the properties of the new combination from the properties of the parts and the way they are arranged. That is true not just when we string words into sentences, but when we string bits of words into complex words. Start with the noun *man*. Combine it with *work*, to produce a new word, *workman*. The scheme for deducing the properties of the new word from its parts is called the right-hand-head rule: take the properties of the rightmost element and copy them up to apply to the whole word. What kind of word is *workman*? It's a noun, because *man*, the rightmost element, is a noun, and the nounhood gets copied up to apply to the whole new word. What does *workman* mean? It's a kind of man, a man who does work: the meaning of *man* is passed upstairs. And what is is the plural of *workman*? It's *workmen*, because the plural of *man* is *men*, and that information, too, gets copied upstairs.

But there is a family of exceptions: headless words, which don't get their features from the rightmost morpheme. In some compound words, for example, the meaning pertains to something that the rightmost noun *has* rather than something the rightmost noun *is*. For example, what is a *lowlife*? A kind of life? No, it is a kind of *person*, namely, a person who has (or leads) a low life. In forming the word, one has to turn off the right-hand-head rule—that is, plug the information pipeline from the root stored in memory to the whole word—in order to prevent the word from meaning a kind of life. And the pipeline is plugged; there is no longer any way for the irregular plural of *life*, *lives*, to percolate up either. That information is sealed in memory, and the regular "add -s" rule steps in as the default. Other examples include *still lifes* (not *still lives*), which is not a kind of life but a kind of painting, and *saberteeth*, not *saberteeth*, because the word refers not to a kind of tooth but to a kind of cat.

Another example showing off this mental machinery comes from verbs that are based on nouns. We say that the artillery *ringed the city*, not *rang*, because the verb comes from a noun: *to ring* in this sense means to form a ring around. To get a noun to turn into a verb, the usual percolation pipeline has to be blocked, because ordinarily the pipeline allows part-of-speech information to be copied from the root to the newly formed word. And that blocked pipeline prevents any irregularity associated with the sound of the verb from applying to the newly formed word. For similar reasons, we say that a politician *grandstanded*, not *grandstood*, because the verb comes from a noun; it means "play to the *grandstand*." Note, by the way, that no machinery has been posited specifically to generate the regularizations; the right-hand-head rule is the standard mechanism to account for morphological composition in general. A single mechanism accounts for morphological composition, for regularizations caused by headless compounds such as *lowlifes*, for regularizations caused by denominal verbs such as *ringed*, and for some half-dozen other grammatical quirks (Marcus et al., 1995; Pinker, 1999).

How can a connectionist model account for these facts? Daugherty et al. (1993) added input nodes representing the degree of semantic distance of the verb from a homophonous noun. From there it is trivial to train the network to have these nodes turn off irregular patterns and turn on the regular one. But these strange nodes are not part of the semantic representation of a verb itself, but an explicit encoding of the verb's relation to the noun that heads it—that is, a crude implementation of morphological structure, wired in to duplicate phenomena that had been discovered and explained by the linguistic structure account. Daugherty et al. tried to motivate the representation with reference to a suggestion by Harris (1992) that speakers regularize denominals to enhance communication (presumably to disambiguate homophones), but as I have pointed out, the evidence runs against this hypothesis: there are hundreds of pairs of ambiguous verbs with irregular verb roots (*blew away* = "wafted," "assassinated", "impressed"), and they do not regularize, and the vast majority of verbs with noun roots are not ambiguous (e.g., *grandstanded*), and they do regularize. A final problem is that Daugherty et al. had to train their model on regular past tenses of denominal verbs homophonous with irregulars (about 5 percent of the training exemplars). But such verbs, though scientifically interesting test cases, are used extremely rarely, and speakers cannot depend on having heard them regularized (Kim et al., 1994).

3. *Childhood overregularization errors.* Children frequently make errors such as *We holded the baby rabbits* and *The alligator goed kerplunk* (Cazden, 1968; Ervin and Miller, 1963). The words-and-rules theory offers a simple explanation: children's memory retrieval is less reliable than adults'. Since children haven't heard *held* and *came* and *went* very often (because they haven't lived as long), they have a weak memory trace for

those forms. Retrieval will be less reliable, and as long as the child has acquired the regular rule, he or she will fill the vacuum by applying the rule, resulting in an error like *comed* or *holded* (Marcus, et al., 1992).

Evidence that weak memory is a factor comes from many sources, summarized in Marcus et al. (1992). For example, Marcus and I found that the more often a child's parent uses an irregular when talking to the child, the less often the child makes an error on it. The theory explains why children, for many months, produce no errors with these forms—at first they say *held* and *came* and *went*, never *holded* and *comed* and *goed*. We proposed that the errors first occur at the point at which the child has acquired the *-ed* rule. Very young children say things like *Yesterday we walk*, leaving out past tense marking altogether. They pass from a stage of leaving out the *-ed* more often than supplying it to a stage of supplying it more often than leaving it out, and the transition is exactly at the point in which the first error like *holded* occurs. This is what we would expect if the child has just figured out that the past tense rule in English is "add *-ed*." Before that, if the child failed to come up with an irregular form, he had no choice but to use it in the infinitive: *Yesterday, he bring;* once he has the rule, he can now fill the gap by overapplying the regular rule, resulting in *bringed*.

In contrast, the connectionist accounts of the transition are incompatible with a number of phenomena. First, the basic assumption of Rumelhart and McClelland's developmental account—that vocabulary growth leads to an increase in the proportion of regular verbs fed into the network—is highly unlikely. Children presumably learn as they listen to the speech coming out of the mouths of their parents, not by scanning their own mental dictionaries and feeding each verb into their network once per pass. That implies that it should be the percentage of *tokens*, not the percentage of *types* (vocabulary items), that must be counted. And the percentage of tokens that are regular remains constant throughout development, because irregular verbs are so high in token frequency that they remain predominant even as the number of regular types increases.

The percentage of regular types does increase as the child's vocabulary expands, of course, because there is a fixed number of irregular verbs in the language and the child will eventually learn them all and thereafter expand his or her vocabulary only by learning regular forms. But the rate of increase in regular vocabulary is *negatively,* not positively, correlated with overregularization in children's speech over time. That is because children's vocabulary spurt, which Rumelhart and McClelland credited for the onset of overregularization, occurs a full year before the first overregularization errors. Plunkett and Marchman (1991, 1993) claimed to have devised a new PDP model that began to overregularize like children without an unrealistic change in the mixture of regular and irregular verbs in the input (see also Marchman and Bates, 1994). But Marcus et al. (1992) and Marcus (1995) have shown that this claim is belied by their own data, and that the developmental curves of the Plunkett-Marchman models are qualitatively different from those of children in several ways.

4. *Neuropsychological dissociations.* A particularly direct test of the words-and-rules theory consists of cases in which the human memory system is directly compromised by neurological damage or disease. Ullman et al. (1997) asked a variety of neurological patients to fill in the blank in items like "Every day I like to (verb); yesterday, I_____." We tested patients with anomia, an impairment in word finding, often associated with damage to the posterior perisylvian region of the left hemisphere; such patients can often produce fluent and largely grammatical speech, suggesting that their mental dictionaries are more impaired than their mental grammars. With such patients, we found that irregular verbs are harder than regulars, which fits the theory that irregulars depend on memory, whereas regulars depend on grammar. We also predicted and observed regularization errors like *swimmed,* which occur for the same reason that children (who also have weaker memory traces) produce such errors: they cannot retrieve *swam* from memory in time. And the patients are relatively unimpaired in doing a *wug*-test (*Today I wug, yesterday I wugged*), because that depends on grammar, which is relatively intact.

Conversely, brain-injured patients with agrammatism (a deficit in stringing words together into grammatical sequences, often associated with damage to anterior perisylvian regions of the left hemisphere) should show the opposite pattern: they should have more trouble with regulars, which depend on grammatical combination, than with irregulars, which depend on memory. They should produce few errors like *swimmed,* and they should have trouble doing the *wug*-test. And that is exactly what happens (Marin, Saffran, and Schwartz, 1976; Ullman, et al., 1997).

These dissociations are part of a growing set of neuropsychological studies showing that the processing of regular forms and the processing of irregular forms take place in different sets of brain areas (Pinker, 1997, 1999).

Double dissociations are difficult to explain in uniform pattern associators because, except for artificially small networks, "lesioning" the networks hurts the irregular forms more than the regular ones (Bullinaria and Chater, 1995). One exception is a simulation of the past tense by Marchman (1993), which seemed to go the other way. But 60 percent of Marchman's "irregular" items were no-change verbs such as *hit-hit*, which use a highly predictable and uniform mapping shared with the regular verbs. This artificial word list, and the fact that the model didn't do well with the regular verbs even *before* it was lesioned, explain the anomalous result. A more recent model by Joanisse and Seidenberg (1999) did try to obtain a double dissociation in a uniform network. But though the authors claimed that double dissociations don't license separate modules, they proceeded to build a model with separate modules. They called one of their modules "semantics," but in fact it was not semantic at all but an extreme version of a lexicon of arbitrary entries: one unit dedicated to each lexical item, with no trace of a meaning representation. (This is a common tactic among connectionists: rhetorically deny the need for a lexicon while in fact enjoying the benefits of having one by giving it the euphemism "semantics.") Moreover, the Joanisse and Seidenberg model failed to duplicate our finding that agrammatic patients have more trouble with regular than with irregular verbs.

5. Crosslinguistic comparisons. Many connectionists have tried to explain away our findings on English by pointing to a possible confound, type frequency: regular verbs are the majority in English. Only about 180 verbs in modern English are irregular, alongside several thousand regular verbs. Since pattern associators generalize the majority pattern most strongly, it is conceivable that a pattern associator that was suitably augmented to handle grammatical structure would have the regular pattern strongly reinforced by the many regular verbs in the input, and would come to generalize it most strongly.

Taking this argument seriously requires yet another act of charity. Pattern associators are driven by tokens rather than types: the models are said to learn in response to actual utterances of verbs, in numbers reflecting their frequencies of usage, rather than in response to vocabulary entries, inputted once for each verb regardless of its frequency of usage. So differences in the sheer number of vocabulary items in a language should not have a dramatic effect, because the irregular forms are high in token frequency and dominate tokens of speech (with different numbers of regular forms rotating in and out of a minority of the conversational slots). Moreover, no pattern-associator model yet proposed has plausibly handled the various grammatical circumstances involving headlessness (*lowlifes, ringed the city,* etc.) in which irregular forms systematically regularize.

But many connectionist researchers have held out the greater type frequency of regular verbs in English as the main loophole by which future pattern associators might account for the psycholinguistic facts reviewed herein (Bybee, 1995; MacWhinney and Leinbach, 1991; Seidenberg, 1992; see Marcus et al., 1995, for quotations). To seal the case for the words-and-rules theory it would be ideal to find a language in which the regular (default) rule applies to a minority of forms in the language. Note that some connectionists, reasoning circularly, treat this prediction as an oxymoron, because they *define* regular as pertaining to the most frequent inflectional form in a language, and irregular to pertain to the less frequent forms. But we are considering a psycholinguistic definition of "regular" as the default operation produced by a rule of grammatical composition and "irregular" as a form that must be stored in memory; the number of words of each kind in the language plays no part in this definition.

One language that displays this profile is German (Marcus et al., 1995). The plural comes in eight forms: four plural suffixes (*-e, -er, -en, -s,* and no suffix), some of which can co-occur with an altered (umlauted) stem vowel. The form that acts most clearly as the default, analogous to English *-s,* is *-s*. German allows us to dissociate grammatical regularity from type frequency (see Marcus et al., 1995, for a far more extensive analysis). In English, *-s* is applied to more than 99 percent of all nouns; in German, *-s* is applied to only about 7 percent of nouns. Despite this difference, the two suffixes behave similarly across different circumstances of generalization. For example, in both languages, the *-s* suffix is applied to unusual-sounding nouns (*ploamphs* in English, *Plaupfs* in German), to names that are homophonous with irregular nouns (*the Julia Childs, die Thomas Manns*), and to many other cases of systematic regularization. Moreover, German-speaking children frequently overregularize the suffix in errors such as *Manns,* analogous to English-speaking children's *mans.* So despite the relatively few nouns in German speech taking an *-s* plural, it shows all the hallmarks of a rule product, showing that the signs of a rule cannot be explained by sheer numbers of vocabulary items.

There is one final escape hatch for the connectionist theory that the generalizability of regular patterns comes from the statistics of regular words in a language. Several connectionist modelers have replied to our arguments about German by saying that it may not be the *number* of regular words that is critical so much as the *scattering* of regular words in phonological space (Forrester and Plunkett, 1994; Hare, Elman, and Daugherty, 1995; Nakisa and Hahn, 1996; Plunkett and Nakisa, 1997). Suppose irregulars fall into tight clusters of similar forms (*sing, ring, spring; grow, throw, blow;* etc.), while regulars are kept out of those clusters but are sprinkled lightly and evenly throughout no man's land (*rhumba'd, out-Gorbachev'd, oinked,* etc.). Then one can design pattern associators that devote some of their units and connections to the no man's land, and they will deal properly with any subsequent strange-sounding word. These models cannot be taken seriously as theories of a human child, because they have the inflections of a language innately wired in, one output node per inflection, and merely learn to select from among them. Moreover, a concatenation operation, allegedly superfluous, is needed to combine the inflection and the stem. And as usual, the problem of rootless and headless words is ignored. But bending over backward even further, we can test the general idea that certain patterns of clustering among regular and irregular sounds are necessary for people to generalize the regular inflection freely.

In any case, Iris Berent has nailed that escape hatch shut (Berent, Pinker, and Shimron, 1999). In Hebrew, regular and irregular nouns live cheek by jowl in the same phonological neighborhoods. Irregular nouns do not carve out their own distinctive sounds, as they do in English. Nonetheless, the regular plural suffixes *-im* and *-ot* behave similarly to

-s in English and German: speakers apply them to unusual-sounding nouns, and to names based on irregular nouns (analogous to our *the Childs, the Manns*).

Summary of Empirical Comparisons between Connectionist and Symbolic Theories of the Past Tense

The preceding comparisons have shown that despite the identical function of regular and irregular inflection, irregulars are avoided, but the regular suffix is applied freely in a variety of circumstances, from *chided* to *ploamphed* to *lowlifes* to anomia, that have nothing in common except a failure of access to information in memory. Crucially, I chose these diverse cases precisely because they are so heterogeneous and exotic. Even if a separate connectionist model were devised that successfully accounted for each of these phenomena in a psychologically plausible way—and that is far from the case—it would be hard to treat the set of models as a psychologically plausible theory of language. Clearly, we don't have separate innate neural mechanisms each designed to generate regular forms in one of these cases. Rather, the repeated appearance of the regular pattern falls out of the simple theory that the rule steps in whenever memory fails, regardless of the reason that memory fails. And that in turn implies that rules and memory are different systems.

Let me mention the remaining arguments for the connectionist approach to the past tense and related phenomena. Occasionally I am asked whether it might be unparsimonious to posit two mechanisms, rather than trying to handle all the phenomena in one. But it is hard to put much stock in the a priori argument that the human mind, or any of its major subsystems, ought to contain exactly one part, and in the case of inflection in particular, parsimony works in favor of the words-and-rules theory. No one (including the connectionists) has ever tried to model simple words and productive sentences in a single mechanism (i.e., uninflected monomorphemes, and novel sentences assembled on the fly). Even in the connectionist literature, the models of the lexicon and the models of sentence processing (e.g., the recurrent network of Elman, 1990) are distinct, and Rumelhart and McClelland and MacWhinney and Leinbach admit in the fine print that a more realistic model than theirs would need a separate lexicon. If one has a mechanism for storing and retrieving words, and one has a separate mechanism for assembling and parsing sentences, then one already has words and rules, exactly the mechanisms needed to handle irregular and regular forms. In other words the words-and-rules distinction is needed for language in general; it was not invented to explain regular and irregular forms per se.

According to a second objection, linguists themselves have shown that the distinction between words and rules is obsolete. After all, there are novel complex words like *unmicrowaveability* that have to be generated by rules, and there are phrasal idioms such as *hit the fan* and *beat around the bush* that have to be memorized like words (di Sciullo and Williams, 1987; Jackendoff, 1997). This argument merely underscores the fact that the word *word* is highly ambiguous, with at least four senses. In the sense intended by the words-and-rules theory, "words" refers to what Di Sciullo and Williams call listemes: language chunks of any size, from morpheme to proverb, that are not fully compositional and therefore have to be memorized. And "rule" is not intended to refer narrowly to a classic rewrite production such as S → NP VP; it is meant to refer more broadly to any productive, combinatorial operations on symbolic structures, including principles, constraints, unification, optimality, and so on. A more accurate (but less euphonious) title for the theory would have been *Listemes and Combinatorial Symbolic Operations*.

A third objection is that I am beating a dead horse by testing the old RM model. All the connectionists agree that it was a poor first attempt; there are new models that do much better and account for each of the phenomena on which the RM model failed.

In fact, fifteen years and twenty-five models later, the RM model is still the best connectionist model of the past tense. For one thing, its supposedly low-tech features—the lack of a hidden layer, and its Wickelfeature representation of the input and output—turn out to make little or no difference. Richard Sproat and Dana Egedi (Egedi and Sproat, 1991) did head-to-head comparisons of the original model and a version with a hidden layer and a state-of-the-art representation and output decoder. The souped-up version had the same problems as the original.

Many connectionist fans are surprised to learn that the RM model isn't any worse than its successors, because the standard doctrine in connectionist modeling is that hidden-layer models are more powerful than perceptrons. But as Marvin Minsky and Seymour Papert (1988) have pointed out, one can compensate for the lack of a hidden layer by beefing up the input representation, and that's what Rumelhart and McClelland, perhaps inadvertently, did. Every word got a "blurred" input representation, in which a smattering of incorrect units were activated for that word together with the correct ones. The blurring was not, however, random noise: the same set of incorrect units got activated for a given word every time it was fed into the network. Moreover, many of the incorrect units code for sound sequences that cannot exist in English words. Thus each set of "blurred" units can serve as a unique code for that lexical item. This was particularly effective because many of the blurred units represented Wickelfeatures that are phonologically impossible, so those units didn't have to do any work in the sound-to-sound mapping, and were therefore available to code individual lexical entries without interference. This compensated nicely for the lack of a hidden layer, which under ordinary training circumstances comes to code indirectly for distinct lexical entries. Once again, traditional linguistic notions (in this case a lexical entry) have to be reintroduced into the pattern-associator models through the back door.

But most important, the RM network remains the only model with empirical content—that is, the only model whose behavior makes a novel and correct prediction about human behavior. Rumelhart and McClelland built a model to compute and generalize the past tense, and the model not only did that but successfully predicted what kinds of irregular forms children find more or less easy, and it also successfully predicted several forms of such errors. (The model made some incorrect predictions as well, for interesting reasons.) That kind of predictive record cannot be ignored, which is why Alan Prince and I noted back in 1988 that the model, at the very least, had to be taken seriously as capturing something about the memory system in which irregular forms are stored.

In contrast, the immediate follow-up models either made empirical predictions that are demonstrably false (see Marcus, 1995, 2000; Kim et al., 1994; Marcus et al., 1995; Berent et al., 1999), or didn't make predictions at all, because they were kluged by hand to mimic a specific phenomenon that our group had previously documented. The modelers seem content to show that some connectionist model or other can mechanically generate some behavior (true enough, in the same sense that some FORTRAN program or other can generate the behavior), without showing that the model is true or even plausible.

One irony of the past tense debate is that it is often framed as a nature-nurture battle, with the connectionists on the side of nurture. But in fact the connectionist models build in innate features specific to the English past tense that would make Jerry Fodor blush—such as a layer of nodes whose only purpose is to generate anomalies such as *ringed the city* and *grandstanded,* or an output layer that consists of exactly one innate node for each inflection in English, with the model merely selecting among them in a multiple-choice task. Indeed, those innate-inflection networks have taken over in the most recent generation of connectionist models of inflection (by Elman, Nakisa and Hahn, and others); Rumelhart and McClelland's more ambitious goal of *computing* the output form has mostly been abandoned without comment. As mentioned earlier, there is a double irony in these models: the English inflections are innate, and since the model only selects the appropriate suffix or vowel change, some unmentioned postprocessor has to apply the suffix or vowel change to the stem to generate the actual form. That postprocessor, of course, is what linguists call a rule—exactly what the models are supposed to be doing without.

I am sometimes asked if I would deny that *any* connectionist model could *ever* handle inflection. Of course not! I am skeptical only of the claim that the current favored style—a single pattern associator—can handle it in a psychologically realistic way. A neural network model consisting of an associative memory (for words, including irregulars) and a hierarchical concatenator (for combinatorial grammar, including regulars) could (if the details were done correctly) handle all the phenomena I discussed, and I would have no problem with it. My objections are aimed not at connectionism, but at the current fashion for denying compositional structure and shoehorning phenomena into a single uniform net. I'm cautiously optimistic about connectionist models that take structure seriously, such as the various proposals by Paul Smolensky (1990), Lokendra Shastri (Shastri, 1999; Shastri and Ajjanagadde, 1993), John Hummel and colleagues (1992, 1997), and others.

A Final Word

As a scientist and a member of an influential intellectual community in the history of cognitive science, Jacques Mehler helped to found the modern field of psycholinguistics that first framed the contrast between rules and associations in language as a problem in empirical psychology and computational modeling. As an editor, he has provided a fair and lively forum for debate, in which both sides of this (and many other) scientific controversies have refined their arguments, clarified the issues, proposed and tested hypotheses, and discovered common ground. It is a great honor to participate in the long-overdue recognition of this outstanding scientist, editor, colleague, and friend.

Notes

1. A possible exception is the model of Cottrell and Plunkett, which uses a recurrent network rather than the more common feedforward pattern associator to model inflection. Their model, however, does not deal with the range of phenomena handled by the feedforward models, and does not represent the kind of model standardly offered by connectionists as their best hypothesis about how inflection works.

2. A model by Gasser and Lee handles both production and comprehension, but it handles only regular forms, no irregulars.

References

Berent, I., Pinker, S., and Shimron, J. (1999). Default nominal inflection in Hebrew: Evidence for mental variables. *Cognition, 72,* 1–44.

Berko, J. (1958). The child's learning of English morphology. *Word, 14,* 150–177.

Bullinaria, J. A., and Chater, N. (1995). Connectionist modeling: Implications for cognitive neuropsychology. *Language and Cognitive Processes, 10,* 227–264.

Bybee, J. L. (1995). Regular morphology and the lexicon. *Language and Cognitive Processes, 10,* 425–455.

Cazden, C. B. (1968). The acquisition of noun and verb inflections. *Child Development, 39,* 433–448.

Daugherty, K. G., MacDonald, M. C., Petersen, A. S., and Seidenberg, M. S. (1993). Why no mere mortal has ever flown out to center field but people often say they do. In *Fifteenth Annual Conference of the Cognitive Science Society.* Mahwah, NJ: Erlbaum.

di Sciullo, A. M., and Williams, E. (1987). *On the definition of word.* Cambridge, MA: MIT Press.

Egedi, D. M., and Sproat, R. W. (1991). *Connectionist networks and natural language morphology.* Murray Hill, NJ: Linguistics Research Department, Lucent Technologies.

Elman, J. L. (1990). Finding structure in time. *Cognitive Science, 14,* 179–211.

Ervin, S. M., and Miller, W. (1963). Language development. In H. W. Stevenson (Ed.), *Child psychology: The sixty-second yearbook of the National Society for the Study of Education.* Part 1. Chicago: University of Chicago Press.

Fodor, J. A., and Pylyshyn, Z. (1988). Connectionism and cognitive architecture: A critical analysis. *Cognition, 28,* 3–71.

Forrester, N., and Plunkett, K. (1994). Learning the Arabic plural: The case for minority default mappings in connectionist networks. In A. Ram and K. Eiselt (Eds.), *Sixteenth Annual Conference of the Cognitive Science Society.* Mahwah, NJ: Erlbaum.

Hare, M., and Elman, J. (1992). A connectionist account of English inflectional morphology: Evidence from language change. In *Fourteenth Annual Conference of the Cognitive Science Society.* Mahwah, NJ: Erlbaum.

Hare, M., Elman, J., and Daugherty, K. (1995). Default generalization in connectionist networks. *Language and Cognitive Processes, 10,* 601–630.

Harris, C. L. (1992). Understanding English past-tense formation: The shared meaning hypothesis. In *Fourteenth Annual Conference of the Cognitive Science Society.* Mahwah, NJ: Erlbaum.

Hummel, J. E., and Biederman, I. (1992). Dynamic binding in a neural network for shape recognition. *Psychological Review, 99,* 480–517.

Hummel, J. E., and Holyoak, K. J. (1997). Distributed representations of structure: A theory of analogical access and mapping. *Psychological Review, 104,* 427–466.

Jackendoff, R. (1997). *The architecture of the language faculty.* Cambridge, MA: MIT Press.

Joanisse, M. F., and Seidenberg, M. S. (1999). Impairments in verb morphology after brain injury: A connectionist model. *Proceedings of the National Academy of Sciences of the United States of America, 96,* 7592–7597.

Kim, J. J., Marcus, G. F., Pinker, S., Hollander, M., and Coppola, M. (1994). Sensitivity of children's inflection to morphological structure. *Journal of Child Language, 21,* 173–209.

Kim, J. J., Pinker, S., Prince, A., and Prasada, S. (1991). Why no mere mortal has ever flown out to center field. *Cognitive Science, 15,* 173–218.

Kiparsky, P. (1982). Lexical phonology and morphology. In I. S. Yang (Ed.), *Linguistics in the morning calm.* Vol. 3–91. Seoul: Hansin.

Lachter, J., and Bever, T. G. (1988). The relation between linguistic structure and associative theories of language learning—A constructive critique of some connectionist learning models. *Cognition, 28,* 195–247.

Lakoff, G. (1987). Connectionist explanations in linguistics: Some thoughts on recent anti-connectionist papers. Unpublished manuscript, posted on the CONNECTIONISTS mailing list.

MacWhinney, B., and Leinbach, J. (1991). Implementations are not conceptualizations: Revising the verb learning model. *Cognition, 40,* 121–157.

Marchman, V. (1993). Constraints on plasticity in a connectionist model of the English past tense. *Journal of Cognitive Neuroscience, 5,* 215–234.

Marchman, V. A., and Bates, E. (1994). Continuity in lexical and morphological development: A test of the critical mass hypothesis. *Journal of Child Language, 21,* 339–366.

Marcus, G. F. (1995). The acquisition of inflection in children and multilayered connectionist networks. *Cognition, 56,* 271–279.

Marcus, G. F. (2001). *The algebraic mind: Reflections on connectionism and cognitive science.* Cambridge, MA: MIT Press.

Marcus, G. F., Brinkmann, U., Clahsen, H., Wiese, R., and Pinker, S. (1995). German inflection: The exception that proves the rule. *Cognitive Psychology, 29,* 189–256.

Marcus, G. F., Pinker, S., Ullman, M., Hollander, M., Rosen, T. J., and Xu, F. (1992). Overregularization in language acquisition. *Monographs of the Society for Research in Child Development, 57.*

Marin, O., Saffran, E. M., and Schwartz, M. F. (1976). Dissociations of language in aphasia: Implications for normal function. In S. R. Harnad, H. S. Steklis, and J. Lancaster (Eds.), *Origin and evolution of language and speech. Annals of the New York Academy of Sciences, 280.*

Miller, G. A. (1967). The psycholinguists. In G.A. Miller (Ed.), *The psychology of communication.* London: Penguin Books.

Minsky, M., and Papert, S. (1988). Epilogue: The new connectionism. In M. Minsky and S. Papert (Eds.), *Perceptrons,* Expanded edition. Cambridge, MA: MIT Press.

Nakisa, R. C., and Hahn, U. (1996). Where defaults don't help: The case of the German plural system. In G. W. Cottrell (Ed.), *Eighteenth Annual Conference of the Cognitive Science Society.* Mahwah, NJ: Erlbaum.

Pinker, S. (1997). Words and rules in the human brain. *Nature, 387,* 547–548.

Pinker, S. (1999). *Words and rules: The ingredients of language.* New York: HarperCollins.

Pinker, S., and Mehler, J. (1988). *Connections and symbols.* Cambridge, MA: MIT Press.

Pinker, S., and Prince, A. (1988). On language and connectionism: Analysis of a parallel distributed processing model of language acquisition. *Cognition, 28,* 73–193.

Plunkett, K., and Marchman, V. (1991). U-shaped learning and frequency effects in a multi-layered perceptron: Implications for child language acquisition. *Cognition, 38,* 43–102.

Plunkett, K., and Marchman, V. (1993). From rote learning to system building. *Cognition, 48,* 21–69.

Plunkett, K., and Nakisa, R. (1997). A connectionist model of the Arab plural system. *Language and Cognitive Processes, 12,* 807–836.

Prasada, S., and Pinker, S. (1993). Generalizations of regular and irregular morphological patterns. *Language and Cognitive Processes, 8,* 1–56.

Rumelhart, D. E., and McClelland, J. L. (1986). On learning the past tenses of English verbs. Implicit rules or parallel distributed processing? In J. L. McClelland, D. E. Rumelhart, and the PDP Research Group (Eds.), *Parallel distributed processing: Explorations in the microstructure of cognition,* Vol. 2: *Psychological and biological models.* Cambridge, MA: MIT Press.

Seidenberg, M. (1992). Connectionism without tears. In S. Davis (Ed.), *Connectionism: Theory and practice.* New York: Oxford University Press.

Selkirk, E. O. (1982). *The syntax of words.* Cambridge, MA: MIT Press.

Shastri, L. (1999). Advances in SHRUTI: A neurally motivated model of relational knowledge representation and rapid inference using temporal synchrony. *Applied Intelligence.*

Shastri, L., and Ajjanagadde, V. (1993). From simple associations to systematic reasoning: A connectionist representation of rules, variables, and dynamic bindings using temporal synchrony. *Behavioral and Brain Sciences, 16,* 417–494.

Shirai, Y. (1997). Is regularization determined by semantics, or grammar, or both? Comments on Kim, Marcus, Pinker, Hollander and Coppola. *Journal of Child Language, 24,* 495–501.

Smolensky, P. (1990). Tensor product variable binding and the representation of symbolic structures in connectionist systems. *Artificial Intelligence, 46,* 159–216.

Sproat, R. (1992). *Morphology and computation.* Cambridge, MA: MIT Press.

Ullman, M., Corkin, S., Coppola, M., Hickok, G., Growdon, J. H., Koroshetz, W. J., and Pinker, S. (1997). A neural dissociation within language: Evidence that the mental dictionary is part of declarative memory, and that grammatical rules are processed by the procedural system. *Journal of Cognitive Neuroscience, 9,* 289–299.

Williams, E. (1981). On the notions "lexically related" and "head of a word." *Linguistic Inquiry, 12,* 245–274.

10

The Roll of the Silly Ball

Anne Cutler, James M. McQueen, Dennis Norris, and A. Somejuan

What is the syllable's role in speech processing? We suggest that the answer to this question championed by Jacques Mehler some twenty years ago (see Mehler, 1981) is wrong. But our purpose is not to flog a dead horse. Rather, we would like to bring Jacques the comforting news that he was less wrong than we might once have thought he was about the syllable's role. On the basis of recent research, it appears that the syllable does have a language-universal part to play in speech processing, just as Jacques argued two decades ago. It is, however, not the role he proposed then.

The experimental study of spoken word recognition is barely three decades old. The syllable played a role in the theoretical debate from the earliest days (see, e.g., Savin and Bever, 1970), and Jacques has been and still is one of the syllable's staunchest advocates. The focus of the debate, as might be expected, has changed somewhat in those thirty years; but the syllable, like its trusty champion, is still an active contender.

Why the Syllable Is Not the Unit of Perception

Below, we argue that the syllable in fact has two roles to play in speech processing. Crosslinguistic evidence suggests that one of these roles is language-specific, whereas the other is language-universal. But neither corresponds to the position adopted by Mehler (1981), which was that "the syllable is probably the output of the segmenting device operating upon the acoustic signal. The syllable is then used to access the lexicon" (p. 342). On this view, syllables are the "units of perception" which provide the interface between the speech signal and the mental lexicon. This theory is no longer tenable. In many experiments carried out in the 1970s and early 1980s which compared monitoring times for phonemes and syllables, syllables were always detected faster than phonemes (e.g., see Savin and Bever, 1970; Segui, Frauenfelder, and Mehler, 1981; reviewed in Norris and Cutler, 1988). This might suggest that, in line with the theory that syllables are the mental representations used for lexical access, syllable perception is primary, and that the detection of phonemes depends on the perception of syllables. As pointed out by Foss and Swinney (1973), however, there is a flaw in this argument: the order of identification of units in a laboratory task need not reflect the order in which the corresponding units are processed during speech perception. More generally, there is no logical necessity that the units which listeners detect in laboratory tasks must correspond directly to representations used in normal lexical access (Cutler et al., 1987). Furthermore, the advantage which the syllable showed in identification times was found to be an experimental artifact. Norris and Cutler (1988) asked listeners to monitor for either phoneme or syllable targets in lists of words and nonwords. The materials contained foils: phonemes or syllables that were phonologically very similar to the targets. In contrast to earlier experiments, listeners were therefore obliged to analyze each stimulus fully. Under these conditions, phoneme targets were detected faster than syllable targets.

Stronger evidence for the syllable's role as the "unit of perception" was presented by Mehler et al. (1981). French listeners were faster to detect the sequence BA, for example, in "balance" than in "balcon" (where the target matches the syllabification of ba.lance), and faster to detect BAL in "balcon" than in "balance" (where the target matches the syllabification of bal.con). Although this crossover interaction could be taken as evidence that the speech signal is classified into syllables prior to lexical access (as Mehler, 1981, and Mehler et al., 1981, argued), there is an alternative explanation for this finding which we prefer. We discuss it later. For the moment, let us simply note that neither English nor Dutch listeners, presented with exactly the same French materials, showed this syllabic effect (Cutler, 1997; Cutler et al., 1983, 1986), and that the effect did not replicate when French listeners were asked to detect similar targets in French nonwords (Frauenfelder and Content, 1999). These findings challenge the hypothesis that the syllable is the universal prelexical unit.

A further major problem for this hypothesis is that lexical access appears to be continuous. The rapidity of spoken word recognition (Marslen-Wilson, 1973; Marslen-Wilson and Welsh, 1978) suggests that lexical access begins as soon as a word's first few sounds are available (Marslen-Wilson, 1987); the initiation of access is not delayed until, say, the end of the first syllable. This observation is incompatible with any model of lexical access in which there is a strict classification of the speech signal into syllables which intervenes between low-level acoustic analysis and the mental lexicon. Even more striking is the repeated demonstration that lexical activation changes with fine-grained changes in acoustic-phonetic information in the speech signal. For example, words are recognized more slowly when they contain subphonemic mismatches than when they do not (Marslen-Wilson and Warren, 1994; McQueen, Norris, and Cutler, 1999; Streeter and Nigro, 1979; Whalen, 1984, 1991).

These studies show that the activation of lexical representations is modulated by changes in the speech signal which are not only subsyllabic but also subphonemic. Marslen-Wilson and Warren (1994) therefore argued that prelexical representations are featural. The subphonemic mismatch data, however, are in fact neutral with respect to the size of prelexical representations (McQueen et al., 1999; Norris, McQueen, and Cutler, 2000). It is true that these data rule out strictly serial models, in which information is only sent on to the lexical level in discrete chunks (whether those chunks are phonemic, syllabic, or whatever). If information cascades continuously up to the lexicon, however, subphonemic changes could modulate lexical activation, whatever the size of the units coding that information. The evidence of continuous uptake of information at the lexical level therefore rules out only the strongest form of the syllabic model, that of strict serial classification into syllables prior to lexical access.

Another argument against the syllable as the unit of perception comes from the analysis of the phonological structure of the world's languages. The syllable structure of some languages is very clear; in others it is not. For example, there is no consensus among English native speakers on the syllabification of English (Treiman and Danis, 1988). There is even no consensus on the syllabification of French by French speakers (Content, Dumay, and Frauenfelder, 2000; Frauenfelder and Content, 1999).

Phonologists (e.g., Kahn, 1980) have argued that some consonants in English (such as single intervocalic consonants following a lax vowel and preceding schwa, like the /l/ in balance) are ambisyllabic, that is, they belong to two syllables. A model of lexical access based on classification into syllables cannot work in a straightforward way in languages with ambisyllabicity. Unless one were willing to posit that the form of prelexical representations varies from language to language, such that only the speakers of languages with fully regular syllable structure perform lexical access on the basis of syllabic units, ambisyllabicity provides another challenge to the hypothesis that syllables are the units of perception. We believe that there are important crosslinguistic differences in speech processing, driven by differences in the phonological structures of languages, but we think it very unlikely that such differences would be so large that spoken word access would be based on different units in different languages.

The strong position advocated by Mehler (1981) can therefore be rejected. Note, however, that we are unable to rule the syllable out completely as a prelexical unit. The nature of the mental representations which provide the key to lexical access remains to be determined: they could be gestural (Liberman and Mattingly, 1985), featural (Marslen-Wilson and Warren, 1994), phonemic (Fowler, 1984; Nearey 1997), or even syllabic. What is clear, however, is that if these representations do prove to be syllabic, they will have to operate in such a way that information can cascade continuously from them to lexical representations, and they would have to operate in a way which would deal with ambisyllabicity. In other words, there can be no strict classification of the speech signal into discrete and indivisible syllables before lexical access.

How the Syllable Is a Possible Word

How, then, can we explain the findings of Mehler et al. (1981)? We will do so in the context of something which was not available in the early 1980s: Shortlist, a computational model of spoken word recognition (Norris, 1994; Norris et al., 1997). Shortlist does not provide an answer to the unit-of-perception question. The input in the current implementation of Shortlist is a string of phonemes. The key aspects of the model's function, however, do not depend on the validity of this assumption. The domain which the model seeks to explain is one step higher in the processing sequence: lexical access itself, and how spoken word recognition is achieved, in particular word recognition in the normal case, continuous speech. This is difficult for two independent reasons. One is that continuous speech lacks reliable cues to word boundaries, so that listeners have to segment the continuous stream of sounds that a speaker produces into the words of the speaker's message. The second reason is the nonuniqueness of most speech: a given word onset is often consistent with many continuations, and many words have other words embedded within them. This further compounds the segmentation problem. Multiple lexical parses of a given input are indeed sometimes possible, for example in the sequence /ðəsɪlɪbɒlzrol/.

The Shortlist model has two main properties. The first is that only the words most consistent with the information in the input (the shortlist of best candidates, hence the model's name) are activated. The second is that the activated candidate words compete with each other (via inhibitory connections). Competition between candidate words provides a solution to the segmentation problem. Words like "silly," "balls," and "roll," along with other words like "sill," "syllabub," and "roe" will be activated along with "syllable" when the listener hears /ðəsɪlɪbɒlzrol/. The speech signal itself can act to disfavor some of these alternative words (such as the mismatch between "syllabub" and the syllable /bɒlz/; see, e.g., Connine et al., 1997, and Marslen-Wilson, Moss, and van Halen, 1996, for discussion of the role of mismatching information in lexical access). Lexical competition, however, provides an efficient mechanism by which alternative candidates can be evaluated relative to each other even when there is no clear information in the signal to favor or disfavor particular alternatives (McQueen et al., 1995). The net result of the competition process is a lexical parse which is the optimal interpretation of the input. The parse "The silly balls roll" is better than "The silly balls roe," for example, since the latter parse leaves the final /l/ unaccounted for. The candidate "roe" will lose in its competition with the candidate "roll" since there is more evidence in favor of "roll."

There is more to speech segmentation than lexical competition, however. While it is true that speech is continuous, and lacks reliable cues to word boundaries, it is not true that the speech signal contains no word-boundary cues. An obvious cue is that provided by silence: before and after an utterance, or in pauses at clause boundaries within an utterance. There is evidence that listeners are sensitive to this cue, and indeed to many others, when they are available in the speech signal (see Norris et al., 1997, for a review).

In Norris et al. (1997), we sought to unify the evidence that listeners use this variety of segmentation cues with a theory of word recognition based on competition. The idea was that the cues indicate the likely locations of word boundaries, and that this information could be used to bias the competition process. What was required, therefore, was a mechanism by which each activated lexical hypothesis could be evaluated relative to the signals provided by the segmentation cues. Clearly, a candidate that is perfectly aligned with a likely word boundary (i.e., a word which begins or ends at the point signaled as a likely boundary cue) should be at an advantage relative to a candidate that is misaligned with a signaled boundary. But what should count as misaligned? The crucial factor in determining the goodness of fit of a lexical parse for a given input sequence is whether the entire input has been accounted for in a plausible way. The alignment constraint we proposed, therefore, was the possible word constraint (PWC). If there is an impossible word between a candidate word and a likely word boundary, then the parse including that word is highly implausible, and the candidate in question is disfavored (in the computational implementation of the PWC in Shortlist, the word's activation level is halved). Since silence at the end of the phrase /ðəsɪlɪbɒlzrol/ indicates a likely word boundary after the final /l/, and "l" by itself is not a possible English word, the candidate "roe" can be penalized ("The silly balls roe l" is a highly implausible utterance). Although, as discussed earlier, competition alone might favor "roll" over "roe," the PWC deals "roe" a deathblow (pushing it well down the ranking of shortlisted candidates).

Norris et al. (1997) presented experimental evidence in support of the PWC. Listeners were asked to spot English words embedded in nonsense sequences. They found it much harder to spot a word such as "apple" in "fapple" than in "vuffapple." According to the PWC, this is because in "fapple" the word "apple" is misaligned with the boundary cued by silence at the beginning of the sequence; between the word and the boundary is a single consonant, "f," which is not a possible English word. In "vuffapple," however, there is a syllable, "vuff," between "apple" and the boundary, and this syllable, though not an actual English word, is a possible English word. The PWC penalty should therefore be applied in the former but not in the latter case. The simulations with Shortlist (also in Norris et al., 1997) showed not only that the PWC is required for the model to be able to fit these data (and other data on segmentation) but also that the model's recognition performance on continuous speech was better with the PWC than without it.

The PWC evaluates activated lexical hypotheses relative to likely word boundary locations no matter how those boundaries are cued. Multiple segmentation cues, when available in the speech stream, can therefore all bias the competition process via the same mechanism. It should be clear that these cues are language-specific. What counts as a permissible phonotactic sequence of phonemes varies from language to language. Vowel harmony, for example, can only provide a segmentation cue in a language with vowel harmony constraints (Suomi, McQueen, and Cutler, 1997). Other cues which must be language-specific are those related to rhythm: different languages have different rhythmic properties. A large body of crosslinguistic research suggests that listeners use the rhythm of their native language in speech segmentation (Cutler and Mehler, 1993; Cutler and Norris, 1988; Otake et al., 1993).

Where the Syllable Plays a Language-Specific Role

With this crosslinguistic perspective, we can now look again at French, and the sequence monitoring results of Mehler et al. (1981). The reason why French listeners are faster to detect BA in "ba.lance" than in "bal.con," and BAL in "bal.con" than in "ba.lance," we suggest, is that in French syllables can serve to assist in speech segmentation for lexical access (see Frauenfelder and Content, 1999, and Content et al., 2000, for a similar suggestion).

The source of the syllable's role in lexical segmentation in French is its role in the rhythmic structure of that language. Likewise, the reason that the syllable apparently plays no role in lexical segmentation by users of some other languages is that rhythm, in those languages, is not syllabically based. In English and Dutch, for instance, rhythm is based on stress;

in Japanese, rhythm is based on a subsyllabic unit called the mora. Rhythmic structure, whatever its (language-specific) nature, provides cues to the likely position of word boundaries, and these cues in turn are exploited by the PWC. Note that these cues need not be deterministic—it does not matter that, for instance, resyllabification in French makes some syllable boundaries not correspond to word boundaries, nor does it matter that some English words begin with unstressed syllables. Lexical access is based on continuous incoming information, and is unaffected by the rhythmic structure. But the rhythmic cues, where they are available, combine with the other available cues to assist the PWC in getting rid of at least some of the activated words.

In other words, this view of lexical processing differs substantially from that taken in the debate on units of perception (see Norris and Cutler, 1985). No strong claim about prelexical access representations is made, nor needs to be made. Information from language rhythm (and from other cues) simply indicates where in the signal likely word boundaries occur. This information is then used by the PWC in the evaluation of lexical hypotheses. The claim that in French syllabic structure marks the locations of likely word boundaries does not depend on any particular assumptions about lexical access units. The reason why BA is easier to detect in ba.lance than in bal.con, therefore, is not that the unit for /ba/ is extracted prior to lexical access given ba.lance (but not given bal.con), but that BA is perfectly aligned with the likely word boundary cued by syllable structure before the /l/ in ba.lance but misaligned with the likely word boundary cued by syllable structure after the /l/ in bal.con.

We prefer this explanation because it has crosslinguistic validity. It explains why speakers of different languages appear to segment their native languages in different ways. What is common across languages is that native language rhythm provides segmentation cues. This account also explains why speakers of different languages also segment non-native languages in different ways. French listeners segment English syllabically, whereas English listeners segment neither English nor French syllabically (Cutler et al., 1983, 1986). Like English listeners, Dutch listeners do not segment French syllabically (Cutler, 1997). French listeners also segment Japanese syllabically, unlike Japanese listeners (Otake et al., 1993). Finally, even French-English bilinguals do not segment their two languages in the same way. Cutler et al. (1989, 1992) found that some bilinguals segmented French but not English syllabically; the other bilinguals showed no evidence of a syllabic strategy in either language; instead, they segmented English (but not French) using a stress-based strategy based on English rhythm. A theory in which the syllable is the universal "unit of perception" cannot account for this crosslinguistic variability.

How the Syllable Also Plays a Language-Universal Role

Jacques might well be disappointed if he were to decide that the syllable might therefore have only a language-specific role to play in speech processing. But this conclusion would be premature. Our recent research has suggested that the syllable also has a language-universal role.

The PWC depends on multiple simultaneous activation of candidate words, and on cues to the location of likely word boundaries, but it also depends on a third factor: a clear definition of what counts as a "possible word," that is, an acceptable portion of speech between a candidate word and a boundary location. In Norris et al. (1997), consonantal portions, like the "f" in "fapple," were compared with syllabic portions, like the "vuff" in "vuffapple." As described above, the syllable here is clearly a possible word of English; the consonant is not. In the implementation of the PWC, therefore, a word is considered to be an acceptable candidate if the stretch of speech between its edge and a likely word boundary contains a vowel (i.e., if it is a syllable) and is considered to be an implausible candidate if the stretch of speech does not contain a vowel.

This simple implementation of the PWC could work across all languages. But does it? Languages differ with respect to what constitutes minimal possible words. Thus, in English, for example, consonant-vowel (CV) syllables in which the vowel is lax (e.g., /væ/, /tɛ/) are not well-formed words. But in French these syllables are well-formed words ("va" and "thé" are in fact both words). If the current implementation of the PWC is correct, CV syllables with lax vowels should be treated as acceptable words in the segmentation of any language. In a word-spotting task, therefore, words should be easier to spot in open-syllable contexts with lax vowels than in single-consonant contexts. If, however, the PWC respects the fact that open syllables with lax vowels violate the phonological constraints of English, word spotting should be as hard in lax-vowel contexts as in consonantal contexts in English. Norris et al. (2001) tested this prediction. English listeners spotted "canal," for example, more rapidly and accurately in a CV context with a lax vowel (zɛkənæl) than in a consonantal context (skənæl).

Another constraint on English words is that weak syllables (containing the reduced vowel schwa) are not well-formed content words. Words from the closed set of functors can have schwa as their only vowel (the, a, of, etc.), but words from the open set of content words all have at least one full vowel. Norris et al. (2000) also tested whether the PWC is sensitive to this constraint on the well-formedness of English words. Listeners could spot targets like "bell" faster in /bɛlʃəf/ than in /bɛlʃ/. This difference was equivalent to the difference between syllabic and consonant contexts with the same set of target words in Norris et al. (1997), where, for example, "bell" was easier to spot in contexts with syllables containing full vowels (/bɛlʃɪɡ/) than in consonantal contexts (/bɛlʃ/).

Syllables with open lax vowels or with schwa therefore do not appear to be treated by the PWC like single consonants. This suggests that there is no correspondence between what constitutes a possible word in a language according to abstract phonological constraints and what constitutes a viable portion of speech in the computation of acceptable lexical parses during continuous speech recognition. Another demonstration of this comes from a word-spotting experiment in Sesotho, a language spoken in southern Africa. Bantu languages like Sesotho have the phonological constraint that any surface realization of a content word must be minimally bisyllabic. Monosyllables are thus not well-formed words in Sesotho. Cutler, Demuth, and McQueen (submitted) compared Sesotho speakers' ability to spot a word like "alafa" (to prescribe) in "halafa" (where the single consonant context "h" is an impossible word) and "ro-alafa" (where the monosyllabic context "ro" is not a word, and not even a well-formed Sesotho word). Listeners were able to spot words in the monosyllabic contexts faster and more accurately than in the consonantal contexts. It would appear that the fact that "ro" is not a possible word in the Sesotho vocabulary does not make "ro" unacceptable as part of the parse "ro + alafa."

All of these results point to the idea that, across the world's languages, possible parses of continuous speech consist of chunks no smaller than a syllable. In other words, the segmentation procedure instantiated by the PWC does not vary from language to language, depending on the well-formedness of words in each language. Instead, the PWC mechanism appears to respect a simple language-universal constraint: if the stretch of speech between a likely word boundary and the edge of a candidate word is a syllable, that candidate word is a viable part of the parse; if the stretch of speech is not syllabic (a single consonant or a string of consonants), the candidate word is not part of a plausible parse, so its activation is reduced. Thus, across many languages, including English and Sesotho, listeners find it harder to spot words in nonsyllabic contexts than in syllabic contexts: in Dutch, "lepel" (spoon) is harder to spot in /blepəl/ than in /səlepəl/. McQueen and Cutler (1998) note that weak syllables are also not possible content words in Dutch; and in Japanese, "ari" (ant) is harder to spot in /rari/ than in /eari/ (McQueen, Otake, and Cutler, 2001). The size or nature of the syllable does not appear to matter: any syllable will do.

The syllable therefore does appear to have a central role to play in speech processing. Syllables are not, as Mehler (1981) suggested, units of perception into which listeners classify spoken input prior to lexical access. Instead, the syllable appears to be the measuring stick against which viable and unviable parses of continuous speech are judged. Syllables form acceptable chunks in the ongoing lexical parse of the speech stream; nonsyllabic sequences do not. La syllabe est morte; vive la syllabe!

Acknowledgments

The order of authors (C, M, N, S) is not arbitrary. We thank Peter Gipson for suggesting the title to us, in the form of a most memorable reprint request, which long adorned the wall of Jacques's office.

References

Connine, C. M., Titone, D., Deelman, T., and Blasko, D. (1997). Similarity mapping in spoken word recognition. *Journal of Memory and Language, 37,* 463–480.

Content, A., Dumay, N., and Frauenfelder, U. H. (2000). The role of syllable structure in lexical segmentation: Helping listeners avoid mondegreens. In A. Cutler, J. M. McQueen, and R. Zondervan (Eds.), *Proceedings of the SWAP Workshop* (pp. 39–42). Nijmegen, Netherlands: Max Planck Institute for Psycholinguistics.

Cutler, A. (1997). The syllable's role in the segmentation of stress languages. *Language and Cognitive Processes, 12,* 839–845.

Cutler, A., and Mehler, J. (1993). The periodicity bias. *Journal of Phonetics, 21,* 103–108.

Cutler, A., and Norris, D. (1988). The role of strong syllables in segmentation for lexical access. *Journal of Experimental Psychology: Human Perception and Performance, 14,* 113–121.

Cutler, A., Demuth, K., and McQueen, J. M. (submitted). Universality versus language-specificity in listening to running speech.

Cutler, A., Mehler, J., Norris, D., and Segui, J. (1983). A language specific comprehension strategy. *Nature, 304,* 159–160.

Cutler, A., Mehler, J., Norris, D., and Segui, J. (1986). The syllable's differing role in the segmentation of French and English. *Journal of Memory and Language, 25,* 385–400.

Cutler, A., Mehler, J., Norris, D., and Segui, J. (1987). Phoneme identification and the lexicon. *Cognitive Psychology, 19,* 141–177.

Cutler, A., Mehler, J., Norris, D., and Segui, J. (1989). Limits on bilingualism. *Nature, 340,* 229–230.

Cutler, A., Mehler, J., Norris, D., and Segui, J. (1992). The monolingual nature of speech segmentation by bilinguals. *Cognitive Psychology, 24,* 381–410.

Foss, D. J., and Swinney, D. A. (1973). On the psychological reality of the phoneme: Perception, identification and consciousness. *Journal of Verbal Learning and Verbal Behavior, 12,* 246–257.

Fowler, C. A. (1984). Segmentation of coarticulated speech in perception. *Perception and Psychophysics, 36,* 359–368.

Frauenfelder, U. H., and Content, A. (1999). The role of the syllable in spoken word recognition: Access or segmentation? In *Proceedings of the Second Journées d'Etudes Linguistique* (pp. 1–8). Nantes, France: Université de Nantes.

Kahn, D. (1980). Syllable-structure specifications in phonological rules. In M. Aronoff and M.-L. Kean (Eds.), *Juncture* (pp. 91–105). Saratoga, CA: Anma Libri.

Liberman, A. M., and Mattingly, I. G. (1985). The motor theory of speech perception revised. *Cognition, 21,* 1–36.

Marslen-Wilson, W. D. (1973). Linguistic structure and speech shadowing at very short latencies. *Nature, 244,* 522–523.

Marslen-Wilson, W. D. (1987). Functional parallelism in spoken word-recognition. *Cognition, 25,* 71–102.

Marslen-Wilson, W., and Warren, P. (1994). Levels of perceptual representation and process in lexical access: Words, phonemes, and features. *Psychological Review, 101,* 653–675.

Marslen-Wilson, W. D., and Welsh, A. (1978). Processing interactions and lexical access during word recognition in continuous speech. *Cognitive Psychology, 10,* 29–63.

Marslen-Wilson, W., Moss, H. E., and van Halen, S. (1996). Perceptual distance and competition in lexical access. *Journal of Experimental Psychology: Human Perception and Performance, 22,* 1376–1392.

McQueen, J. M., and Cutler, A. (1998). Spotting (different types of) words in (different types of) context. In *Proceedings of the Fifth International Conference on Spoken Language Processing.* Vol. 6 (pp. 2791–2794). Sydney: Australian Speech Science and Technology Association.

McQueen, J. M., Cutler, A., Briscoe, T., and Norris, D. (1995). Models of continuous speech recognition and the contents of the vocabulary. *Language and Cognitive Processes, 10,* 309–331.

McQueen, J. M., Norris, D., and Cutler, A. (1999). Lexical influence in phonetic decision making: Evidence from subcategorical mismatches. *Journal of Experimental Psychology: Human Perception and Performance, 25,* 1363–1389.

McQueen, J. M., Otake, T., and Cutler, A. (2001). Rhythmic cues and possible-word constraints in Japanese speech segmentation. *Journal of Memory and Language, 45,* in press.

Mehler, J. (1981). The role of syllables in speech processing: Infant and adult data. *Philosophical Transactions of the Royal Society of London. Series B: Biological Sciences, 295,* 333–352.

Mehler, J., Dommergues, J.-Y., Frauenfelder, U. H., and Segui, J. (1981). The syllable's role in speech segmentation. *Journal of Verbal Learning and Verbal Behavior, 20,* 298–305.

Nearey, T. (1997). Speech perception as pattern recognition. *Journal of the Acoustical Society of America, 101,* 3241–3254.

Norris, D. G. (1994). Shortlist: A connectionist model of continuous speech recognition. *Cognition, 52,* 189–234.

Norris, D., and Cutler, A. (1985). Juncture detection. *Linguistics, 23,* 689–705.

Norris, D., and Cutler, A. (1988). The relative accessibility of phonemes and syllables. *Perception and Psychophysics, 43,* 541–550.

Norris, D., McQueen, J. M., Cutler, A., Butterfield, S. and Kearns, R. (2001). Language-universal constraints on speech segmentation. *Language and Cognitive Processes, 16,* in press.

Norris, D., McQueen, J. M., Cutler, A., and Butterfield, S. (1997) The possible-word constraint in the segmentation of continuous speech. *Cognitive Psychology, 34,* 191–243.

Norris, D., McQueen, J. M., and Cutler, A. (2000). Merging information in speech recognition: Feedback is never necessary. *Behavioral and Brain Sciences, 23,* 299–325.

Otake, T., Hatano, G., Cutler, A., and Mehler, J. (1993). Mora or syllable? Speech segmentation in Japanese. *Journal of Memory and Language, 32,* 258–278.

Savin, H. B., and Bever, T. G. (1970). The non-perceptual reality of the phoneme. *Journal of Verbal Learning and Verbal Behavior, 9,* 295–302.

Segui, J., Frauenfelder, U., and Mehler, J. (1981). Phoneme monitoring, syllable monitoring and lexical access. *British Journal of Psychology, 72,* 471–477.

Streeter, L. A., and Nigro, G. N. (1979). The role of medial consonant transitions in word perception. *Journal of the Acoustical Society of America, 65,* 1533–1541.

Suomi, K., McQueen, J. M., and Cutler, A. (1997). Vowel harmony and speech segmentation in Finnish. *Journal of Memory and Language, 36,* 422–444.

Treiman, R., and Danis, C. (1988). Syllabification of intervocalic consonants. *Journal of Memory and Language, 27,* 87–104.

Whalen, D. H. (1984). Subcategorical phonetic mismatches slow phonetic judgments. *Perception & Psychophysics, 35,* 49–64.

Whalen, D. H. (1991). Subcategorical phonetic mismatches and lexical access. *Perception & Psychophysics, 50,* 351–360.

11

Phonotactic Constraints Shape Speech Perception: Implications for Sublexical and Lexical Processing

Juan Segui, Ulricht Frauenfelder, and Pierre Hallé

The way in which we perceive speech is determined by the phonological structure of our native language. Though this conclusion may appear today as somewhat trivial, it is one of the more important advances in modern psycholinguistics, both from a theoretical and a methodological point of view. Indeed, language-specific specialization is now understood as necessary for humans to process their native language(s) with the fluidity and the fluency that have long been considered as a puzzling performance. Far from being a limitation, narrowing down speech perception to the native categories is required to cope with the variability inherent to the speech signal, and to benefit from "perceptual constancy." In recent decades, acquisition data have accumulated showing that language-specific specialization is largely guided by innate learning programs and is almost achieved by the end of the first year, both for speech perception *and production*. (This latter aspect has perhaps received less attention than it deserves.) It is interesting to note that Polivanov (1931) had already had this insight seventy years ago when he stated that "le phonème et les autres représentations phonologiques élémentaires de notre langue maternelle ... se trouvent si étroitement liées avec notre activité perceptive que, même en percevant des mots (ou phrases) d'une langue avec un système phonologique tout différent, nous sommes enclins à décomposer ces mots en des représentations phonologiques propre à notre langue maternelle" [the phoneme and the other phonemic representations for our native language ... are so intricately linked to our perceptual activity that even when we hear words (or sentences) from a language with an utterly different phonemic system, we tend to analyze these words in terms of our native language phonemic representations]. Other scholars from this early period shared similar assumptions about cross-language perception (Troubetzkoy, 1939/1969; Sapir, 1921, 1939).

It follows from Polivanov's observations that researchers cannot understand how listeners process speech without referring to listeners' native language. Indeed, it is by relating the perception performance of listeners to the phonological structure of their language that considerable progress has been made. In the last three decades, researchers have been able to characterize the evolution of the perceptual capacities of infants as well as the speech processing mechanisms in adults. Jacques Mehler is one of the first psycholinguists to have understood the importance of establishing this relation. By working within this tradition, he has made major empirical contributions in research on both infant and adults.

This chapter presents some of our recent research that fits well into this tradition. It examines the role of language-specific phonotactic constraints in speech processing. The experiments provide a striking demonstration of a new kind of perceptual assimilation that depends upon the listeners' knowledge of the phonotactic constraints in their language. To set the stage for this presentation, we first summarize briefly some of the major findings that have shown the impact of a listener's phonological knowledge on language acquisition and use.

The Influence of the Native Phonological System on Speech Perception

Important findings in infant language literature have emerged from research on both the perception of individual segments and that of segment sequences. Young infants appear to be equipped at birth with "universal" capacities for discriminating speech segments. Thus, for example, it has been shown that infants were generally able to discriminate nonnative speech contrasts during their first months of life (Streeter, 1976; Trehub 1976). This capacity is declining by nine months or before. By ten to twelve months, infants have reorganized their perceptual system and can no longer discriminate consonant contrasts not present in the language they are learning (Werker and Tees, 1984). This loss of phonological contrasts that are not functional in the language environment reflects the emergence of *phonological deafness* as it is observed in adults. Examples of phonological deafness of adults to foreign contrasts are abundantly documented in the literature (e.g., Dupoux et al., 1997; Polka, 1991; Pallier, Bosch, and Sebastián-Gallés, 1997). It is important to note that adults are not deaf to all nonnative language contrasts: Their ability to discriminate nonnative language contrasts depends upon the articulatory-phonetic differences between native and nonnative phonemic categories. For example, Best, McRoberts, and Sithole (1988) observed that English-speaking listeners are able to discriminate Zulu contrasts between clicks, which are all quite distant from any English sound. The Perceptual Assimilation Model (PAM) proposed by Best (1994) explains the differential sensitivity to foreign language contrasts by appealing to the notion of phonological perceptual assimilation. According to this model, two sounds that can be assimilated to the same segment of the native language are difficult to discriminate. A good illustration of such perceptual assimilation can be found in the well-known case of Japanese listeners who have great difficulties in discriminating the American English /r/-/l/ contrast (Goto, 1971; Mochizuki, 1981). Indeed, Japanese listeners assimilate both English phonemes to the single Japanese phoneme /r/. Note that French listeners also have some trouble with the American English /r/, which they tend to assimilate to /w/ (Hallé, Best, and Levitt, 1999).

Another important aspect of infants' speech development is their increasing sensitivity to recurrent patterns in general (Safran, Aslin, and Newport, 1996), and more specifically to sound sequences or phonotactic regularities in the language they are acquiring. It has been shown that nine-month-old Dutch infants are sensitive to phonotactically legal onset and offset clusters in Dutch words (Friederici and Wessels, 1993; see also Jusczyk et al., 1993). Moreover, Jusczyk, Luce, and Charles-Luce (1994) have shown that nine-month-old American infants prefer listening to monosyllabic nonwords with a high rather than a low phonotactic probability.

These data suggest that by the end of the first year, children have acquired the phonological knowledge that is relevant to the possible and likely combinations of sounds in their language. In the next section, we turn to the influence on adult speech perception of this kind of phonological knowledge, that is, the knowledge of phonotactic constraints.

198   Segui, Frauenfelder & Hallé

The Role of Phonotactic Constraints in Adult Speech Perception

Polivanov claimed that the native language affects not only the way in which individual sounds are processed but also the way in which sequences of sounds are perceived. He illustrated this point by observing that when Japanese native speakers are presented with an English word like "drama," they say that they hear /do.ra.ma/. This latter sequence is consistent with the phonotactic rules of Japanese, which does not allow syllable-initial clusters. Here then, we have an example of a transformation of a sequence that is illegal into one that is permitted in the language. This phonotactic effect in Japanese has recently been studied experimentally (Dupoux et al., 1999). When Japanese participants were presented with pairs of nonsense words such as /ebzo/ and /ebuzo/, they failed to perceive the difference. They reported hearing the vowel /u/ in /ebzo/ even though there was not any acoustic correlate for this vowel in the signal. By contrast, French listeners did not hear /u/ in /ebzo/. The authors interpreted their results in terms of an extension of Best's PAM.

These findings and other similar ones suggest that listeners interpret a nonnative illegal input sequence by assimilating this sequence to sound patterns that exist in their own language. We shall call this transformation process *phonotactic assimilation*. Three types of phonotactic assimilation may be distinguished: (1) the listener "ignores" individual phonemes (or stress patterns: Dupoux et al., 1997) that are present in the signal; (2) the listener "perceives" illusory phonemes that have no acoustic correlate in the signal; (3) the listener "transforms" one phoneme in another.

The objective of this chapter is to present some recent findings that shed light on how phonotactically illegal sequences of native language sounds are assimilated to legal sequences according to the third type of assimilation. We will investigate the processing of /tl/ and /dl/ clusters that are illegal in French in word-initial position, although they are attested in this position in other languages. These sequences represent an accidental gap within the set of possible obstruent + liquid (OBLI) clusters in French. As we will show, /dl/ and /tl/ are perceptually assimilated to legal OBLI clusters by changing the place of articulation of the initial stop consonants. Our intention is first to demonstrate the existence of such phonotactic assimilation and, second, to uncover its time course and its perceptual cost. Finally, we examine the consequences of this phonotactic assimilation for lexical access. As it will appear, the outcome of these studies also bear on the general issue of how word forms are represented and how they are matched to the lexical input forms.

Experimental Studies

Massaro and Cohen (1983) conducted the first experiments on the perception of illegal consonant clusters in English. Here, participants were presented with synthetic speech stimuli beginning with an obstruent consonant followed by a "liquid" taken from an /r/-/l/ continuum. The category boundary in the /r/-/l/ continuum was shifted, relative to a neutral situation, such that more /r/s than /l/s were identified after /t/ and conversely, more /l/s than /r/s were identified after /s/. These results suggest the existence of a phonotactic assimilation process which biases listeners to perceive word-initial legal /tr/ and /sl/ rather than illegal /tl/ or /sr/. More recently, Pitt (1998) replicated these findings and further showed they were better explainable by phonotactic constraints than by cooccurrence frequencies.

In our first experiment, we wanted to discover what French listeners hear when they are presented with items beginning with the illegal cluster /dl/ or /tl/. A simple and straightforward approach to answering this question is to let participants freely transcribe what they hear when presented with such clusters. Nonwords with an initial /tl/ ('tlabdo,' 'tlabod,' 'tlobad,' and 'tlobda'), and with its voiced counterpart /dl/ ('dlapto,' 'dlapot,' 'dlopat,' and 'dlopta') were used as test stimuli. The outcome of this experiment was clear-cut: the /dl/ and /tl/ items were transcribed with a /gl/ or /kl/ cluster 85.4 percent of the time, and with a /dl/ or /tl/ cluster only 13.2 percent of the time. This finding was confirmed in a second experiment in which participants were given a forced choice identification task on the initial plosive consonant. Again, the results obtained showed a very substantial dental-to-velar shift for the items with an illegal cluster and suggested that some kind of phonotactic assimilation occurred for /dl/- and /tl/-initial items.

In order to eliminate alternative explanations for this shift in terms of the acoustic-phonetic properties of the illegal clusters, we conducted some acoustic analyses of the speech signal and performed two *phonemic gating experiments*. In this variant of the gating paradigm, participants are asked to transcribe exactly what they hear, rather than to guess words. The results showed that the initial stops of /dl/- and /tl/-items were first identified as being dental, not velar, in the early gates (first 60 ms after release burst). So it is unlikely that there were cues to velar articulation in the speech signal in the portion corresponding to the stop onset. This was confirmed by the acoustic analyses. However, at the point where the /l/ and the following vowel were first identified, subjects revised their dental responses and began to produce velar responses. This assimilation process for the initial stop consonant (dental to velar) thus depends upon the following phonological context. Hence rather than showing misperception of the illegal cluster's first constituent, these experiments provide a striking example of phonotactic assimilation.

Having demonstrated the existence of a phonotactic assimilation process in offline tasks, we then asked whether the process is also observed using a more online task such as the generalized phoneme monitoring task (Frauenfelder and Segui, 1989). The same /dl/- and /tl/-items as in the previous experiments were used. When the participants were instructed to detect /d/ and /t/ targets, they most often failed to do so (69 percent misses, on average). In contrast, when participants were instructed to detect velar targets in the same items, they did so (erroneously) 80 percent of the time. These findings suggest that listeners categorized the initial phoneme of the illegal sequence as a velar rather than a dental in real time. An additional finding of interest was the longer reaction times (RTs) to the assimilated target (/k/ in tlabod) than to the actual target (/k/ in klabod). Thus, we have shown that these illegal clusters induce phonotactic assimilation and require more processing time than legal clusters.

In more recent experiments, we explored whether the additional processing load associated with the phonotactic assimilation of the initial consonant cluster also affects the processing of later portions of the speech input. One way to measure later-occurring effects of a word-initial illegal cluster on subsequent processing is to have participants detect phonemes occurring later in the test items. In case the additional load induced by an illegal cluster is not confined to the processing of that cluster, phonemes that occur later in the speech sequence should be detected more slowly in illegal than in phonetically matched legal nonwords.

In this experiment, participants detected the target that occurred as the fourth phoneme in the item (e.g., /b/) either as the coda of the first syllable (/b/ in /tlobda/) or as the onset of the second syllable (/b/ in /tlobad/). This manipulation of the syllabic position of the target phoneme additionally allowed us to test whether the processing cost is confined to the syllable containing the illegal cluster or whether it propagates downstream to the initial phoneme of the second syllable.

The results showed that the presence of an illegal cluster in word-initial position slowed down the detection latencies to a phoneme located in either the first or the second syllable. Furthermore, the syllabic target position affected miss rate in illegal items: more misses occurred for phoneme targets in the first than in the second syllable. Therefore, the processing load due to the presence of a phonotactic violation appears to depend upon the syllabic position of the target.

It is possible that there were some acoustic-phonetic differences in the production of the target phonemes following legal and illegal clusters due to a difficulty the speaker may have encountered in producing the illegal stimuli. To control for this possibility, we conducted a further experiment in which the initial consonant-consonant-vowel (CCV) part of legal and illegal stimuli were cross-spliced. If the effect observed in the previous experiment was due to the presence of a phonotactic violation (rather than to reduced acoustic saliency of the target), then the same pattern of results should be observed. This was indeed the case. RTs were longer for the targets following illegal sequences. Moreover, both RT and miss rate data showed a significant effect of syllabic position, with faster and more precise detection of the target in the second than in the first syllable.

These experiments showed that detection of the word-medial stops was delayed by the presence of a word-initial illegal cluster /dl/ or /tl/. The increased difficulty in detecting the word-medial stops in illegal items was not due to phonetic-acoustic or prosodic differences compared to the legal items. It is important to note that the greater processing cost observed in the detection of the coda of the first syllable has a structural rather than a temporal origin since in both conditions being compared the target phoneme always occurred in the same serial position. These results suggest that the syllable plays a role in the perceptual integration process (Hallé, Segui, and Frauenfelder, in preparation).

The final series of experiments use lexical decision and cross-modal priming paradigms (Hallé, Segui, and Frauenfelder, submitted) to evaluate the effect of the phonotactic assimilation process at the lexical level. In particular, an auditory lexical decision experiment compared the responses to nonwords which either could or could not be assimilated to existent words. For instance, a nonword like "dlaïeul"—beginning with the illegal consonant cluster /dl/—can be assimilated to the word "glaïeul," whereas the legal nonword "droseille"—beginning with the legal consonant cluster /dr/—should not be assimilated to the word "groseille." Note that these two nonwords differ from the base words from which they are derived by the same, single phonetic feature (namely, dental vs. velar place of articulation), and thus are equally distant phonologically from their respective base words. The results showed that more "word" responses were observed for the phonotactically assimilated items (68.3 percent) than for the legal nonwords (19.6 percent), which served as a control condition.

A more subtle test was performed using the cross-modal repetition priming paradigm. Here participants heard either the base words or the derived nonwords ("dlaïeul" or "droseille") and made a lexical decision on the base words, presented visually ("glaïeul" or "groseille"). The findings indicated that RTs to the words like "glaïeul" were similar whether the prime was the real word (repetition condition) or the phonotactically assimilated nonword (471 ms vs. 496 ms). In contrast, a large RT difference was obtained for the case of phonotactically nonassimilated nonwords (469 ms vs. 552 ms). Taken together, these experimental findings demonstrate the effect of phonotactic assimilation at the lexical level. In addition, they also suggest that the amount of lexical preactivation produced by the prior presentation of a phonologically related nonword does not depend solely upon the phonological distance between prime and target measured in terms of distinctive features but is also determined by the phonotactic properties of the prime. Indeed, given that prime-target pairs always differed by the same phonological distance, the differential amount of lexical preactivation observed can only be related to the legal vs. illegal nature of the initial cluster.

Discussion

In this chapter we have presented the main findings of a series of experiments that explored a case of *phonotactic assimilation*. We have demonstrated a simple yet robust phenomenon: When listeners are presented with the illegal consonant cluster /dl/ or /tl/ they report hearing the legal cluster /gl/ or /kl/, respectively. The observed effect seems to be very robust and unrelated to the acoustic properties of the signal. In fact, a gating experiment showed that listeners initially perceived the first consonant of the cluster correctly and only somewhat later, when they were able to identify the second consonant of the cluster, began to revise their responses and report the phonotactically legal cluster /gl/ or /kl/.

Experiments using a phoneme monitoring paradigm confirm the on-line nature of the assimilation process and furthermore reveal a processing cost associated with the phonotactic assimilation. This processing cost was observed not only in the detection latencies of the initial member of the illegal cluster but also for phoneme targets located at later positions in the carrier items. Moreover, the size of the perceptual cost as reflected by both miss rates and RTs was greater for phonemes belonging to the syllable containing the illegal cluster than for phonemes located in the following syllable.

Finally, we studied the effect of this type of phonotactic violation in lexical access. Phonotactic assimilation was found to operate for words just like it does for nonwords, so that altered forms obtained by changing, for example, /gl/ to /dl/, are readily mapped to the lexical forms from which they are derived. We return to this point later.

Overall, the experimental results presented in this chapter confirm the role of phonotactic constraints in speech perception. We interpreted the results in terms of an extended version of PAM according to which an illegal sequence phonologically similar to a legal sequence is assimilated to that legal sequence. Differently said, listeners hallucinate another phoneme than that which actually occurs in the sequence, illustrating the third type of phonotactic assimilation described above. It is important to note that these experiments compared the perceptual processing of legal and illegal items within the same language. Therefore, the case of perceptual assimilation we found is not explicitly addressed in PAM, which in its current version focuses on cross-language perception. How could PAM deal with the case of within-language perceptual assimilation that we found? PAM makes predictions about the type of assimilation pattern and the level of discrimination performance that should be observed for a nonnative contrast, according to both the phonetic and structural relations between native and nonnative phonemic categories. Yet, we can consider that /dl/ and /tl/ are foreign to the phonology of French. In the terminology of Best's model, the fact that French listeners hear /dl/ as /gl/ is most readily interpreted as a case of category goodness (CG) assimilation: /gl/ is indeed a good exemplar of /gl/, while /dl/ is probably heard as an atypical exemplar of /gl/. (The reason why /dl/ assimilates to /gl/ rather than to /bl/ is to be found in the phonetic-acoustic properties of the sounds involved; see Hallé et al., 1998.) That the phonotactic assimilation we found is related to the CG type rather than, for example, the single category (SC) type of assimilation could be tested crosslinguistically using categorization and discrimination tasks. For example, modern Hebrew allows both velar and dental stop + /l/ clusters. Given the abundance of evidence for crosslinguistic deafnesses, we expect that French listeners would have difficulty in discriminating the Hebrew /dl/-/gl/ and /kl/-/tl/ contrasts. More specifically, we predict that French listeners would categorize the dental stops of these clusters as velar.

In addition to contributing to a more complete account of perceptual assimilation in speech perception, our data also shed some light on word recognition. Indeed, the pattern of data obtained in the cross-modal experiment is informative about the mapping process underlying word recognition. It is generally assumed that a lexical item is recognized by means of a mapping from the speech signal onto its representation. The activation level of any given lexical entry is mainly a function of its "vertical similarity" to the input (Connine, 1994; Connine, Blasko, and Titone, 1993), that is, the goodness of fit between the speech input and a particular lexical representation. (The activation level also depends on the possible competition with phonological neighbors: this modulates what Connine, 1994, called "horizontal similarity.") Previous experiments indicated that vertical similarity takes into account two main factors: the amount of deviation between input and lexical forms in terms of the number of mismatching distinctive features, and the serial position of the mismatching phonemes (see, e.g., Marslen-Wilson and Zwitserlood, 1989). The results obtained in our research showed that the goodness of fit also depends on the phonological context immediately following the altered phoneme, and in particular, on the possibility that this context induces "phonotactic assimilation." In terms of phonetic features and serial position, "dlaïeul" differs from "glaïeul" in the same way as "droseille" differs from "groseille." Yet, only the former nonword strongly activates the related word. In other words, goodness of fit must also take into account *contextual similarity* and language-specific phonotactic knowledge.

To the extent that our cross-modal repetition priming data reflect real-time processes, the differential priming effects for "dlaïeul" vs. "droseille" suggest that lexical access proceeds through a stage of perceptual integration into sublexical units larger than phonemes or phonetic features. Our results are not compatible with the notion of a direct mapping of a sequence of phonemes onto lexical forms. A mapping from a structured array of phonetic features extracted from the input is equally unlikely for the same reasons. (Also, such a mapping is not consistent with the processing discontinuities observed in the phonemic gating data at a sublexical level.) Our data thus point to the mandatory perceptual integration of the speech flow into sublexical units that are at least of the size of complex syllable onsets. Speech seems to be processed with an extreme ease and fluidity when it is free from possible distortions, and this might suggest a continuous, direct kind of processing which can be based on very fine-grained units of perception. Our data, on the other hand, strongly suggest that speech sounds also are integrated into coarser-grained units such as syllable (complex) onsets. In our view, these two notions are not necessarily contradictory. We suggest that different analyses of the speech input are performed in parallel and, as it were, compete to produce a plausible outcome as quickly as possible. We therefore do not claim a precise size for the coarse-grained units which were evidenced in our data. Could they be the size of the plain syllable? Although this is prima facie not in line with recent propositions that dismiss the syllable as a processing unit (Cutler et al., chapter 10 in this book; Frauenfelder and Content, 1999), our findings of a greater sensitivity to the processing load induced by phonotactic violations within than across syllables suggest that the syllable still is a good candidate for a prelexical unit of speech integration.

Acknowledgments

This research was supported by a grant from the Fonds National de la Recherche Suisse (FNRS 1113-049698.96) and a BQR grant from the René Descartes University, Paris V.

References

Best, C. (1994). The emergence of native-language phonological influence in infants: A perceptual assimilation model. In J. Goodman and H. Nusbaum (Eds.), *The development of speech perception: The transition from speech sounds to spoken words* (pp. 167–224). Cambridge, MA: MIT Press.

Best, C., McRoberts, G., and Sithole, N. (1988). Examination of perceptual reorganization for nonnative speech contrasts: Zulu click discrimination by English-speaking adults and infants. *Journal of Experimental Psychology: Human Perception and Performance, 14,* 345–360.

Connine, C. (1994). Horizontal and vertical similarity in spoken word recognition. In C. Clifton, Jr., L. Frazier, and K. Rayner (Eds.), *Perspectives on sentence processing* (pp. 107–120). Hillsdale, NJ: Erlbaum.

Connine, C., Blasko, D., and Titone, D. (1993). Do the beginnings of spoken words have a special status in auditory word recognition? *Journal of Memory and Language, 32,* 193–210.

Dupoux, E., Kakehi, Y., Hirose, C., Pallier, C., and Mehler, J. (1999). Epenthetic vowels in Japanese: A perceptual illusion? *Journal of Experimental Psychology: Human Perception and Performance, 25,* 1–11.

Dupoux, E., Pallier, C., Sebastián, N., and Mehler, J. (1997). A destressing deafness in French? *Journal of Memory and Language, 36,* 406–421.

Frauenfelder, U., and Segui, J. (1989). Phoneme monitoring and lexical processing: Evidence for associative context effects. *Memory and Cognition, 17,* 134–140.

Frauenfelder, U., and Content, A. (1999). The role of the syllable in spoken word recognition: Access or segmentation? In *Proceedings of Deuxièmes Journées d'Etudes Linguistiques* (pp. 1–8). Nantes, France: Université de Nantes.

Friederici, A., and Wessels, J. (1993). Phonotactic knowledge of word boundaries and its use in infant speech perception. *Perception and Psychophysics, 54,* 287–295.

Goto, H. (1971). Auditory perception by normal Japanese adults of the sounds "L" and "R." *Neuropsychologia, 9,* 317–323.

Hallé, P., Best, C., and Levitt, A. (1999). Phonetic vs. phonological influences on French listeners' perception of American English approximants. *Journal of Phonetics, 27,* 281–306.

Hallé, P., Segui, J., and Frauenfelder, U. (in preparation). Perceiving /dl/ as /gl/: What does it cost?

Hallé, P., Segui, J., and Frauenfelder, U. (submitted). The role of phonotactic constraints in word recognition.

Hallé, P., Segui, J., Frauenfelder, U., and Meunier, C. (1998). Processing of illegal consonant clusters: A case of perceptual assimilation? *Journal of Experimental Psychology: Human Perception and Performance, 24,* 592–608.

Jusczyk, P., Friederici, A., Wessels, J., Svenkerud, V., and Jusczyk, A. (1993). Infants' sensitivity to the sound patterns in the native language. *Journal of Memory and Language, 32,* 402–420.

Jusczyk, P., Luce, P., and Charles-Luce, J. (1994). Infants' sensitivity to phonotactic patterns in the native language. *Journal of Memory and Language, 33,* 630–645.

Marslen-Wilson, W., and Zwitserlood, P. (1989). Accessing spoken words: On the importance of word onsets. *Journal of Experimental Psychology: Human Perception and Performance, 15,* 576–585.

Massaro, D., and Cohen, M. (1983). Phonological context in speech perception. *Perception and Psychophysics, 34,* 338–348.

Mochizuki, M. (1981). The identification of /r/ and /l/ in natural and synthesized speech. *Journal of Phonetics, 9,* 283–303.

Pallier, C., Bosch, L., and Sebastián-Gallés, N. (1997). A limit on behavioral plasticity in speech perception. *Cognition, 64,* B9, B17.

Pitt, M. (1998). Phonological processes and the perception of phonotactically illegal consonant clusters. *Perception and Psychophysics, 60,* 941–951.

Polivanov, E. (1931). La perception des sons d'une langue étrangère. *Travaux du Cercle Linguistique de Prague, 4,* 79–96.

Polka, L. (1991). Cross-language speech perception in adults: Phonemic, phonetic, and acoustic contributions. *Journal of the Acoustical Society of America, 89,* 2961–2977.

Saffran, J., Aslin, R., and Newport, E. (1996). Statistical learning by 8-month-old children. *Science, 274,* 1926–1928.

Sapir, E. (1921). *Language.* New York: Harcourt Brace.

Sapir, E. (1939). *Language: An introduction to the study of speech.* New York: Harcourt Brace.

Streeter, L. (1976). Language perception of 2-month-old infants shows effect of both innate mechanisms and experience. *Nature, 259,* 39–41.

Trehub, S. (1976). The discrimination of foreign speech contrasts by infants and adults. *Child Development, 47,* 466–472.

Troubetzkoy, N. S. (1939/1969). *Principles of phonology.* Translated by C.A.M. Baltaxe. Berkeley: University of California Press.

Werker, J., and Tees, R. (1984). Cross-language speech perception: Evidence for perceptual reorganization during the first year of life. *Infant Behavior and Development, 7,* 49–63.

12

A Crosslinguistic Investigation of Determiner Production

Alfonso Caramazza, Michele Miozzo, Albert Costa, Niels Schiller, and F.-Xavier Alario

Research by Jacques Mehler and his collaborators (e.g., Cutler et al., 1989, 1992; Mehler et al., 1993, 1996) has shown that the processing routines (prelexical segmentation, word segmentation, etc.) that are engaged in speech perception are not identical from one language to the other, but are finely tuned to the specific properties of the native language. In contrast, current models of speech production have emphasized the language-universal aspects of the process; that is, the details of the processing routines (processing levels, units, time course of the computations) are always thought to be identical across languages. Is speech production special in that it uses exclusively language-universal procedures, or is there as much language-specific tuning in speech production as has been found in speech perception? Here, we argue that the crosslinguistic investigation of noun phrase (NP) production reveals language-specific differences in the procedures used to select determiners and other closed-class words.

There is a clear difference in the kind of information that is used to select open- and closed-class words in language production. The selection of open-class words, such as nouns and verbs, depends primarily on their individual meanings. For example, in sentence 1a, the selection of "rose" depends only on its meaning (to fulfill the speaker's intention to communicate the proposition: *is on* rose, table). Quite a different process seems to be at play in the selection of closed-class words, such as determiners and pronouns. The selection of the latter types of words depends largely on properties of other words in the sentence. This is clearly illustrated in sentences 1b–d for determiners and in sentences 1e–g for demonstratives and possessives.

1a. *The* rose is on the table/*The* roses are on the table
1b. *La*$_{fem}$ rosa è sul tavolo/*Le*$_{fem}$ rose sono sul tavolo [Italian, The rose is on the table/The roses are on the table]
1c. *Il*$_{masc}$ tulipano è nel vaso/*I*$_{masc}$ tulipani sono nel vaso [Italian, The tulip is in the vase/The tulips are in the vase]
1d. *La*$_{fem}$ tulipe est dans le vase/*Les*$_{fem}$ tulipes sont dans le vase [French, The tulip is in the vase/The tulips are in the vase]
1e. *My* rose is in the vase/*My* tulip is in the vase
1f. *Esta*$_{fem}$ rosa está en el jarron/*Este*$_{masc}$ tulipán está en el jarrón [Spanish, The rose is in the vase/This tulip is in the vase]
1g. *Ma*$_{fem}$ rose est dans le vase/*Mon*$_{masc}$ chapeau est sur la table [French, My rose is in the vase/My hat is on the table]

In sentence 1b, the Italian (definite) determiner *la* is used because the noun "rosa" is feminine and singular; in sentence 1c, the determiner *il* is used with the masculine, singular noun "fiore." The masculine and feminine plural determiner forms are, respectively, *i* and *le*, as shown in sentences 1b and 1c. These examples illustrate that the selection of determiners depends on a property of the nouns they are associated with: their grammatical gender. Similarly, the selection of the French possessive adjectives *mon* and *ma* in sentence 1g depends on the gender of the head noun of the phrase. A similar process can be seen in the case of the Spanish demonstratives *esta* and *este*. What implications follow from these facts for the process of NP production? One obvious implication is that the selection of a determiner (or demonstrative pronoun or possessive adjective) can only take place *after* the head noun of the NP has been selected and its gender feature becomes available to guide the selection of the proper form of the determiner. This much is obvious.[1] What is not clear is *how* the selection takes place. In fact, the broader question of how closed-class words are selected remains controversial. Are the principles that govern the selection of closed-class words the same as those for open-class words (e.g., Dell, 1990; Stemberger, 1984), or are closed-class words selected by a special set of processes (e.g., Garrett, 1980)?

The issue of whether or not the selection of closed-class words involves special processes has received much attention in the literature. There is little doubt that in certain conditions closed-class words behave differently from open-class words. For example, speech errors seem to be distributed differently for open- and closed-class words (Garrett, 1992). And there is the well-known phenomenon of disproportionate difficulty in the production of closed-class words relative to open-class words in certain types of brain-damaged patients (e.g., Berndt and Caramazza, 1980; Kean, 1977). However, it has proved rather difficult to demonstrate that the observed differences between word classes is due to their categorical status as closed- and open-class words as opposed to some other characteristic (e.g., frequency, concreteness, etc.) that distinguishes between them. It is not our intention here to attempt to resolve this complex issue. Rather, we have the more modest goal of trying to ascertain how determiner selection might differ from open-class word selection across languages. Any insight we might gain into such differences could help us formulate both general and language-specific principles in the selection of closed- and open-class words in language production.

Grammatical Gender and Determiner Selection

One approach that could be used to address the processes that underlie determiner selection is to investigate the time course of determiner production in NPs. A promising experimental paradigm for this purpose is the picture-word interference task, a variant of the classical Stroop task (Klein, 1964; for a review, see McLeod, 1991). This experimental paradigm has been used successfully to address various issues concerning the dynamics of activation at different levels of lexical access (e.g., see Glaser and Glaser, 1989; Schriefers, Meyer, and Levelt, 1990; Starreveld and La Heij, 1996).

In the picture-word interference naming task, subjects are required to name a picture while ignoring a distractor word that is printed on (or near) the picture (see, e.g., Glaser, 1992). The relationship between the distractor word and the picture has been shown to affect the reaction time (RT) to name the picture. For example, two major effects are the semantic interference and the phonological facilitation effect. Picture naming is slower if the distractor word is semantically related to the picture, for example, the word "car" superimposed on the picture of a bus, relative to a baseline condition defined by an unrelated word (e.g., see Lupker, 1979). However, picture naming is faster when the distractor word and the picture name are phonologically related, for example, the word "bar" superimposed on the picture of a bus (e.g., see Briggs and Underwood, 1982; Lupker, 1982). These interference and facilitation effects are assumed to reflect processes at different levels of lexical access. The semantic interference effect is commonly thought to reflect competition at the level of lexical node selection, and the phonological facilitation effect is thought to reflect priming of the phonological content of the lexical node selected for production. Therefore the investigation of these effects could reveal properties of the lexical access system.

Schriefers (1993) extended the use of the picture-word interference paradigm to investigate the mechanisms that control the selection of a word's grammatical features. Instead of manipulating the semantic relatedness or the phonological relatedness between distractor word and picture name he varied whether or not the two words were grammatically congruent. Specifically, he varied the gender relatedness between distractor word and picture name: the two words could either have the same or different genders. Schriefers reasoned that if grammatical feature selection functions with principles similar to those involved in the selection of lexical nodes and phonological segments (i.e., graded activation and selection competition), the manipulation of gender relatedness should produce measurable effects. We will refer to effects of gender relatedness as "gender congruency" effects, without commitment to a specific claim about the locus of this effect within the lexical access process.

In a seminal set of experiments, Schriefers asked Dutch speakers to produce NPs (e.g., "the red table") in response to colored pictures. In Dutch, determiners are marked for gender: *de* is used for common (com) gender nouns (e.g., *de* tafel, the table, com), and *het* is used for neuter (neu) gender nouns (e.g., *het* boek, the book, neu). Thus, speakers would produce either a *de* + Adj + N phrase or a *het* + Adj + N phrase. RTs in naming pictures associated with words of the same gender or different genders were compared. The results showed that naming latencies were longer when targets and distractors had different genders (see also La Heij et al., 1998, and van Berkum, 1997). Schriefers interpreted this gender congruency effect as arising at the level of gender feature selection. He argued that the distractor word activates its gender feature, which interferes with the selection of the target word's gender feature when there is a mismatch between the two. On this interpretation, the selection of a gender feature is a competitive process that is dependent on its relative level of activation and is not simply an automatic consequence of selecting a lexical node.

However, the results reported by Schriefers do not unambiguosly imply gender feature competition. They could alternatively reflect competition between determiners (*de* vs. *het*) instead of competition between gender features. That is, the selection of the target word's determiner could be slower when a different determiner is activated by the distractor word. This outcome would be possible if the determiner of the distractor word is activated even though the distractor lexical node is not selected for production.

In a recent series of experiments, Schiller and Caramazza (submitted) tested these alternative accounts of the gender congruency effect. They exploited an interesting property of the Dutch and German determiner systems. In these languages, determiners are gender-marked in the singular but not in the plural. We have seen that, in Dutch, the determiners *de* and *het* are selected, respectively, for common and neuter singular nouns. However, plural NPs take the determiner *de* irrespective of gender (e.g., de tafel/de tafels; het boek/de boeken). If the gender congruency effect reflects the selection of the noun's gender feature, we should observe interference in the gender incongruent condition in the production of both singular and plural NPs. However, if the gender congruency effect reflects competition between determiners, we should find interference only in the production of singular NPs and not in the production of plural NPs, since in the latter case the same determiner is produced irrespective of gender. A similar set of predictions can be made for German NP production. In German, different determiners are selected for masculine (m.), feminine (f.), and neuter (n.) nouns when they are used in the singular (e.g., in the nominative case, the determiners are, respectively, *der, die,* and *das,* as in *der* Tisch ["the table," m.], *die* Wand ["the wall," f.], and *das* Buch ["the book," n.]). However, like Dutch, the same determiner is used for all genders in the plural (in the nominative case it is the determiner *die,* as in *die* Tische ["the tables," m.], *die* Wände ["the walls," f.], *die* Bücher ["the books," n.]). Therefore, the same pattern of results should be observed in German and Dutch. Namely, either the gender congruency effect is obtained for singular and plural NPs (competition in the selection of the gender feature) in both languages or the effect is only obtained for singular NPs (competition in the selection of the determiner) in both languages.

The results of several experiments were clear-cut: a gender congruency effect was observed in the production of singular NPs with both Dutch and German speakers (see also Schriefers and Teruel, 2000) but not in the production of plural NPs. This interaction of gender congruency by number (plural vs. singular) was robust across different stimulus onset asynchronies (SOAs) for the distractor word in relation to the picture (see figure 12.1). Furthermore, the absence of a gender congruency effect for plural NPs was observed in the context of strong and reliable semantic interference and phonological facilitation effects, signature effects of target-distractor interaction at early and late stages of processing, respectively. The latter pattern of results cannot be dismissed as the consequence of an inability to reveal effects at early or late stages of selection in lexical

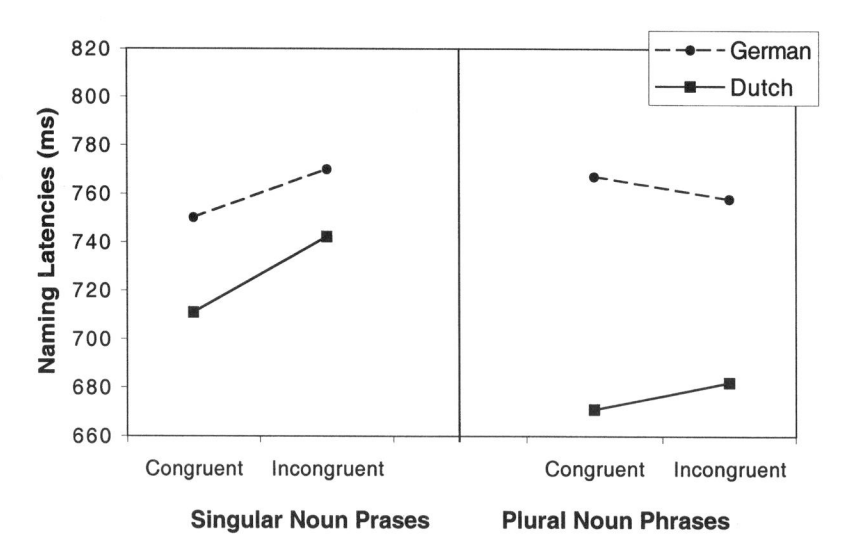

Figure 12.1
Picture naming latencies obtained in German and Dutch for singular and plural noun phrases. Picture word-pairs were of the same gender (congruent) or of different gender (incongruent). Data from Schiller and Caramazza (submitted).

access. Thus, the results with plural NPs in Dutch and German indicate that the gender congruency effect is actually a determiner interference effect, and therefore we can conclude that the selection of grammatical features of words is a noncompetitive, automatic consequence of the selection of a lexical node.

The hypothesis that the gender congruency effect reflects interference between competing determiners receives further support from the results of picture-word interference experiments in which subjects are required to produce only nouns (without determiners). If the gender congruency effect were the result of competition for selection between determiners as opposed to competition for selection between gender features, we should *not* observe interference when the noun alone is produced (since there is no selection process for determiners in this case). As mentioned earlier, La Heij et al. (1998) found a gender congruency effect in Dutch when participants were required to produce singular, determiner-NPs. However, they failed to find a congruency effect when subjects were required to name nouns without a determiner. The latter result has also been obtained with Italian speakers (Miozzo and Caramazza, unpublished): a strong semantic interference effect was observed but there was no trace of a gender congruency effect (see figure 12.2). In other words, gender incongruency does not result in interference in the picture-word naming task unless determiners must be produced along with the pictures' names. These results are consistent with the determiner interference hypothesis and problematic for the gender feature interference hypothesis.[2]

The results of NP production with the picture-word naming task that we have reviewed thus far suggest that determiners are selected in roughly the same fashion as nouns. The process could work as follows: a bundle of features (e.g., definite, singular, masculine) activates all determiners that are associated with that bundle of features (e.g., all definite determiners, all singular determiners, all masculine determiners), and the determiner node with the highest activation level is selected for further processing.[3] Of course, determiner selection differs from noun selection in that some of the features (e.g., gender) used to activate determiners are provided by other lexical items in a sentence (indirect election). However, the dynamics of activation and the process of selection appear to be the same across word classes (e.g., see Dell, 1990; Stemberger, 1984).

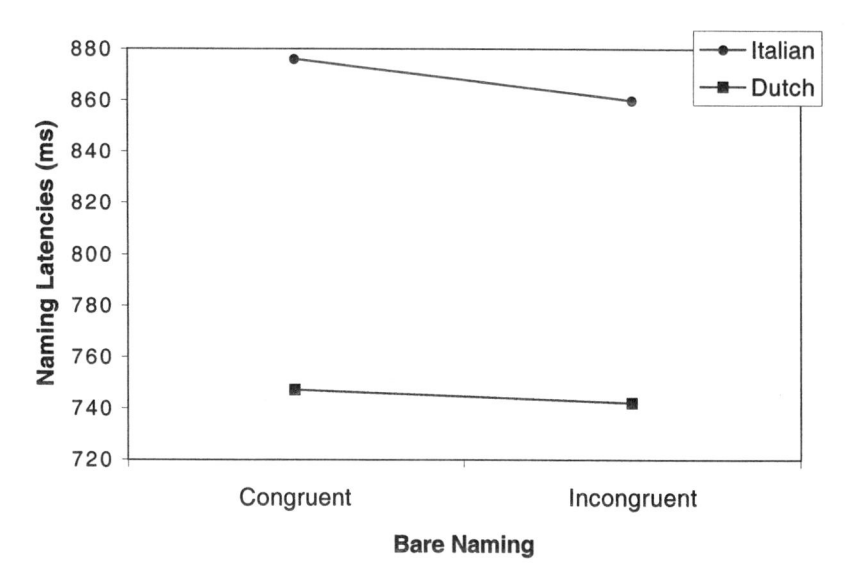

Figure 12.2
Picture naming latencies obtained in Italian and Dutch for picture-word pairs that were of the same gender (congruent) or of different gender (incongruent). Italian data from Miozzo and Caramazza (unpublished); Dutch data from La Heij et al. (1998).

Another implication that could be drawn from the results reviewed thus far is that determiner selection happens independently of phonological processes. That is, conceptual information that leads to the selection of the feature [+/− definite] and information from the head noun that specifies the features [number] and [gender] jointly comprise the "lemma" for a determiner,[4] which then activates a specific determiner form (Levelt, 1989). This process is not informed by considerations of the phonological properties of other words in the NP. However, these conclusions, which we have derived from research with Dutch and German speakers, are not consistent with the facts of NP production in other languages.

The selection of determiners in Italian is more complex than in Dutch or German. Consider the following Italian NPs:

2a. Il_{masc} treno/i_{masc} treni [the train/the trains]
2b. Lo_{masc} sgabello/gli_{masc} sgabelli [the stool/the stools]
2c. La_{fem} forchetta/le_{fem} forchette [the fork/the forks]

2d. *Il*~masc~ piccolo treno [the small train]
2e. *Il*~masc~ piccolo sgabello [the small stool]
2f. *La*~fem~ piccola forchetta [the small fork]
2g. *Il*~masc~ treno piccolo [literally, the train small]
2h. *Lo*~masc~ sgabello piccolo [literally, the stool small]
2i. *La*~fem~ forchetta piccola [literally, the fork small]

Several facts are immediately apparent from a consideration of the determiner forms (and adjectives) used in the phrases 2a–i. Determiners and adjectives are marked for number and gender. Feminine nouns take the definite article *la* (plural: *le*), while masculine nouns take either *il* or *lo* (plural: *i* and *gli,* respectively). A crucial fact for present purposes is that the selection of the masculine article is dictated by *phonological* characteristics of the onset of the word that immediately follows it: the determiners *lo* and *gli* are selected if the next word starts with a vowel, with a consonant cluster of the form "s + consonant" or "gn," or with an affricate; the determiners *il* and *i* are selected for all the remaining cases. Note also that since Italian allows adjectives to occupy both prenominal and postnominal NP positions, the relevant phonological context for determiner selection is not specified until the major constituents of the phrase are ordered. It is only at this point that the phonological context relevant to determiner selection (the onset of the noun or adjective) becomes available. In other words, in order to produce the proper determiner, Italian speakers must access not only grammatical information (e.g., the gender of the noun) but also phonological information about the onset of the word that follows it (compare *lo* sgabello/*il* piccolo sgabello/*lo* sgabello piccolo). This fact has two related implications for determiner production. One implication is that the selection of a determiner form is based on a mixture of phrasal (number), lexical (gender), and phonological features. The other implication is that determiner selection occurs very late in the process of NP production: the point at which the phonological forms of the noun and adjectives are ordered and inserted into a *phonological phrase.* Miozzo and Caramazza (1999) refer to languages with these properties as *late selection languages.* These properties of the structure of gender in Italian have specific consequences for the gender congruency effect in the picture-word naming paradigm.

We have argued that the gender congruency effects observed with Dutch and German speakers reflect determiner-selection interference. The interference is caused by the involuntary activation of the determiner of the distractor word which competes for selection with the determiner of the target word. However, if our analysis of the determiner-selection process in Italian is correct, we should not expect a gender congruency effect in this language. This absence is due to the fact that in Italian determiners are selected so late in the production process that the activation of potentially competing information has long dissipated and hence cannot interfere with the selection of the target determiner (see Miozzo and Caramazza, 1999, for a more detailed explanation). In various tests of this hypothesis we repeatedly failed to observe a gender congruency effect despite the fact that we obtained strong and reliable semantic interference and phonological facilitation effects in the same experiments (see figure 12.3).

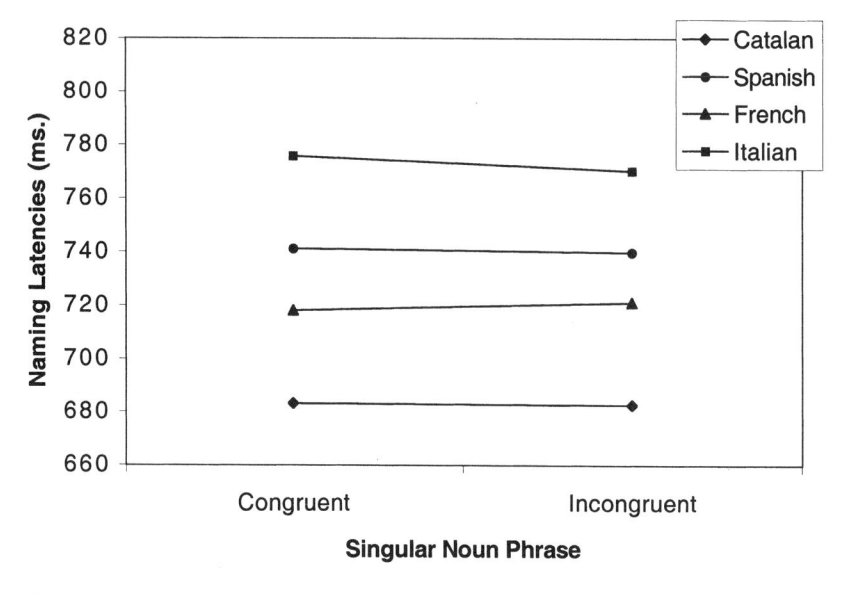

Figure 12.3
Picture naming latencies obtained in Catalan, Spanish, French, and Italian for singular noun phrases. Picture word-pairs were of the same gender (congruent) or of different gender (incongruent). Data from Costa et al. (1999), Alario and Caramazza (submitted), Miozzo and Caramazza (1999).

The effects observed in Italian generalize to other Romance languages with qualitatively similar gender systems: Catalan, French, and Spanish. In Catalan, for example, the form of masculine determiners depends on the phonological properties of the word that follows it. Thus, the determiner *el* is used when the following word begins with a consonant (*el got*, the glass; *el meu ull*, (literally) the my eye, but *l'* is used when the following word begins with a vowel (*l'ull*, the eye). Similarly, in Spanish, feminine singular nouns take the determiner *la* except when followed by a word beginning with a stressed /a/ (about 0.5 percent occurrence overall), where the determiner *el* is used instead (*el agua*, the water). Similar rules apply in French. Thus, for these languages the determiner form is specified by a complex interaction between grammatical and phonological properties. In various tests of the gender congruency effect with Spanish and Catalan speakers (Costa et al., 1999) and French speakers (Alario and Caramazza, submitted), we systematically failed to find a congruency effect (see figure 12.3). Furthermore, the absence of a gender congruency effect in Italian and Spanish is also observed when distractors are presented at different SOAs (Miozzo, Costa, and Caramazza, submitted). Thus we can conclude that the gender congruency effect is not observed in late-selection languages.

It is also possible to test directly whether we can observe determiner interference independently of gender congruency. This test is possible because some languages have multiple determiners for a given gender (e.g., the Italian masculine determiners *il* and *lo*). We can investigate therefore whether a mismatch between same-gender determiners for the target and the distractor word leads to production interference. For example, is the production of "il tavolo" (the table, m.) slower when paired with the word "struzzo" (ostrich, m.), which takes the determiner *lo*, than when paired with the word "nastro" (ribbon, m.), which takes the determiner *il*? In an experiment designed to answer this question, we found no trace of determiner interference (Miozzo and Caramazza, 1999). This result further strengthens our conclusion that the selection of determiners occurs so late in the process of NP production that potentially conflicting information from the distractor word that is relevant to determiner selection has already dissipated.[5]

Is Determiner Selection Special?

Languages differ in the degree to which the syntactic and phonological levels in sentence processing interact in determining the selection of closed-class words. For example, the determiner systems in Dutch and German are such that once the gender and number (and case) of the head noun of a phrase are selected, the phonological form of the determiner is fully specified and can be selected for production. These languages exemplify the possibility of a clear separation between syntactic and phonological processes in the selection of closed-class words. In contrast, the form of determiners in Italian, Spanish, Catalan, and French depends not only on grammatical features of the head noun of the NP but also on local phonological context (the onset of the word that immediately follows the determiner). Because of this property of the determiner system, the selection of determiners can only take place very late in the process of NP production. In these languages we do not have a clear separation between syntactic and phonological information in the process of determiner selection. This is also true for other closed-class words such as, for example, demonstrative pronouns and possessives in French (*ce* chapeau, *cet* arbre [this hat, this tree, both masculine] or *ma* bicyclette, *mon* étoile [my bike, my star, both feminine]). Instead it appears that grammatical and phonological information interact in selecting closed-class words, suggesting a close link between grammatical and phonological information during lexical access (Caramazza, 1997).

We have reviewed research that explored the implications of these cross-language differences for the process of determiner selection in NP production. This research has generated the following facts.

a. In picture-word naming experiments with Dutch and German speakers there is a reliable effect of gender congruency in the production of singular NPs.

b. Dutch and German speakers do not show a gender congruency effect in producing plural NPs; nor is there a gender congruency effect in the production of bare nouns by Dutch speakers.

c. No gender or determiner congruency effects are obtained in NP production with Italian, Spanish, Catalan, or French speakers.

We have argued that facts (a) and (b) jointly show that the gender congruency effect reflects selection competition between determiners and not between gender features (contra Schriefers, 1993). This is an important conclusion because it establishes that the selection of a word's syntactic features is not a competitive process like the selection of lexical nodes but instead appears to be an automatic, discrete process: the selection of a word's lexical node automatically makes available that word's syntactic features.

We have also argued that fact (c) shows that determiner competition is a property of only those languages in which determiners can be selected so early in the process of NP production that activation from *distractor* words is still strong enough to interfere with selection of the target's determiner. A corollary of this conclusion is that for languages such as Italian, determiner selection is a very late process, occurring at the point where the phonological form of an NP is assembled. This implies that, although we do not expect determiner interference in these languages, we do expect effects of phonological interference on determiner selection. That is, since the selection of a determiner involves "inspecting" the onset of the word that follows it, any uncertainty about that phonological parameter could interfere with the selection decision. This expectation can be tested by manipulating the phonological context for determiner production.

Consider the pairs of phrases "il treno/il piccolo treno" (the train/the small train) and "lo sgabello/il piccolo sgabello" (the stool/the small stool). In the phrase "il piccolo treno," both the adjective's and the noun's onsets are consistent with the selection of the determiner *il*. However, for the phrase "il piccolo sgabello," the onsets of the adjective and noun provide conflicting information for the selection of the determiner: the adjective's onset requires the determiner *il*, whereas the noun's onset requires the determiner *lo*. The conflicting phonological information within the NP could interfere with the selection of the determiner. We tested this possibility by comparing RTs to produce phonologically consistent (il piccolo treno) and phonologically inconsistent (il piccolo sgabello) NPs, relative to their respective baseline phrases (il treno and lo sgabello). The results showed that phonologically inconsistent NPs are named more slowly than phonologically consistent NPs (Miozzo and Caramazza,

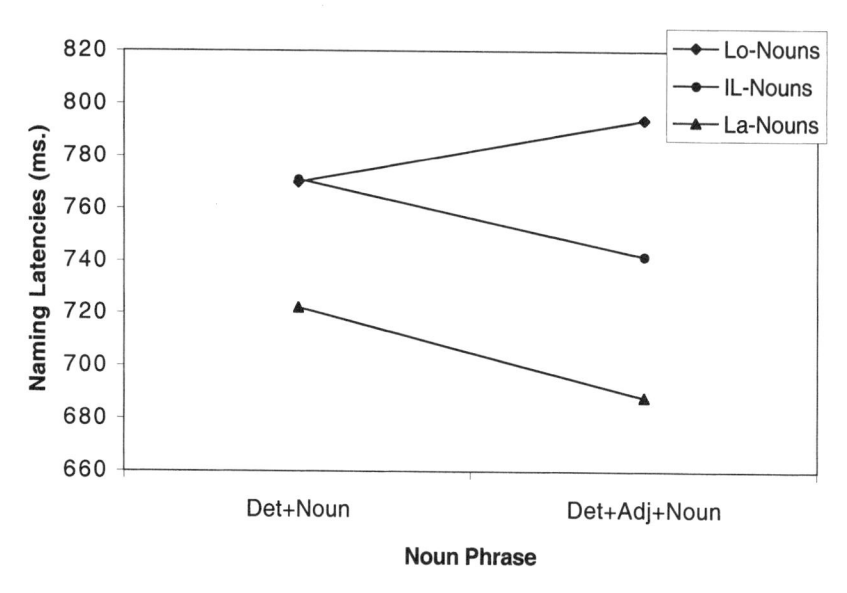

Figure 12.4
Naming latencies in a task in which Italian speakers produced noun phrases formed by determiner + noun or determiner + adjective + noun. Targets (picture names) were *lo*-nouns (masculine nouns, determiner *lo*), *il*-nouns (masculine nouns, determiner *il*) or *la*-nouns (feminine nouns, determiner *la*). Data from Miozzo and Caramazza (1999).

1999). In other words, there is a substantial interference effect induced by conflicting phonological information at the level of noun-adjective ordering within an NP (see figure 12.4). We have recently replicated this result in French (Alario and Caramazza, submitted). These results confirm that determiner selection is sensitive to phonological context, defined by the lexical items in an NP.

Finally, the pattern of results across languages allows us to address the principles that determine the point at which closed-class words are selected for production. Miozzo and Caramazza (1999) speculated that the point at which determiners are selected in a language is defined by a "temporal optimization" principle: prepare phonological material for production at the earliest possible stage of processing. However, the results from Spanish are inconsistent with this principle. In Spanish, the feminine (definite) determiners depend on the phonological context of production, but the masculine (definite) determiners *el* and *los* are fully specified once number and gender are selected. If the temporal optimization principle were to define the point at which determiners are selected, we would be forced to conclude that Spanish masculine determiners are selected early, as appears to be the case for Dutch and German. We would then expect to observe a gender congruency effect in the production of masculine NPs. However, as already noted, there was no trace of gender congruency effects in the production of NPs (Costa et al., 1999) by Spanish speakers, either for feminine or for masculine determiners. This finding implies that it is not the "temporal optimization" principle that determines the point at which closed-class words are selected, but something more like a "maximum consistency" or "highest common denominator" principle. The procedure that is adopted maximizes consistency of the selection process across words (morphemes) of a particular type. Thus, for example, determiner selection in a given language occurs at the same point for *all* determiners, even though some of them could be selected earlier. In the case of Spanish, even though masculine determiners could be selected early, they are nevertheless selected at the same late point as feminine determiners.

Our cross-language investigation of NP production has shown that determiner selection *is* special, at least in some respects and for some languages. For languages such as Italian, Catalan, Spanish, and French, determiner selection involves a highly interactive process across types of information. Grammatical and phonological information (but perhaps also conceptual information, depending on how we construe such features as [+definite]) interact in specifying the form of determiners and other closed-class words. This process is clearly different from the procedure involved in the selection of open-class words, where only semantic (and perhaps grammatical) factors are considered. The process of determiner selection is also special in that, unlike open-class word selection, it can vary across languages. Languages vary in the degree of interactivity between the types of information (conceptual, grammatical, and phonological) that are necessary for the selection of determiners and other closed-class words. Variation along this dimension determines the point in the process of NP production where closed-class words can be selected.

Our research has shown that this point is defined by the "maximum consistency" principle.

Acknowledgments

The work reported here was supported in part by NIH grant DC 04542 to Alfonso Caramazza. Albert Costa was supported by a postdoctoral fellowship from the Spanish government (Fulbright program). F.-Xavier Alario was supported by a postdoctoral fellowship from the Fyssen foundation. We thank Kimiko Domoto-Reilly for editorial assistance.

Notes

1. Levelt (1989) has called the process of determiner selection "indirect election" because it depends on the features of other lexical items in the sentence.

2. Theories that assume the existence of competition in the selection of grammatical features could account for the bare noun results if they assumed that the selection of the gender feature only occurs when this information is required for the utterance (e.g., see Levelt, Roelofs, and Meyer, 1999). Accounts of this type would have to invoke some mechanism that decides which features of a word are to be selected. No explicit proposal of such a mechanism has been offered.

3. If it were assumed that the set of grammatical features directly activates a phonological lexical node (as opposed to a determiner lemma node), we would then also have to assume that this process operates via the principle of cascaded activation. This assumption would be necessary since, by hypothesis, the lexical nodes of distractor words are not selected and therefore would not send activation down to the next level of processing.

4. For a discussion of the processes involved in the selection of the feature [number], see Bock, Nicol, and Cooper Cutting (1999) and Eberhard (1997).

5. An interesting case that seems to lie between the early-selection and the late-selection languages is that of English. In English, the form of some closed-class words also varies as a function of the phonological context of the utterance. For instance, the form of the indefinite article depends on the phonology of the word that follows it (if the word starts with a consonant, the determiner is *a* [e.g., "a pear"]; otherwise, it is *an* [e.g., an angel]). In principle, the application of this rule could be implemented as a two-step process: early selection of determiner form (*a*) followed by a very late output process that modifies this form if needed. This modification rule would be defined solely in phonological terms and would leave English an early-selection language. On the other hand, one may argue that the processes of selection of this determiner form in English parallel those postulated for "late-selection languages." That is, determiner form selection is not car-

ried out before the phonological properties of the following word is computed. This remains an empirical question.

References

Alario, F.-X., and Caramazza, A. (submitted). The production of determiners: Evidence from French.

Berndt, R. S., and Caramazza, A. (1980). A redefinition of the syndrome of Broca's aphasia: Implications for a neuropsychological model of language. *Applied Psycholinguistics, 1,* 225–278.

Bock, K., Nicol, J., and Cooper Cutting, J. (1999). The ties that bind: Creating number agreement in speech. *Journal of Memory and Language, 40,* 330–346.

Briggs, P., and Underwood, G. (1982). Phonological coding in good and poor readers. *Journal of Experimental Child Psychology, 34,* 93–112.

Caramazza, A. (1997). How many levels of processing are there in lexical access? *Cognitive Neuropsychology, 14,* 177–208.

Costa, A., Sebastian-Galles, N., Miozzo, M., and Caramazza, A. (1999). The gender congruity effect: Evidence from Spanish and Catalan. *Language and Cognitive Processes, 14,* 381–391.

Cutler, A., Mehler, J., Norris, D., and Segui, J. (1989). Limits on bilingualism. *Nature, 340,* 229–230.

Cutler, A., Mehler, J., Norris, D., and Segui, J. (1992). The monolingual nature of speech segmentation by bilinguals. *Cognitive Psychology, 24,* 381–410.

Dell, G. S. (1990). Effects of frequency and vocabulary type on phonological speech errors. *Language and Cognitive Processes, 5,* 313–349.

Eberhard, K. M. (1997). The marked effect of number on subject-verb agreement. *Journal of Memory and Language, 36,* 147–164.

Garrett, M. F. (1980). Levels of processing in sentence production. In B. Butterworth (Ed.), *Language Production.* Vol. 1: *Speech and Talk* (pp. 177–220). London: Academic Press.

Garrett, M. F. (1992). Disorders of lexical selection. *Cognition, 42,* 143–180.

Glaser, W. R. (1992). Picture naming. *Cognition, 42,* 61–106.

Glaser, W. R., and Glaser, M. O. (1989). Context effects in Stroop-like word and picture processing. *Journal of Experimental Psychology: General, 118,* 13–42.

Kean, M.-L. (1977). The linguistic interpretation of aphasic syndromes: Agrammatism and Broca's aphasia, an example. *Cognition, 5,* 9–46.

Klein, G. S. (1964). Semantic power measured through the interference of words with color-naming. *American Journal of Psychology, 77,* 576–588.

La Heij, W., Mark, P., Sander, J., and Willeboordse, E. (1998). The gender-congruency effect in picture-word tasks. *Psychological Research, 61,* 209–219.

Levelt, W. J. M. (1989). *Speaking: From intention to articulation.* Cambridge, MA: MIT Press.

Levelt, W. J. M., Roelofs, A., and Meyer, A. S. (1999). A theory of lexical access in speech production. *Behavioral and Brain Sciences, 22,* 1–75.

Lupker, S. J. (1979). The semantic nature of responses competition in the picture-word interference task. *Memory and Cognition, 7,* 485–495.

Lupker, S. J. (1982). The role of phonetic and orthographic similarity in picture-word interference. *Canadian Journal of Psychology, 36,* 349–367.

MacLeod, C. M. (1991). Half a century of research on the Stroop effect: An integrative review. *Psychological Bulletin, 109,* 163–203.

Mehler, J., Dupoux, E., Nazzi, T., and Dehaene-Lambertz, G. (1996). Coping with linguistic diversity: The infant's viewpoint. In James L. Morgan, and K. Demuth (Ed.), *Signal to syntax: Bootstrapping from speech to grammar in early acquisition* (pp. 101–116). Mahwah, NJ: Erlbaum.

Mehler, J., Sebastian, N., Altmann, G., Dupoux, E., Christophe, A., and Pallier, C. (1993). Understanding compressed sentences: The role of rhythm and meaning. In P. Tallal, A. M. Galaburda, R. R. Llinás, and C. Von Euler (Eds.), *Temporal information processing in the nervous system: Special reference to dyslexia and dysphasia.* Annals of the New York Academy of Sciences, 682, 272–282.

Miozzo, M., and Caramazza, A. (1999). The selection of lexical-syntactic features in noun phrase production: Evidence from the picture-word interference paradigm. *Journal of Experimental Psychology: Learning Memory, and Cognition, 25,* 907–922.

Miozzo, M., Costa, A., and Caramazza, A. (submitted). The lack of gender congruity in Romance languages: A matter of SOA?

Schiller, N. O., and Caramazza, A. (submitted). Gender or determiner selection interference? Evidence from noun phrase production in German and Dutch.

Schriefers, H. (1993). Syntactic processes in the production of noun phrases. *Journal of Experimental Psychology: Learning, Memory, and Cognition, 19,* 841–850.

Schriefers, H., and Teruel, E. (2000). Grammatical gender in noun phrase production: The gender interference effect in German. *Journal of Experimental Psychology: Learning, Memory, and Cognition, 26,* 1368–1377.

Schriefers, H., Meyer, A. S., and Levelt, W. J. M. (1990). Exploring the time course of lexical access in language production: Picture-word interference studies. *Journal of Memory and Language, 29,* 86–102.

Starreveld, P. A., and La Heij, W. (1996). Time-course analysis of semantic and orthographic context effects in picture naming. *Journal of Experimental Psychology: Learning Memory, and Cognition, 22,* 896–918.

Stemberger, J. P. (1984). Structural errors in normal and agrammatic speech. *Cognitive Neuropsychology, 1,* 281–313.

van Berkum, J. J. A. (1997). Syntactic processes in speech production: The retrieval of grammatical gender. *Cognition, 64,* 115–152.

13

Now You See It, Now You Don't: Frequency Effects in Language Production

Merrill Garrett

When it comes to studying lexical processing, there is no variable in the experimental language scientist's armamentarium more useful than lexical frequency. If you want to know whether the experiment is working, check the frequency effects. That claim and formula apply with force to language recognition and comprehension studies. Matters are not so tidy for the corresponding language production questions. It is not that frequency effects are absent, but rather that their effects are less robust, and in some respects inconsistent. In this chapter, I explore some aspects of these conditions.

Over the past decade or two, study of language production processes has become an increasingly significant factor in the larger spectrum of psycholinguistic activity. Most of this is directly attributable to the blossoming of experimental work that has greatly expanded the scope of an inquiry that was formerly based largely on observation of naturally occurring speech error, hesitation phenomena, and closely related self-correction systems ("repair"). To a substantial extent, the findings that have emerged from wide-ranging experimental efforts have converged in productive ways with the findings from the natural observational data. There are, however, some interesting exceptions to this convergence, and evidence bearing on the role of lexical frequency in language production processes is one of them. It's a pretty puzzle that I don't have an answer for, but posing the questions may suggest some possibilities. It is useful to begin with some of the things that seem well established and then look at the frequency issues from that vantage point.

Issues in Lexical Retrieval

Looking at the history of claims about lexical recovery for production best begins with the classic work by Fay and Cutler (1977) in which they argued for two types of retrieval operations on the basis of the distribution of word substitution errors observed for spontaneous speech Their analysis of their error corpus has since been several times affirmed for other corpora. The central feature of their observations was a sharp dissociation in the phonological similarity metric for words that displayed a clear semantic relation and those lacking it. Examples of these types are readily available in various published papers as well as the appendix to the Fay and Cutler paper. Here are some memorable ones:

- Jimmy Stewart on the *Tonight Show* in response to queries about how it was to ride an elephant in a recent movie (elephants apparently have very rough hides that make bare-legged riding attire distinctly less than pleasant): "People don't know! —people who've never sat on an envelope!" (target was "elephant").
- From Fromkin (1973): "... white, Anglo-Saxon prostitute" (target was "protestant").
- Gerald Ford on the Presidency: "It's a single four-year sentence . . . (pause) . . . you see I have no illusions about the job" (target was "four-year term").
- From Fromkin (1973): "I should stand on my nose" (target was "on my head").

Pairs like *envelope/elephant, prostitute/protestant, mushroom/mustache, garlic/gargle, married/measured*, and so on, typify the semantically unrelated substitutions that Fay and Cutler dubbed "malapropism errors." They prove to have a qualitatively and quantitatively different overlap in segmental composition for target and intrusion words than the substitutions with a clear semantic relation, as seen in pairs like *sentence/term, nose/head, wife/husband, watch/listen, later/earlier*, and so on. Fay and Cutler proposed a semantically guided and a phonologically guided set of retrieval operations. The former provides an account of meaning-related word substitutions and the latter an account of form-based substitutions.

How does frequency of occurrence enter this picture? The natural prediction to make for word substitutions would be that target and intrusion should show an asymmetry. Comparing the frequency of occurrence for target word and intrusion word, one would predict a *higher frequency-value for the intruding word* on average given normal assumptions about competition between representations varying in strength of activation. The intruding word with greater activation based on frequency would have a compensatory advantage to help overcome effects of semantic or form mismatches with the target representation. Note that this claim must be separately applied to the two error classes, semantically based and form-based. Tests of target-intrusion frequencies in substitutions lend some support to the view that semantically based and form-based lexical errors differ in frequency-sensitivity. There is little evidence for a target-intrusion frequency difference in semantic substitutions, but there are some indications of such an effect in form-based errors. The details of this are left for consideration in a later section, and merely stipulated here as preamble to a summary of various experimental findings that lead to the same general picture as that suggested by the error data. The experimental effects, however, are considerably more robust than the word-substitution effects; this contrast is one I want to emphasize, and so note it here at the outset.

A paper by Levelt, Roelofs, and Meyer (1999) comprehensively lays out major empirical claims about the complex of processing systems underlying lexical retrieval in language production. That complex rests on a variety of experimental and observational pillars that fit together in a compelling way to suggest a system of staged processing that maps from conceptual systems to articulation. The way stations in the processing path distinguish semantic, syntactic, and morphological variables, as well as various aspects of phonological and phonetic form. Frequency variables play a role in performance at sublexical levels in the system the authors describe, but I will not attend to those matters; it is the lexical levels that I want to focus on. The lexical processing levels in their model readily assimilate the contrast drawn by Fay and Cutler for word substitutions and other evaluations of error data, my own among them (e.g., see Garrett, 1980).

The scheme laid out by Levelt et al. is an elaboration with some variation on the architecture described in Levelt (1989), in which a distinction between *lemma* and *word form* ("lexeme") is drawn. Lemma structures provide the means for linking conceptual representations to the syntactic constraints associated with individual lexical items. In Levelt's earlier version, lemmas included both semantic and syntactic information; in the later elaboration, these are represented separately. That change is not immediately germane to the frequency issue, however, since the locus of lexical frequency effects is argued to reside at the word-form level—roughly as construed in the contrast drawn by Fay and Cutler for their account of word substitution. This conclusion about word frequency is experimentally based on a series of studies reported in Jescheniak and Levelt (1994). The experimental approach used picture naming derived from early work by Oldfield and Wingfield (1965), complemented by an ingenious study using a translation task to extend the range of claims about frequency contributions to word production.

Briefly, the facts are these. Conceptual-level contributions were controlled in the picture naming experiments reported by Jescheniak and Levelt, as were times for phonological integration of naming targets. For the architecture of the model being tested, the remaining level of potential control for the observed lexical frequency effects in picture naming is the layer *between* semantic and syntactic control (lexical concept and lemma levels) and sublexical phonological planning: the word-form (or "lexeme") level. And there is direct evidence for frequency control of lexical retrieval at this level from the translation task. In this task, fluent bilingual subjects (Dutch-English) were cued for a naming target by a word presented in their second language (English), and they were required to provide its translation in their native language (Dutch). Low-frequency Dutch translation targets that were homophonic with high-frequency Dutch words were compared with low-frequency targets lacking a homophonic twin. Even after recognition times for the English cue words were subtracted from translation times, the homophonic targets were responded to faster than their nonhomophonic counterparts. More compellingly, response times for the low-frequency homophone were comparable to the high-frequency member of the homophonic pair. In short, the low-frequency homophone seemed to inherit the high-frequency performance profile of its twin. This effect was robust and was coupled with ancillary evidence to make a prima facie case that frequency is allocated to individual entries in the word-form inventory and that homophones are represented by a single entry in that inventory.

This outcome not only has strong implications for production processing per se, it also provides counterindication for the hypothesis that production and comprehension systems overlap at the word-form level. This latter implication follows from results of word-recognition experiments that seem best understood in terms of independent frequency records for homophones. So, for example, Forster and Bednall's (1976) results in speeded judgments of acceptability for elementary phrases containing homophones seem best explained if the access codes for homophonic lexical items are distinct and frequency-ordered. Thus, for comprehension judgments, the form representations for homophones are not collapsed, and frequency predictions based on their relative occurrence rates are borne out in some detail. If the same form entry were associated with both terms of the ambiguity, access to one should give access to the other, but in Forster and Bednall, this was not the case. In similar vein, other recent lexical ambiguity research seems to best fit with ranked-order access to the senses of the test forms in terms of relative frequency of occurrence. So, this experimentally based frequency account for lexical access in production is powerfully indicated and it is production-specific.

Jescheniak and Levelt's results are also readily linked to some other findings in production studies of error phenomena. In particular, an experimental test by Dell (1991) for the induction of sublexical speech errors (e.g., sound exchanges) has a very similar cast to their result. In Dell's experiment, induced error rates for low-frequency open-class words approximated the error rates for high-frequency closed-class homophones. In short, the low-frequency members of the homophone pair seemed to inherit the insulation of the high-frequency twin vis-à-vis susceptibility to the "disintegration" required for sound error. This is coupled in Dell's work with additional experimental verification of frequency as a factor in susceptibility to sound error. This fits overall with the more general proposition that word frequency and sound error incidence in natural error corpora are inversely correlated (e.g., see Stemberger and MacWhinney, 1986). Other interesting indications of frequency effects in production are to be found in Bock's work (1987) on lexical retrieval in sentence generation tasks. Moreover, aphasia patients show frequency-correlated failures for sound error (e.g., see Nickels 1995). And finally, one may note that failures of lexical retrieval in tip-of-tongue (TOT) states seem to reflect some frequency influence (lower-frequency elements having greater likelihood of such interruption in retrieval efficiency), and there is some reason to assign this effect to the form retrieval stage of processing (e.g., see Vigliocco, Antonini, and Garrett, 1997).

The Frequency Dilemmas

All in all, one might suppose that the frequency results summarized above present an acceptably well-ordered picture. Speech errors, experimental studies of lexical processing in sentences, picture naming experiments, translation experiments, aphasia patterns, and TOT studies all indicate a staged retrieval architecture in which semantic or conceptual processes access abstract lexical representations that are semantically and syntactically active (lexical concepts, lemmas), followed by access of word-form representations, and it is this latter class of representations for which frequency is a contributing factor. So what am I grumbling about? What is wrong with this picture? In brief: some aspects of the naturally occurring speech error data do not fit this picture as well as one might wish. The following facts do not settle comfortably into place:

• Asymmetries for targets and intrusions in word substitutions are weak at best. Given the strength of the experimental effects reported by Jescheniak and Levelt (1994), a more decisive outcome ought to be identifiable.

• The weak target-intrusion asymmetries in word substitutions contrast quite perplexingly with a strong correlation between the frequencies of target and intrusion words that applies to both semantically and word-form–based error types of substitution. That correlation fits neither the Jescheniak and Levelt results, nor the selectivity of the target-intrusion results for form based errors.

• There is no compelling frequency effect in word and morpheme exchanges. Word substitutions are the most obvious candidates for display of frequency influences. But word and morpheme exchanges can also be distinguished in terms of lemma and word-form processes, and are likewise interesting candidates for frequency effects in systems that account for error in terms of activation-based timing failures.

Put briefly, the places in language generation for which lexical frequency effects should have a natural expression in speech error patterns show pallid or inconsistent effects. There are some rationalizations that might be examined as candidates to account for limited effects of lexical frequency, given suitable details of production architecture, but nothing that seems very satisfactory emerges from this exercise. I will pursue it nevertheless.

The distinction between semantically, syntactically, and phonologically controlled processes is a central matter in this enterprise. It is a distinction strongly indicated in the Jescheniak and Levelt results and similarly for word-substitution errors. In order to bring experimental and natural error data patterns into correspondence, the regularities of the latter should be consistent across several tests of frequency influence with the implications of the experimental findings. At first blush, this looks manageable. The evidence is that semantically based word substitutions do not show even the weak frequency bias that has been reported for form-based substitutions. Several reports support a claim for dissociation of frequency in semantic and form errors in word substitutions. Hotopf (1980) reported on target-intrusion frequency relations in his own corpus of semantically related substitutions, as well as those in the Merringer and Meyer (1895) German corpus, and found no evidence for an asymmetry in target-intrusion pairs. Del Viso, Igoa, and Garcia-Albea (1991) contrasted both semantically and form-based substitutions in a substantial Spanish error corpus (n = 555). They reported a selective effect for frequency: target-intrusion pairs showed a significant asymmetry for form-based errors, but not for meaning-based errors. Raymond and Bell (1996) reported an analysis of a corpus of word errors (n = 401) in which they likewise distinguished frequency effects for word form–based errors and semantically based errors. The latter showed no bias in the relation between target and intrusion word frequencies; the former did, but weakly. It must be borne in mind that of the studies that did show a target-intrusion asymmetry effect, none were large, and in Raymond and Bell's results, not general. They reported that such effects were limited to polysyllabic words. Moreover, other recent analysis of substantial sets of word substitutions showed no significant effect favoring higher-frequency intrusion words for either semantically or word-form–based error types (Harley and McAndrew, 1995; Silverberg 1998). In short, the pattern of dissociation of frequency effects for semantic- and form-driven aspects of lexical selection that is powerfully indicated by the experimental work is only dimly discernible in the data on naturally occurring word substitution errors when target and intrusion word frequencies are compared.

One may be inclined to say at this point, Well, error data is after all messy and full of observational glitches, so some slop in the outcomes is hardly surprising. There is some justice in this, but also good reason not to take much comfort from that line. First of all, there are a number of samples at this point, and some are substantial in size and come from different observers. The second and more telling point is this: there *is* an effect of frequency in the word-substitution errors for these same corpora. And here we find a strong effect. It is a correlation of target and intrusion word frequencies. The error words in a target-intrusion pair, both semantically and word-form–based error types, come from similar frequency domains (Hotopf, 1980; Harley and McAndrew, 1995; Raymond and Bell, 1996; Silverberg, 1998). The correlations are substantial in all reports. Raymond and Bell's report is detailed and shows a correlation, as do the other reports, for both form- and meaning-based pairs. They report a robust relation between target and intrusion frequencies after partialing out correlated variables of form ($r = .48$ for form-based errors; $r = .70$ for meaning-based errors). The relation accounts for a considerable portion of variation in what is acknowledged to be an intrinsically messy class of data. The effect is, if anything, stronger in the semantic set than in the form-based set.

What does this pattern mean? It means that frequency predicts quite well the simultaneous availability of target and intrusion words. One natural interpretation of this is that frequency controls access to lexical records, and only when the competing words are of roughly comparable availability does a competition ensue. Now trouble starts. As a departing assumption, the idea that frequency controls simultaneous accessibility for the target and intrusion runs right into a couple of contradictions.

The assumption runs afoul of the Jescheniak and Levelt results, since the correlation applies to both semantically and word-form–based error retrieval, and the (very robust) experimental results indicate only the latter should be so affected. Second, if frequency controls availability (as the correlation suggests), then it should also lead to a strong target-intrusion asymmetry. But, as already noted, the competition is not reliably resolved in favor of the higher-frequency member of the competing pair. The effect is not present at all for semantic substitutions and the target-intrusion asymmetries are weak and nonuniform for word-form–based error types. I see no obvious way to encompass these results in a single explanatory scheme.

One might be tempted to appeal to a postretrieval mechanism in the substitution process. Raymond and Bell's (1996) discussion of this raises the idea in terms of the "timing of lexical insertion," suggesting that frequency affects when words are available as encoding proceeds; those with similar windows of availability will interact, and this could lead to a correlation of the sort observed. This general line has some plausibility since there is a strong grammatical category constraint on word substitution, and so the insertion of elements in phrasal environments is a plausible source of constraint. But several problems with this come immediately to mind. It does not escape the inconsistency with target-intrusion results noted earlier. If frequency dictates when words become available, why does the same mechanism not influence the competition between the candidates? To address this, one must assume that, for whatever reason, initial retrieval is strongly frequency-governed (hence the correlation), but insertion into the phrasal frame only weakly so, and only in the case of the word-form–based substitutions. The line of argument requires that insertion processes are form-driven, or at least form-sensitive, and hence a frequency effect for lexical insertion can arise when the competitors are of similar form.

There is an independent index of the plausibility of such a postretrieval contribution to the substitution patterns that can be got by looking at processing in word and morpheme exchanges. Word exchanges might be expected to show a bias in favor of higher-frequency targets displacing lower-frequency targets for the first slot of the two involved in an exchange. This is modeled on the same idea as the word substitutions. The later-occurring word must displace the earlier-ordered one, and higher frequency might plausibly be expected to contribute to the necessary activation boost for premature selection. Certainly, in models for which activation levels of yet-to-be-uttered sentence elements are a determining factor in exchange error, this is a plausible expectation.

There are indications of contributions of form-based variables to some word and morpheme exchange processes (e.g., Dell and Reich, 1981; del Viso et al., 1991), though they are more clearly discernible for length variation than for segmental structure. Still, the effects in the word substitutions are weak, so perhaps the limited scope of form variables in word exchanges fits the bill. I note in passing that exchanges may be distinguished in terms of stage of processing, roughly by distinguishing exchanges that are between phrases for elements of corresponding grammatical category ("early stage planning") from those that are within phrase exchanges of differing grammatical category. The former on my account are lemma-level processes and the latter are word-form–based (e.g., see Garrett, 1980). Looking at these different classes of exchange (from early to late) for frequency effects provides progressively stronger grounds for claims of the relevance of form-based insertion processes.

Alas, there is, so far as I know, no support for any of the frequency-based expectations regarding exchange errors of various sorts among words and stems. My analysis of substantial sets of exchanges does not indicate reliable frequency differences between first- and second-ordered word or morpheme exchange elements (Garrett, 2000). So, on the movement error index of the problem, lexical frequencies have minimal impact once retrieval is accomplished. Explanation of the correlation of target and intrusion by appeal to a distinction between retrieval and insertion processes does not seem promising on this evidence. Moreover, note that even were this enterprise to have had a different outcome, the theoretical problem of integrating experimental and natural error data would remain acute, though of different form. The notion that retrieval accounts for the strong correlations between target and intrusion frequencies applies to both semantically and form-based pairs, and hence is incompatible with the selective experimental effects reported in Jescheniak and Levelt.

The obvious line of response to the frequency inconsistencies at this point (maybe even sooner) is just to give up the idea that the frequency correlation for target and intrusion in semantically related pairs is retrieval-based: adopt a two-factor tack that assigns frequency a causal role in form retrieval (and hence fits the experimental constraints), but explains the correlation for semantic pairs by appeal to other mechanisms that happen to be correlated with frequency. Raymond and Bell (1996) raise a version of this idea. They note that other cognitive systems might cluster items with common frequency, and selection factors for such systems might adventitiously produce a correlation for semantic processing (or, somewhat less obviously, for word form). If this could be established, it would avoid the troubles that stem from the need to use frequency as a determinant of accessibility for the competing words of semantic origin.

The trouble with this line is that persuasive general accounts of why word sets that *are likely to be co-candidates in some communicatively relevant selection process* might display a frequency correlation are hard to come by. This kind of idea has been suggested for associatively related target-intrusion pairs (Levelt, 1989) on grounds that associates tend to come from roughly similar frequency domains. Sensible as the salient fact about association is, some difficulties for its application to the instant case arise. For one thing, associatively related pairs are a significant component of semantically related word substitution error sets but not dominant. So it is not immediately clear that the correlational evidence could be accounted for—though that analysis remains to be done at this point. An analysis that I have done, however, is not encouraging for this general line. Consider cases in which the words that compete in semantic selection processes are organized into semantic fields and subfields. This is a quite general feature of substitutions. If items closely related semantically within such fields tended to be clustered in frequency strata, it would be an example of the condition required for explaining the correlation of semantically related substitutions as the byproduct of a communicatively relevant process in the production system. My examination of the frequency effects in semantic fields for word substitutions (e.g., body parts, clothing terms, color names, etc.) does not afford any comfort for this idea (Garrett, 1993). The distribution of substitutions from the fields does not seem to strongly reflect frequency clusters. Competing terms have closer semantic relations to each other than to other members of the semantic field, but may differ substantially in frequency. This evaluation rests on a relatively small set of observations, and so is certainly not decisive. But, as it stands, the indication for the general argument is not positive.

One is left at this somewhat inconclusive point with a problem that has these necessary components if experimental and natural error data are to be reconciled.

• Find a solution for the correlation of semantic pairs in word substitutions that does not rely on frequency-governed semantic retrieval. This is a precondition set by the experimental results.

• Find a solution for the contrast in strength between the correlational effect in form-based substitutions (strong) and the target-intrusion asymmetry (very weak). This is required for coherence of the overall theoretical account. If some filtering effect of lexical insertion processes on the products of word selection is a part of this account, it needs an independent defense.

• Find a solution for the contrast in strength of effect for form-based word substitutions (very weak) and the experimental effects as indexed by the translation task (strong). If form-based word substitutions are to be treated as retrieval failures from the word form inventory that the experimental tasks are presumed to tap, this is a problem that has to be dealt with.

Raymond and Bell's (1996) sensible discussion of relations between frequency and other form variables is instructive in this context. They report that length, phonetic composition at different serial positions, stress, and grammatical category are in one or another instance covariates with different relations to frequency in the set of substitution errors they analyzed. The overall picture is impressively intricate, though the intricacies do not suggest a ready solution to the conflicts I have been discussing. It appears that a satisfactory account of frequency effects in production will require a range of new observations. The obvious avenues of explanation are still underspecified and underconstrained by available data.

References

Bock, J. K. (1987). Coordinating words and syntax in speech plans. In A. Ellis (Ed.), *Progress in the psychology of language*. Vol. 3 (pp. 337–390). London: Erlbaum.

Dell, G. (1991). Effects of frequency and vocabulary type on phonological speech errors. *Language and Cognitive Processes, 5,* 313–349.

Dell, G., and Reich, P. (1981). Stages in sentence production: An analysis of speech error data. *Journal of Verbal Learning & Verbal Behavior, 20,* 611–629.

del Viso, S., Igoa, J., and Garcia-Albea, J. (1991). On the autonomy of phonological encoding: Evidence from slips of the tongue in Spanish. *Journal of Psycholinguistic Research, 20,* 161–185.

Fay, D., and Cutler, A. (1977). Malapropisms and the structure of the mental lexicon. *Linguistic Inquiry, 8,* 505–520.

Forster, K., and Bednall, E. (1976). Terminating and exhaustive research in lexical access. *Memory and Cognition, 4,* 53–61.

Fromkin, V. A. (1973). Slips of the tongue. *Scientific American, 229,* 110–117.

Garrett, M. (1980). Levels of processing in sentence production. In B. Butterworth (Ed.), *Language production:* Vol 1. *Speech and talk* (pp. 177–220). London: Academic Press.

Garrett, M. (1993). Lexical retrieval processes: Semantic field effects. In E. Kittay and A. Lehrer (Eds.), *Frames, fields, and contrasts.* Hillsdale, NJ: Erlbaum.

Garrett, M. (2000). Frequency variables in lexical processing for sentence production: Error patterns and non-patterns. Presented at the workshop on language production, Saarbrücken, Germany, 5 September 2000.

Harley, T. A., and McAndrew, S. B. G. (1995). Interactive models of lexicalisation: Some constraints from speech error, picture naming, and neuropsychological data. In J. P. Levy, D. Bairaktaris, J. A. Bullinaria, and P. Cairns (Eds.), *Connectionist models of memory and language* (pp. 311–331). London: UCL Press.

Hotopf, W. (1980). Semantic similarity as a factor in whole-word slips of the tongue. In V. Fromkin (Ed.), *Errors in linguistic performance: Slips of the tongue, ear, pen, and hand.* New York: Academic Press.

Jescheniak, J., and Levelt, W. J. M. (1994). Word frequency effects in speech production: Retrieval of syntactic information and of phonological form. *Journal of Experimental Psychology: Language, Memory, and Cognition, 20,* 824–843.

Levelt, W. J. M. (1989). *Speaking: From intention to articulation.* Cambridge, MA: MIT Press.

Levelt, W. J. M., Roelofs, A., and Meyer, A. (1999). A theory of lexical access in speech production. *Behavioral and Brain Sciences, 22,* 1–75.

Merringer, R., and Meyer, K. (1895). *Versprechen und Verlesen: Eine Psychologisch-linguistische Studie.* Stuttgart: Groschene Verlagsbuchhandlung.

Nickels, L. (1995). Getting it right? Using aphasic errors to evaluate theoretical models of spoken word recognition. *Language and Cognitive Processes, 10,* 13–45.

Oldfield, R., and Wingfield, A. (1965). Response latencies in naming objects. *Quarterly Journal of Experimental Psychology, 17,* 273–281.

Raymond, W., and Bell, A. (1996). Frequency effects in malapropisms. Presented at Acoustical Society of America, Honolulu, December 1996.

Silverberg, N. (1998). *How do word meanings connect to word forms?* PhD dissertation, University of Arizona, Tucson.

Stemberger, J. (1995). An interactive activation model of language production. In A. Ellis (Ed.), *Progress in the psychology of language*. Mahwah, NJ: Erlbaum.

Stemberger, J. P., and MacWhinney, B. (1986). Form-oriented inflectional errors in language processing. *Cognitive Psychology, 18,* 329–354.

Vigliocco, G., Antonini, T., and Garrett, M. (1997). Grammatical gender is on the tip of Italian tongues. *Psychological Science, 8,* 314–317.

14

Relations between Speech Production and Speech Perception: Some Behavioral and Neurological Observations

Willem J. M. Levelt

One Agent, Two Modalities

There is a famous book that never appeared: Bever and Weksel (shelved). It contained chapters by several young Turks in the budding new psycholinguistics community of the mid-1960s. Jacques Mehler's chapter (coauthored with Harris Savin) was entitled "Language Users." A normal language user "is capable of producing and understanding an infinite number of sentences that he has never heard before. The central problem for the psychologist studying language is to explain this fact—to describe the abilities that underlie this infinity of possible performances and state precisely how these abilities, together with the various details . . . of a given situation, determine any particular performance." There is no hesitation here about the psycholinguist's core business: it is to explain our abilities to produce and to understand language. Indeed, the chapter's purpose was to review the available research findings on these abilities and it contains, correspondingly, a section on the listener and another section on the speaker.

This balance was quickly lost in the further history of psycholinguistics. With the happy and important exceptions of speech error and speech pausing research, the study of language use was factually reduced to studying language understanding. For example, Philip Johnson-Laird opened his review of experimental psycholinguistics in the 1974 *Annual Review of Psychology* with the statement: "The fundamental problem of psycholinguistics is simple to formulate: what happens if we understand sentences?" And he added, "Most of the other problems would be halfway solved if only we had the answer to this question." One major other problem is, of course, How do we produce sentences? It is, however, by no means obvious how solving the issue of sentence or utterance understanding would halfway solve the problem of sentence or utterance production.

The optimism here is probably based on the belief that utterance production is roughly utterance understanding in reverse. In fact, that has long been a tacit belief among psycholinguists in spite of serious arguments to the contrary, such as these: "An ideal delivery in production requires completeness and well-formedness at all linguistic levels involved. The pragmatics should be precisely tuned to the discourse situation. The words, phrases, sentences should be accurate and precise renditions of the information to be expressed. Syntax and morphology have to be complete and well-formed and the same holds for the segmental and suprasegmental phonology of the utterance. Finally, the phonetic realization has to conform to the standards of intelligibility, rate, formality of the speech environment" (Levelt, 1996, p. x). These representations are generated completely and on the fly, that is, incrementally. This multilevel linguistic completeness and well-formedness is in no way required for successful utterance understanding. "Almost every utterance that we encounter is multiply ambiguous, phonetically (*I scream*), lexically (*the organ was removed*), syntactically (*I enjoy visiting colleagues*), semantically (*there are two tables with four chairs here*) or otherwise. As listeners we hardly notice this. We typically do not compute all well-formed parses of an utterance, even though ambiguities can produce momentary ripples of comprehension. Parsing is hardly ever complete. Rather, we go straight to the one most likely interpretation, given the discourse situation" (Levelt, 1996, p. x). In other words, the aims of the two systems are deeply different: attaining completeness is a core target of production; ambiguity is hardly ever a problem. Attaining uniqueness in the face of massive ambiguity is a core target of speech perception; completeness of parsing should definitely be avoided—it would make the system explode.

Meanwhile, the unbalance of comprehension vs. production perspectives in psycholinguistics has been somewhat redressed. After three decades, the title (and the content) of Herbert Clark's 1996 treatise *Using Language* returns us to the language user who is as much a speaker as a listener. It is not in detail Mehler and Savin's language user; that one was mainly concerned with relating surface phonetic and underlying semantic representations, but these are still core ingredients of a language user in Clark's sense, a participant in intentional, joint action.

It is, in my opinion, a major theoretical and empirical challenge to reconcile the unicity of the language user as an agent with the fundamental duality of linguistic processing, speaking, and understanding. In the following, I first consider some perception-production relations from the perspective of our speech production model, which has been taking shape over the years. I then change the perspective to cognitive neuroscience, not only because the recent neuroimaging literature often provides additional support for the existing theoretical notions but also because it provides new, additional challenges for how we conceive of the production-perception relations.

Speech Perception in a Model of Production

The theory of production proposed in Levelt (1989) and the further modeling of its lexical access component in Levelt, Roelofs, and Meyer (1999) postulate three types of relation between the productive and receptive mechanisms of the language user. They reside in the "perceptual loop," in the concept-lemma system, and in connections at the form level. I consider them in turn.

The Perceptual Loop
The global architecture of the speaker as proposed in Levelt (1989) is essentially a feedforward system. Utterances are incrementally produced going through stages of conceptual preparation, grammatical encoding, phonological encoding, and articulation. The perceptual component in this architecture consists of a dual-feedback loop. As a speaker you are normally hearing your own overt speech and that will be parsed just as any other-produced speech you hear or overhear. The one crucial difference is that the attribution of the speech is to self. That being the case, the speaker may use it for self-monitoring. As in any complex motor action, speaking involves some degree of output control. If the self-perceived speech is disruptively deviant from the intended delivery, you

may self-interrupt and make a repair (see Levelt, 1983, for details of this self-monitoring mechanism). As a speaker you can, in addition, monitor some internal, covert representation of the utterance being prepared. The 1989 model identified this "internal speech" as the phonetic code that serves as input to the articulatory mechanism, that is, the stuff that you can temporarily store in your "articulatory buffer" (Morton, 1970). Experimental evidence obtained by Wheeldon and Levelt (1994), however, supports the notion that the code is more abstract. Effective self-monitoring is not wiped out when the articulatory buffer is filled with nuisance materials. The more likely object of internal self-monitoring is the self-generated phonological representation, a string of syllabified and prosodified phonological words. Both the external and the internal feedback loops feed into the language user's normal speech-understanding system. The model is maximally parsimonious in that it does not require any reduplication of mechanisms (as is often the case with alternative models of self-monitoring).

Shared Lemmas and Lexical Concepts
The 1989 model follows Kempen and Hoenkamp (1987) in defining *lemmas* as the smallest units of grammatical encoding. These lemmas were semantic and syntactic entities. Retrieving lemmas for content words from the mental lexicon involved matching a conceptual structure in the input message to the semantic structure of the lemma. A major observation at that stage of theory formation was that all existing theories of lexical selection in speech production were deeply flawed. They all ran into the "hyperonym problem": the speaker would select an item's superordinates instead of, or in addition to, the target itself. This was, essentially, due to the fact that if a word's critical semantic features are all activated or selected, then the critical features of any of its superordinates (or "hyperonyms") are necessarily also activated or selected; they form, after all, a subset of the target word's critical features. Roelofs (1992) solved this problem by splitting up the lemma into a "lexical concept" and a "syntactic lemma" (or "lemma" for short). If the terminal elements of the speaker's message are "whole" lexical concepts, lexical selection is, essentially, a one-to-one mapping of lexical concepts to syntactic lemmas. Lexical concepts are no longer sets or bundles of features. Their semantics

is handled by the relational, labeled network that connects them. The hyperonym problem vanishes, and a more realistic issue arises: What happens if more than a single lexical concept, in particular semantically related concepts, are (co-)activated during conceptual preparation of the message? Multiple activation of lemmas will be the result and there is a selection problem that should be solved "in real time." The computational model, now called WEAVER (Roelofs, 1992, 1997; see figure 14.1 for a fragment), handles this issue in detail and is meanwhile supported by a plethora of chronometric experimental data. The (syntactic) lemma is the unit of grammatical encoding. Lemmas are, essentially, lexical

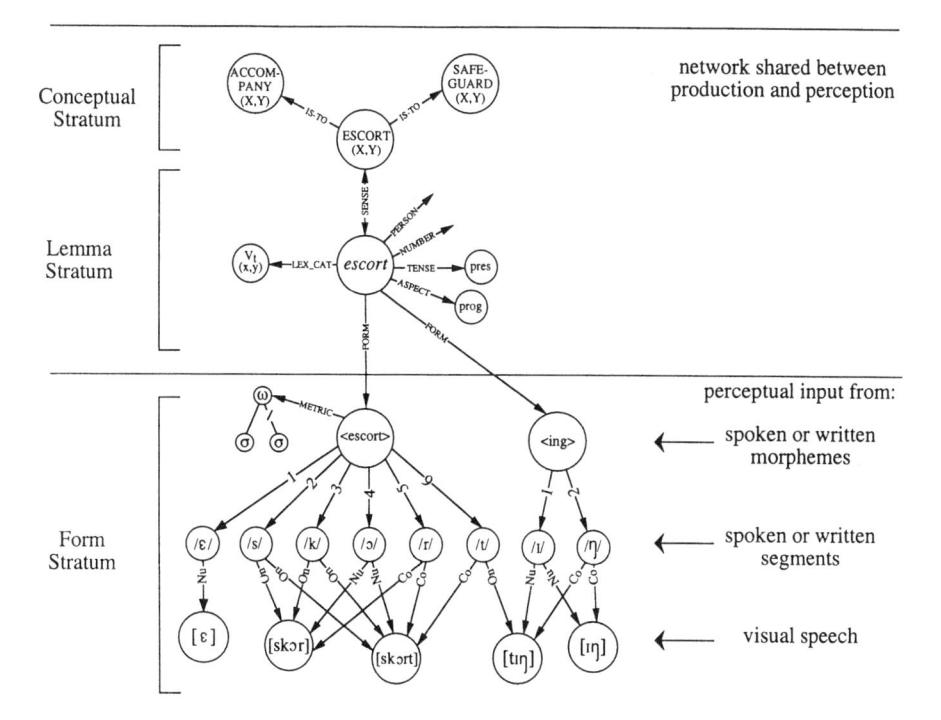

Figure 14.1
Fragment of the WEAVER lexical network (for lexical item *escort*), displaying the input connections from the perceptual network. The top half of the network is shared between perception and production. The bottom half of the network is specific to production, but is three-way–sensitive to input from the perceptual system. The perceptual form network is not shown. It produces perceptual input to the lemma stratum. Adapted from Levelt et al., 1999.

syntactic trees, called "lexical frames" by Kempen (2000; see also Vosse and Kempen, 2000). Kempen models grammatical encoding as a unification of incrementally selected lexical frames. The output is a surface syntactic tree for the utterance as a whole.

A standard method in word production research is the picture-word interference paradigm. The subject is presented with a picture to be named. At the same time, a distractor word is presented, either auditorily or printed in the picture. A semantically related distractor can affect the response latency. In the WEAVER model this is due to activation of the corresponding lemma. Levelt et al. (1999) proposed to account for this effect of perception on production by assuming that lemmas are shared between speech perception and production (leaving undecided how in detail orthographic input affects the corresponding lemma in the speech network). The obvious further step was to claim that the lexical networks for spoken word perception and spoken word production are shared from the lemma level upward. This rather drastic theoretical merger has an inevitable consequence. The feedforward from lexical concepts to lemmas, required for production, is now complemented by feedback from lemmas to lexical concepts, required for spoken word understanding, that is, concept-to-lemma connections are bilaterally activating. We are, as yet, not aware of empirical counterevidence to this proposition.

It would contribute to the aesthetics of this grand unification to make the further claim that grammatical encoding and grammatical decoding are one. This is exactly the step taken by Kempen (2000; see also Vosse and Kempen, 2000). In both encoding and decoding, lexical frames are incrementally unified. In the production case, the frames (lemmas) are conceptually selected (see above). In the perceptual case, the lexical frames are selected on the basis of a recognized phonological code. But then, in both modalities, the selected lexical frames unify incrementally, growing a syntactic tree, "from left to right." Although this is a well-argued and attractive proposal, its empirical consequences need further scrutiny. For instance, it should be impossible for the language user to simultaneously encode and decode an utterance; that would mix up the lemmas in the unification space. Can you parse your interlocutor's utterance while you are speaking yourself?

Connections at the Form Level

Levelt et al. (1991) proposed that lemma-to-lexeme (=word form) connections are unilateral. Upon selection of the lemma, activation spreads to the form node, but there is no feedback. Both the selection condition and the nonfeedback claim have become controversial issues, leading to ever-more sophisticated experiments (see Levelt et al., 1999, and Levelt, 1999, for reviews). If the nonfeedback claim is correct, then the two networks cannot share the input and output form level nodes. In other words, phonological codes for perception and for production are not identical. Dell (1986) proposed a feedback mechanism to explain unmistakable facts of speech error distributions. There is usually a lexical bias in speech errors (they result in real words a bit more often than random segment changes would predict). And there is a statistical preponderance of mixed errors (such as *cat* for *rat*), where the error is both semantically and phonologically related to the target. But these properties can also be handled in terms of the internal loop monitoring mechanism discussed above. A real word and a word in the correct semantic field have a better chance of slipping through the monitor than a nonword or an odd word (Levelt, 1989). Another reason for keeping the perception and production form nodes apart are reports in the aphasiological literature of selective losses of word perception vs. word production (cf. Caplan, 1992).

Still, the form networks for perception and production must be connected in some way. Given the evidence from picture-word interference experiments, Levelt et al. (1999) made the following two assumptions: First, a distractor word, whether spoken or written, affects the corresponding morpheme node (which represents the phonological code) in the production network; the details of the perceptual mechanism were left unspecified (see figure 14.1). Second, active phonological segments in the perceptual network can also affect the corresponding segment nodes in the production lexicon. Again, the precise perceptual mechanism was left unspecified (see figure 14.1). A possible further specification may be achieved in terms of the merge model of Norris, McQueen, and Cutler (Meyer and Levelt, 2000). In short, we assume the existence of close connections at the form level, though without sharing of segmental or morphological nodes. What this means in neuroarchitectonic terms is a fascinating issue, to which I return below.

Still entirely open are potential relations at the phonetic level. Gestural scores in the production model, whether for segments, syllables, or whole words, are abstract motor representations. Levelt and Wheeldon (1994) proposed the existence of a "syllabary," a repository of gestural scores for high-frequency syllables. English (just as Dutch or German) speakers do about 80 percent of their talking with no more than 500 different syllables. One would expect such highly overlearned motor actions to be appropriately stored in the frontal region of the brain. In our model (Levelt et al., 1999), these syllabic gestures are accessed on the fly, as phonological encoding proceeds. As soon as a selected lemma has activated its phonological code, the activated phonological segments are incrementally packaged into phonological syllables, often ignoring lexical boundaries (as in *I'll-sen-dit*). Each such composed syllable accesses "its" gestural score, which can be almost immediately executed by the articulatory system. We have, so far, not made any assumptions about potential relations between retrieving gestural scores and perceptual representations of syllables or other sublexical units. I presently return to that issue.

Some Neurophysiological Observations

Mehler, Morton, and Jusczyk concluded their extensive paper "On Reducing Language to Biology" (1984) as follows: "We have argued that if a mapping between psychological processing and neurophysiological structures is possible, it will only come about after the key theoretical constructs are established for each level of explanation. In the interim, there are restricted circumstances in which neurophysiological observations can play a role in the development of psychological models. But these circumstances require the consideration of such evidence, not only in terms of the physiological organization of the brain, but also in terms of its functional organization" (p. 111). This statement has gained new force in the present era of cognitive neuroimaging. What emerges as relevant and lasting contributions after a decade of exploring the new tools are those studies that were theoretically driven. They are the studies where the experimental and control tasks are derived from an explicit theory of the underlying cognitive process. Where this was not the case, as in most neuroimaging studies of word production, a post hoc theoretically driven meta-analysis can still reveal patterns in the data that were not apparent in any of the individual studies themselves (Indefrey and Levelt, 2000; Levelt and Indefrey, 2000).

Since 1984 major strides have been made in unraveling the functional organization of both speech production and perception, in particular as far as the production and perception of spoken words are concerned. In addition, as discussed in the previous section, the issues about the relations between these two processing modalities can now be stated in explicit theoretical terms. In other words, the "interim" condition in the conclusion of Mehler et al. is sufficiently satisfied to scan the recent neuroimaging evidence for suggestions that can further the modeling of production-perception relations. Here are a few such "neurophysiological observations":

Self-Monitoring

According to the perceptual loop hypothesis, reviewed above, both the external and the internal feedback are processed by the language user's normal speech-understanding system. This hypothesis is controversial in the literature (MacKay, 1992; Levelt 1992; Hartsuiker and Kolk, 2001), not only for its own sake but also because of its role in accounting for lexical bias and mixed errors in the speech error literature. McGuire, Silbersweig, and Frith (1996) have provided PET data that strongly support the notion of the speech perceptual system being involved in self-monitoring. If self-generated speech is fed back in a distorted fashion (pitch-transformed), there is increased bilateral activation of lateral temporal cortex (BA 21/22), and in particular of the left superior temporal sulcus (see figure 14.2). A highly similar activation of temporal areas is obtained when not the own voice but an alien voice is fed back to the speaker. These and other findings lead to the conclusion "that i) self- and externally-generated speech are processed in similar regions of temporal cortex, and ii) the monitoring of self-generated speech involves the temporal cortex bilaterally, and engages areas concerned with the processing of speech which has been generated externally" (McGuire et al., 1996, p. 101).

However, these data only concern the external loop. Is the auditory perceptual system also involved when the subject speaks "silently," not

planum temporale
- speech perception
- monitoring internal speech
- subvocal word production
- lip reading

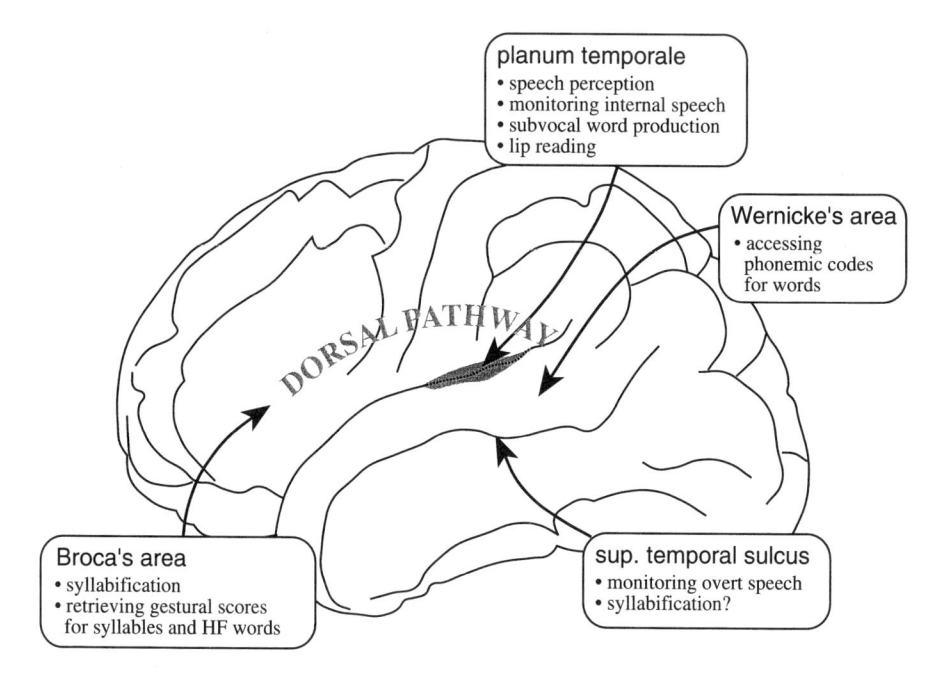

Wernicke's area
- accessing phonemic codes for words

DORSAL PATHWAY

Broca's area
- syllabification
- retrieving gestural scores for syllables and HF words

sup. temporal sulcus
- monitoring overt speech
- syllabification?

Figure 14.2
Schematic lateral view of the left hemisphere with pointers to regions discussed in the text.

producing an overt auditory signal? Paus et al. (1996) have provided relevant evidence on this point. In their PET experiment the subject whispered a string of syllables, with rate of production as the independent variable (30 through 150 syllables per minute). The authors observed a concomitant increase of activation in left auditory cortex, affecting two regions in particular: (1) an auditory region on the planum temporale, just posterior to Heschl's gyrus, and (2) an auditory region in the caudal portion of the Sylvian fissure. Hence, these regions might qualify as candidates for the reception of the speaker's internal feedback. Finally, Levelt and Indefrey (2000) discuss the possibility that the midsuperior temporal gyrus activation that is obtained even in nonword production tasks (as in nonword reading) may reflect overt or covert self-monitoring.

The Indefrey and Levelt meta-analysis has not provided the final answer with respect to the localization of rapid syllabification. As discussed below, the imaging evidence points to both Broca's region and the mid-superior temporal gyrus (see figure 14.2). We do know, however, that internal self-monitoring concerns a syllabified phonological representation (Wheeldon and Levelt, 1995). If that representation is created in Broca's area, it must be fed back to the relevant areas in the superior temporal gyrus where the monitoring takes place. Which anatomical pathway could be involved here? There is less of an anatomical problem if phonological syllabification itself involves the mid or posterior regions of the superior temporal gyrus (or as suggested by Hickok and Poeppel, 2000, some inferior parietal region).

Phonological Codes
The meta-analysis by Indefrey and Levelt (2000) strongly supports Wernicke's original notion that retrieving a word's phonological code for production involves what has since been called Wernicke's area. The neuroimaging data show, in particular, that if word production tasks (which all involve retrieving the words' phonological codes) are compared to nonword reading tasks (which do not require access to a phonological code), the critical difference in activation concerns Wernicke's area. In an MEG study of picture naming, Levelt et al. (1998) also showed dipoles in Wernicke's area, more precisely in the supratemporal plane vicinity, becoming active during a time interval in which phonological access is achieved. As discussed above, phonological codes in the production system can be perceptually primed, both by auditory and visual word or segment/letter stimuli. This may well involve part of Wernicke's area. Which part? Zattore et al. (1992) and Calvert et al. (1997) have shown the involvement of the left posterior supratemporal plane (STP), or planum temporale, in the perception of speech, and even in active lip reading. Recently, Hickok et al. (2000), in an fMRI study of picture naming, showed this area also to be involved in the subvocal production of words. If this is the region where the supposed linkage between the perceptual and production systems is established, then it is important to note that the relevant perceptual output is quite abstract. It may as well result from speech input as from reading the lips. Can it also be the output of seeing a printed word? If so, we are probably dealing with a phonemic level of representation.

A review by Hickok and Poeppel (2000), combining several sources of patient and imaging data, suggests that the auditory-to-motor interface involves a "dorsal pathway" from the just-mentioned region in the left posterior superior temporal gyrus through the inferior parietal lobe toward the frontal motor speech areas (see figure 14.2). They make the explicit suggestion that the auditory-motor interface operations have their site in the left inferior parietal lobe. Although this issue is far from solved, one must be careful not to confuse auditory phonemic codes and phonemic codes for production. They are clearly linked, behaviorally and in terms of functional anatomy, but they are not identical.

Syllabification and Phonetic Encoding
The above-mentioned meta-analysis of word production studies (Indefrey and Levelt, 2000) pointed to the left posterior inferior frontal lobe and the left midsuperior temporal gyrus as being involved in phonological encoding, which is largely rapid syllabification in word production tasks. Not surprisingly, the same study showed bilateral, mostly ventral sensorimotor area involvement in actual articulation. Some speculation should be allowed in a Festschrift. The mentioned inferior frontal lobe involvement may extend beyond strict phonological encoding and also include accessing the gestural scores for successive syllables. In other words, the region would somehow store, retrieve, and concatenate our overlearned articulatory patterns for syllables and other high-frequency articulatory units, such as whole bi- or trisyllabic high-frequency words. More specifically, one would conjecture the involvement of Broca's area, which is, in a way, premotor area. It is, after all, the human homologue of premotor area F5 in the macaque (Broca's area is, however, different from F5 in terms of its cytoarchitecture; it contains a layer IV, which is absent in our and the macaque's premotor cortex; K. Zilles, personal communication). Having mentioned F5, a discussion of mirror neurons is unavoidable (see Gallese and Goldman, 1999, for a review). If (the larger) Broca's region is involved in rapidly accessing syllabic gestures, and if it shares with F5 its mirror-neuron character, one would expect the area to represent *perceived* articulatory gestures as well, that is, what one sees when looking at a speaking face. As early as 1984, Mehler et al. discussed such a possibility in connection with Ojemann's (1983) observation that the

(larger) cortical area "has common properties of speech perception and generation of motor output." These common properties, according to Ojemann, may serve functions "described by the motor theory of speech perception" (cf. Liberman, 1996).

If this speculation is of any value, one should be able to prime the production of a syllable by having a speaker look at a speaking face (on the monitor) that produces the same target syllable. That experiment, with all its controls, was recently run by Kerzel and Bekkering (2000), with the predicted result. Still to be shown is that the effect involves Broca's area or immediately neighboring regions. If so, we are back to Wernicke in a new guise. Wernicke located the auditory word images in the posterior superior temporal area and the motor word images in Broca's area. He supposed the existence of a connection between the two areas, now known to be the arcuate fascicle. That connection is, according to Wernicke, involved in spoken word repetition: the auditory word image activates the corresponding motor word image. Our chronometric work on phonological encoding suggests that an "auditory image" can affect the activation of individual segments in the same or another word's phonological code, that is, the code on which word production is based. As mentioned above, retrieving this detailed, segmented code seems to involve Wernicke's area.

In the WEAVER model, these phonological segments directly affect the activation state of all gestural scores in which they participate. These whole "motor images" for high-frequency syllables are now suggested to be located in the Broca or premotor region. If they are indeed "mirror images," they represent both our own overlearned articulatory gestures and the ones we perceive on the face of our interlocutors. However, they are not auditory images. The auditory-phonemic link between spoken word perception and word production can stay restricted to Wernicke's area, the midsuperior temporal gyrus, the planum temporale, and maybe the inferior parietal region, as long as no evidence to the contrary appears.

Finally, it should be noted that among these syllabic scores are all the articulatory scores for high-frequency monosyllabic *words*. It is a minor step to suppose that there are also whole stored scores for high-frequency combinations of syllables, in particular of disyllabic or even trisyllabic high-frequency words. If, as Wernicke supposed, these motor images can also be directly activated by "object images," that is, conceptually, spoken word production is to some extent possible without input from Wernicke's area. However, the fine-tuning of phonological word encoding requires access to detailed phonological codes. Precise syllabification without paraphasias, resyllabification in context (such as *liaison* in French), and correct stress assignment will always depend on accurate phonemic input from our repository of phonological codes, which, I suggest, involves Wernicke's area. Indefrey et al. (1998) provide fMRI evidence that their assembly involves left lateralized premotor cortex. Phonotactic assembly and the ultimate access to stored articulatory patterns are somehow handled by posterior inferior frontal areas.

Acknowledgments

I am grateful to Gregory Hickok for an e-mail exchange on the functions of Wernicke's area, and to Peter Indefrey for many helpful comments on a draft version of this paper.

References

Bever, T., and Weksel, W. (Eds.) (shelved). *The structure and psychology of language.*

Calvert, G. A., Bullmore, E. T., Brammer, M. J., Campbell, R., Williams, S. C., McGuire, P. K., Woodruff, P. W. R., Iversen, S. D., and David, A. S. (1997). Activation of auditory cortex during silent lipreading. *Science, 276,* 593–596.

Caplan, D. (1992). *Language: Structure, processing, and disorders.* Cambridge, MA: MIT Press.

Clark, H. H. (1996). *Using language.* Cambridge, UK: Cambridge University Press.

Dell, G. (1986). A spreading-activation theory of retrieval in sentence production. *Psychological Review, 93,* 283–321.

Gallese, V., and Goldman, A. (1998). Mirror neurons and the simulation theory of mind-reading. *Trends in Cognitive Sciences, 2,* 493–501.

Hartsuiker, R., and Kolk, H. (2001). Error monitoring in speech production: A computational test of the perceptual loop theory. *Cognitive Psychology, 42,* 113–157.

Hickok, G., and Poeppel, D. (2000). Towards a functional anatomy of speech perception. *Trends in Cognitive Sciences, 4,* 131–138.

Hickok, G., Erhard, P., Kassubek, J., Helms-Tillery, A. K., Naeve-Velguth, S., Strupp, J. P., Strick, P. L., and Ugurbil, K. (2000). An fMRI study of the role of left posterior superior temporal gyrus in speech production: Implications for the explanation of conduction aphasia. *Neuroscience Letters, 287*, 156–160.

Indefrey, P., and Levelt, W. J. M. (2000). The neural correlates of language production. In M. Gazzaniga (Ed.), *The new cognitive neurosciences*. Cambridge, MA: MIT Press.

Indefrey, P., Gruber, O., Brown, C., Hagoort, P., Posse, S., and Kleinschmidt, A. (1998). Lexicality and not syllable frequency determines lateralized premotor activation during the pronunciation of word-like stimuli—An fMRI study. *NeuroImage, 7*, 54.

Johnson-Laird, P. N. (1974). Experimental psycholinguistics. *Annual Review of Psychology, 25*, 135–160.

Kempen, G. (2000). Human grammatical coding. Unpublished manuscript, Leiden University, Leiden, Netherlands.

Kempen, G., and Hoenkamp, E. (1987). An incremental procedural grammar for sentence formulation. *Cognitive Science, 11*, 201–258.

Kerzel, D., and Bekkering, H. (2000). Motor activation from visible speech: Evidence from stimulus-response compatibility. *Journal of Experimental Psychology: Human Perception and Performance, 26-2*, 634–647.

Levelt, W. J. M. (1983). Monitoring and self-repair in speech. *Cognition, 14*, 41–104.

Levelt, W. J. M. (1989). *Speaking: From intention to articulation*. Cambridge, MA: MIT Press.

Levelt, W. J. M. (1992). The perceptual loop theory not disconfirmed: A reply to MacKay. *Consciousness and Cognition, 1*, 226–230.

Levelt, W. J. M. (1996). Foreword. In T. Dijkstra, and K. de Smedt (Eds.), *Computational psycholinguistics*. London: Taylor & Francis.

Levelt, W. J. M. (1999). Models of word production. *Trends in Cognitive Sciences, 3*, 223–232.

Levelt, W. J. M., and Indefrey, P. (2000). The speaking mind/brain: Where do spoken words come from? In A. Marantz, Y. Miyashita, and W. O'Neil (Eds.), *Image, language, brain*. Cambridge, MA: MIT Press.

Levelt, W. J. M., Praamstra, P., Meyer, A. S., Salmelin, R., and Kiesela, P. (1998). An MEG study of picture naming. *Journal of Cognitive Neuroscience, 10*, 553–567.

Levelt, W. J. M., Roelofs, A., and Meyer, A. S. (1999). A theory of lexical access in speech production. *Behavioral and Brain Sciences, 22*, 1–38.

Levelt, W. J. M., Schriefers, H., Vorberg, D., Meyer, A. S., Pechmann, T., and Havinga, J. (1991). The time course of lexical access in speech production: A study of picture naming. *Psychological Review, 98*, 122–142.

Levelt, W. J. M., and Wheeldon, L. (1994). Do speakers have access to a mental syllabary? *Cognition, 50,* 239–269.

Liberman, A. (1996). *Speech: A special code.* Cambridge, MA: MIT Press.

McGuire, P. K., Silbersweig, D. A., and Frith, C. D. (1996). Functional anatomy of verbal self-monitoring. *Brain, 119,* 101–111.

MacKay, D. G. (1992). Awareness and error detection: New theories and research paradigms. *Consciousness and Cognition, 1,* 199–225.

Mehler, J., and Savin, H. (shelved). In T. Bever, and W. Weksel (Eds.), The structure and psychology of language.

Mehler, J., Morton, J., and Jusczyk, P. W. (1984). On reducing language to biology. *Cognitive Neuropsychology, 1,* 83–116.

Meyer, A. S., and Levelt, W. J. M. (2000). Merging speech perception and production. Commentary on target paper by Norris, D., McQueen, J. M., and Cutler, A. *Behavioral and Brain Sciences, 23,* 339–340.

Morton, J. (1970). A functional model for memory. In D. A. Norman (Ed.), *Models of human memory.* New York: Academic Press.

Ojemann, G. A. (1983). Brain organization for language from the perspective of electrical stimulation mapping. *Behavioral and Brain Sciences, 6,* 189–206.

Paus, T., Perry, D. W., Zatorre, R. J., Worsley, K. J., and Evans, A. C. (1996). Modulation of cerebral blood flow in the human auditory cortex during speech: Role of motor-to-sensory discharches. *European Journal of Neuroscience, 8,* 2236–2246.

Roelofs, A. (1992). A spreading-activation theory of lemma retrieval in speaking. *Cognition, 42,* 107–142.

Roelofs, A. (1997). The WEAVER model of word-form encoding in speech production. *Cognition, 64,* 249–284.

Vosse, T., and Kempen, G. (2000). Syntactic structure assembly in human parsing: A computational model based on competitive inhibition and a lexicalist grammar. *Cognition, 75,* 105–143.

Wheeldon, L. R., and Levelt, W. J. M. (1995). Monitoring the time course of phonological encoding. *Journal of Memory and Language, 34,* 311–334.

Zattore, R. J., Evans, A. C., Meyer, E., and Gjedde, A. (1992). Lateralization of phonetic and pitch discrimination in speech processing. *Science, 256,* 846–849.

IV
Development

How to Study Development

Anne Christophe

Jacques Mehler has always been concerned about the proper way to study development. He has long been convinced that it is entirely useless to *describe* stages of development without looking for *explanations* of how babies turn into adults. Today, Jacques's "crusade" against a purely descriptive approach to development has hardly abated. Just a few months ago he wrote the following cutting sentences: "Consider scientists interested in the problem of physical growth: it is obviously useless to measure each growing child, regardless of the virtues of the yardstick. Eventually the news that children get taller with age reaches the crowds as does the fact that this process tapers off after puberty. Why then, one may ask, should anyone pursue such observations? We do not know. . . . Whether we measure the expansion of the lexicon, memory span, attention span, the ability to solve logical puzzles, the facts are similar: children generally get better with age. In the absence of an *explanation* of the observed phenomena, this kind of study does not contribute data of great value" (Mehler and Christophe, in press). Such explanations have to include biological and cognitive factors, not just behavioral descriptions. John Morton and Uta Frith address precisely the problem of how to frame such explanations of development (chapter 15): they describe the "causal modeling" framework, specifically designed as a notation for theories of pathological cognitive development. They illustrate it with dyslexia, a syndrome whose definition was, and sometimes still is, purely descriptive, namely "a discrepancy between mental age and reading age." They emphasize the fact that this "definition" offers no hint of an explanation for the disorder, and as a consequence, may conflate several quite different disorders under the same label. In addition, people may suffer from the disorder, as defined biologically or cognitively, without exhibiting the standard behavioral symptoms. The causal modeling framework allows them to express such ideas explicitly and coherently.

Jacques reacted so much against the idea that the mere description of stages in development was interesting, that for years he claimed that the only interesting objects of study were newborns (as exemplifying the initial state), and adults (the stable state)—see the reminiscing of Peter Jusczyk (chapter 20), who has probably contributed more than anybody else to our database of facts about language development in the first year of life. Even today, when Jacques is ready to admit that studying organisms at intermediate ages may have an interest, he never loses an opportunity to ask us the following question, "What more will you know when you will know at what age this particular feature is acquired?" And the answer is this: in isolation, this piece of information is quite useless indeed (e.g., babies know about xxx at 6 vs. 9 months of life). However, when compared with other such results, it can be used to constrain theories of acquisition. Renée Baillargeon (chapter 19) provides an elegant illustration of this research strategy for the acquisition of object properties. Carefully designed cross-sectional experiments allow her to distinguish between two patterns of results, one reflecting infants' innate expectations for the continuity and solidity of objects, and the other reflecting the operation of a learning mechanism specifically designed for identifying relevant properties of objects in different situations. Another example of this research strategy comes from the acquisition of phonology, where it is useful to know which properties are acquired before a receptive lexicon is compiled, and which are acquired after: in both cases quite different acquisition procedures apply, and the learning path is different (see, e.g., Peperkamp and Dupoux, in press). Peter Jusczyk emphasizes a second domain in which comparing acquisition schedules would be fruitful, even though he regretfully notes that these types of data are mostly absent from the literature: this domain is the crosslinguistic comparison of language development. Indeed, since languages have different characteristics, babies acquiring them may well use different strategies—which would surface as different acquisition schedules. Finally, a third domain of application of this comparative strategy is elegantly demonstrated by Núria Sebastián Gallés and Laura Bosch (chapter 21): they set out to compare the development of monolingual and bilingual babies. In addition to providing constraints on language acquisition procedures, this line of research holds the promise of bringing answers to the vexing problem of what happens in a bilingual speaker's head—a subject which has always fascinated Jacques (himself a fluent heptalingual speaker).

Jacques's insistence on the study of the initial state derived from his conviction that one should concentrate on solving the logical problem of acquisition: how to turn a baby into an adult? This implied studying both infants and adults simultaneously. Cutting across traditional discipline boundaries and administrative partitioning, Jacques managed to create a laboratory devoted to the study of both infants and adults, most often by the same people. The fruitfulness of this research strategy is beautifully illustrated by Elizabeth Spelke and Susan Hespos (chapter 18): they set out to explain babies' strange abilities to succeed in object permanence tasks when the measure is looking, but fail when the measure is reaching. They propose that reaching for an object demands more precise representations than just looking at it, because one has to take into account factors like the size and orientation of the object, rather than just its position. However, such an explanation may sound ad hoc if not supported by independent facts. This converging evidence comes from the study of adults: they, too, show better performance in looking tasks than in reaching tasks, when part of an object's trajectory is hidden. Spelke and Hespos thus conclude that infants and adults are alike in many respects: both expect objects to continue existing when out of sight; both perform better when an object remains visible; the only difference between infants and adults is the precision of object representations, which increases over time.

This chapter also illustrates the *continuity/discontinuity* debate, which is still very alive in theories of acquisition. Thus, the shift in reaching behavior may either be attributed to a conceptual change (namely, that infants initially assume that hidden objects cease to exist), or be considered as the consequence of a quantitative change which influences performance (Spelke and Hespos propose an increase in the precision of representations). The first hypothesis assumes a discontinuity between infants' and adults' conceptual systems, while the second assumes that infants and adults share the same basic cognitive architecture (including the representation of hidden objects). Rochel Gelman and Sara Cordes (chapter 16) investigate number representation abilities in humans and animals, and they also argue in favor of continuity. Susan Carey (chapter 17) illustrates the other side of the debate: she reviews the emergence of the ability to represent integers in human children, and argues that this constitutes an instance of a real conceptual discontinuity.

This debate has been going on for a while, and is likely to go on for some time because of the very nature of the debate. First, the experimental evidence needed to decide between alternatives implies a good understanding of both the initial and the stable state, as well as the processes which lead from one to the other, for each domain in which this question is posed. Second, the debate is asymmetrical, in that everybody agrees that continuity is a good model in some instances (e.g., some parts of language acquisition, presumably); the disagreement arises because some people believe that continuity is the *only* possible model of development (for logical reasons), whereas others believe that discontinuities can *sometimes* arise. As a consequence, the debate is bound to go on until a convincing instance of a conceptual discontinuity can be found (if it can), or until we find other ways of conceptualizing development.

Theoretical debate is a good way of making research advance, and this is why Jacques has always favored it as an editor of *Cognition*—however vehemently he may protest against views he does not share, when speaking in his own name.

References

Mehler, J., and Christophe, A. (in press). Reflections on development. In J. Mehler and L. Bonati (Eds.), *Developmental cognitive science*. Cambridge, MA: MIT Press.

Peperkamp, S., and Dupoux, E. (in press). Coping with phonological variation in early lexical acquisition. In I. Lasser (Ed.), *The process of language acquisition*. Berlin: Peter Lang Verlag.

15

Why We Need Cognition: Cause and Developmental Disorder

John Morton and Uta Frith

Why We Cannot Rely on Behavior

With some notable exceptions, the study of developmental disorders has stayed outside mainstream cognitive science. This is, we believe, because of a lack of a framework that allows us to express and compare different theories in a precise way. As classical cognitive science has benefited from boxological notations of information processing, we believe that a similar notational device, which we have called "causal modeling," is useful to cast theories of developmental disorders (Frith, Morton, and Leslie, 1991; Morton and Frith, 1993a,b, 1995). Developmental disorders are, in a way, more complex than standard problems in cognitive psychology. For instance, to model word recognition, you can refer to a single level of description, the information-processing, or cognitive level (see, e.g., Morton, 1969). However, to model adequately a developmental disorder like dyslexia, you need to refer to several levels of description: biology, cognition, and behavior. The reason for this is that, to capture the essence of the disorder, there has to be a developmental component, and this requires the notion of cause. For example, theories of dyslexia say that it has a genetic origin which leads to abnormalities in the brain. These abnormalities lead to cognitive deficits which, in turn, create the observable behavior.

The richness of such an approach may be contrasted with the mainstream clinical approach, centered on the need for agreed-upon diagnosis, and typified by the technical manuals, one published by the American Psychiatric Association (APA), another by the World Health Organization (WHO). The *Diagnostic and Statistical Manual of Mental Disorders*

(DSM), published by the APA, is now in its fourth revised edition; the *International Classification of Diseases* (ICD) is now in its tenth edition (1992). Extensive work goes into the revisions and this is a continuing process. Better procedures and better ways of describing the diagnostically critical features are constantly sought. However, there are certain underlying limitations. We quote here from ICD-10:

> These descriptions and guidelines carry no theoretical implications and they do not pretend to be comprehensive statements about the current state of knowledge of the disorders. They are simply a set of symptoms and comments that have been agreed, by a large number of advisors and consultants in many different countries, to be a reasonable basis for defining the limits of categories in the classification of mental disorders (1992, p. 2).

Let us take a specific example, DSM III-R (1987, pp. 38–39) specifies the following criteria for autistic disorder:

- qualitative impairment in social interaction;
- qualitative impairment in verbal and non verbal communication, and in imaginative activity;
- markedly restricted repertoire of activities and interests.

There is no mention of the nature of the disorder, only a list of symptoms upon which the diagnosis should be based.

Cognition, we argue, has to be introduced as a level of description that plays an essential role in bridging the gap between brain and behavior. The mapping between brain and behavior is chaotic, unless governed by principled hypotheses which are most sharply and parsimoniously expressed in terms of mental concepts. Our work over the years on both autism and dyslexia has shown how it is possible to simplify a mass of apparently unconnected observations from both biological and behavioral sources, by postulating cognitive deficits.

This general point has been made many times over the last forty or so years during the establishment of functionalist cognitive science in the context of adult mental processes. What is different in the case of developmental disorder is that, although the cognitive level is still necessary, it is not sufficient to describe what is going on. All serious hypotheses concerning developmental disorders attempt to establish causal relations between brain and behavior. Many include cognitive function explicitly, though not systematically, and some theories are confused between the cognitive and behavioral levels of description. Thus, test results are sometimes referred to as being cognitive constructs when, in fact, they represent behavior (see Morton and Frith, 1993a,b, 1995, for examples). What we are reviewing in this chapter is our attempt to introduce an acceptable common framework for the expression of such theories.

Example of a Causal Model

Over the past few years, we have concerned ourselves with establishing the relationships among biological, cognitive, behavioral, and environmental factors in developmental disorders (Frith et al., 1991; Morton and Frith, 1993a,b, 1995). We have done this through a framework—causal modeling—which is neutral with respect to any particular theory. The framework can represent our own causal theories or anyone else's equally well. Indeed, in line with the preceding discussion, the only theoretical point intrinsic to the framework is the necessity of the cognitive level of representation. The framework recognizes the three levels of description mentioned above: brain, cognition, behavior. It also introduces the environment, which can have an effect at any of these three levels. Particular defects are conventionally represented in a box,[1] and located within one of the descriptive levels. If a theory specifies that one defect causes the emergence of another defect during development, the two boxes will be related by an arrow. So, in our framework, arrows represent causality. The point is best made with an illustration.

Figure 15.1 is taken from the British Psychological Society (1999). This diagram represents what might be called the familiar canonical phonological deficit view of dyslexia. In the diagram, the arrows represent statements of the kind "X causes Y" or "X leads to Y" in the theory which is being represented. Note that you don't have to agree with the underlying theory to understand what the notation is intended to accomplish. The first claim of the theory is that dyslexia is a genetic disorder.[2] This contrasts with the traditional, behavioral, view whereby dyslexia was defined simply as a discrepancy between mental age and reading age. We enlarge later on the limitations of the latter viewpoint. The theory claims

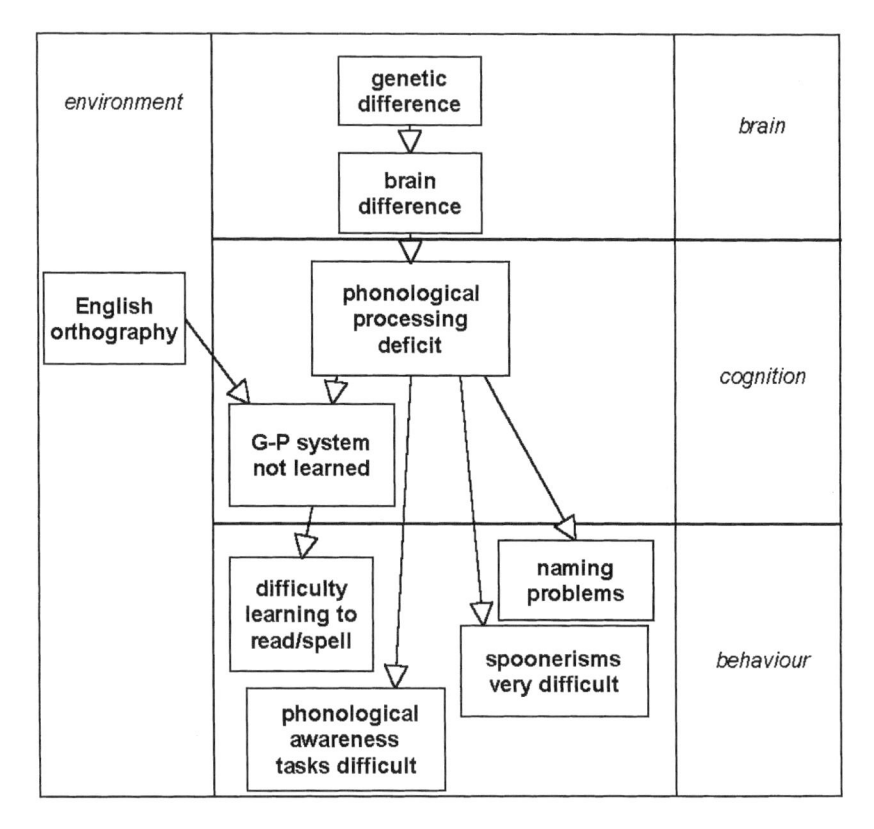

Figure 15.1
The canonical phonological deficit theory of dyslexia, from the British Psychological Society (1999), the arrows showing the causal relations among three levels of description—brain, cognition, and behavior.

that there is a subset of the population which is different genetically from the rest in a particular (as yet unspecified) regard. Over the course of early development, this genetic difference leads to a difference in brain structure. The difference in brain structure, in turn, leads to a deficit in phonological processing which is revealed, in terms of behavior, through problems in naming. In the laboratory, we also find difficulty with phonological awareness tasks and with generating spoonerisms. By default, the theory says that these factors will be found in all environments and cultures. Additionally, the theory says that within the English-speaking culture there will be a further manifest problem—difficulty in learning to read. This problem arises because of the nature of English orthography. Specifically, the theory claims that English orthography, in conjunction with the phonological processing deficit arising from the genetic difference, makes it difficult to set up a grapheme-phoneme mapping system.

Note that there are more indirect claims made by the theory which become very clear from the causal model. The most important claim is that in the presence of a more transparent orthography than English, there will be no reading difficulty. This can be deduced from the figure where "English orthography" and "phonological processing deficit" are both required to cause "G-P system not learned".[3] Now, while this claim might be slightly exaggerated, it is certainly the case that dyslexia, as indexed by manifest difficulty in learning to read, is rare in Italy, where the orthography is much more transparent and consistent than English (Lindgren, De Renzi, and Richman, 1985). However, unless the gene pool in Italy is unexpectedly different from that in the United Kingdom, the theory in figure 15.1 leads to the prediction that there will be Italians with the same genetic difference who will have problems in naming and difficulty with phonological tasks and with generating spoonerisms. Since the problems in naming would not have been particularly noticeable, this group of people would not have been diagnosed as having a developmental disorder, nor would they be aware of their problem.

Current work suggests that such a group has been identified. Perhaps the same proportion of the Italian population have the same genetic condition as the English dyslexic population, but without necessarily manifesting any significant reading problem. The reason for this is that Italian orthography lacks the extremely complex context-sensitive mapping which characterizes English orthography. Apparently, such complexity in orthographic-phonological mapping precipitates the manifestation of the particular genetic configuration, known as dyslexia, in reading behavior. In terms of figure 15.1, the Italian group would be identical to the English apart from the environmental link from the orthography. Without that, the phonological deficit does not lead to a serious difficulty in learning to read in many of the affected individuals. We could, then, view Italian culture as a therapeutic factor! One point of this illustration is that without a cognitive definition of dyslexia, such a phenomenon could not have been predicted, and, if it had been stumbled upon, could not have been explained. The irrelevance of the discrepancy definition of dyslexia (a gap between reading age and mental age) to a complete understanding of the condition should be apparent.

Sublevels of Description in the Brain
In figure 15.1, the theory goes from gene to brain state without explicating the latter. But cognitive neuroscience has produced examples where brain states have already been defined in some detail. Gallese (in press) has made distinctions among a number of levels in the biological domain within which he embeds his work, which is anchored in brain physiology. These levels are:

- membrane proteins
- receptors
- neurons and synapses
- neural assemblies
- cortical circuits
- behavior/cognition

Gallese discussed the possible independence of these levels, pointing out that while complexity at the receptor level is not relevant to behavior/cognition, action potentials, arising from activity in neurons, can have such meaning. Assemblies of neurons are even more likely to have meaning. Functional imaging studies currently relate to cortical circuits. We can note that a full causal model of developmental disorder could start with a problem at the membrane protein level (preceded by a genetic disorder, perhaps) and move down through the five biological levels before leading to a cognitive problem.

Relations between the Cognitive and Biological Levels

In figure 15.1, one can see that there are causal arrows that remain within a descriptive level, and causal arrows that appear to cross levels, that is, go from one descriptive level to another. Closer examination of the shift of level from brain to cognition in the causal chain reveals a particular relationship between the levels. To appreciate this we can start with the causal model sketched in figure 15.2. We have no problem in saying that

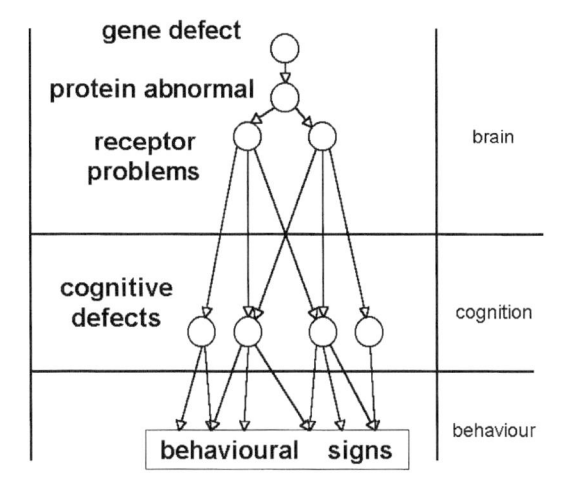

Figure 15.2
Sketch of a generic causal model of a developmental disorder of biological origin. For reasons of simplicity, we omit the environmental factors here.

the gene defect *causes* the various cognitive defects and behavioral signs. This is a very common type of claim in the area of developmental disorder. Equally, it is straightforward to say that receptor problems (assuming that they are specified) *cause* a cognitive defect—such as, in autism, the lack of a theory of mind mechanism (Frith et al., 1991). However, it is clear that there is an explanatory gap between these two elements which would have to be filled before the theory could be considered satisfactory. To start with, using Gallese's scheme, outlined above, the receptor problem could lead to a defect at the level of neurons and synapses, which could, in turn, lead developmentally to an abnormality best described in terms of neural assemblies, *n*. All of this could be specified causally. However, given that we have the specification of a deficit described in terms of *n*, it could be that there is no further causal statement that could be made at the level of brain physiology.

Let us now look at the cognitive level, still using the example of autism. The lack of a theory of mind mechanism, in the case of autism, has been attributed in one version of the theory to the absence of a computational device called an expression raiser (EXPRAIS) (Leslie, 1987). Morton (1986) supposed EXPRAIS to be a cognitive primitive—that is, EXPRAIS

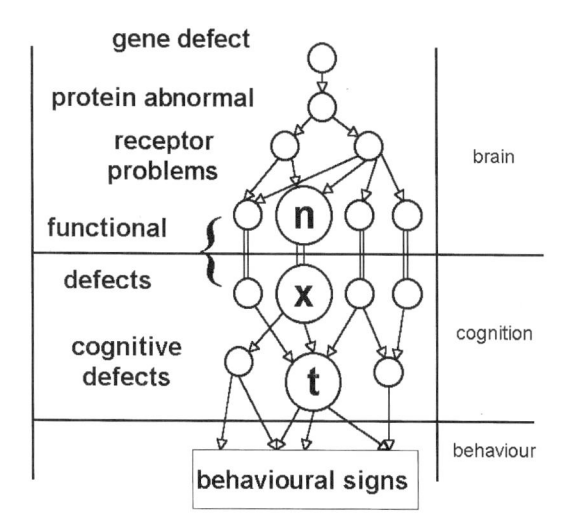

Figure 15.3
An expansion of the model in figure 15.2, showing the equivalence relation between elements at the biological and cognitive levels. n is the neural equivalent to x (EXPRAIS), while t represents the theory of mind mechanism.

is supposed to be irreducible. What this means is that there is no other *cognitive* element above it in the causal chain accounting for its absence in this particular theory of autism.[4]

In figure 15.3, we have brought together the considerations of the previous two paragraphs. In this figure, x represents EXPRAIS and n is the deficit described in terms of neural assemblies. t represents the theory of mind mechanism. From this figure, we might say that the neural deficit (n) causes a deficit in the theory of mind mechanism. Equally, we might say that a deficit in x causes the deficit in t. The odd relationship is that between n and x. To start with, the deficit in n, in the context of this model, is *defined* by the deficit in x. If n was abnormal from the point of view of neuroanatomy it would still only count as a relevant deficit in the theory if it no longer performed the function x. Secondly, the only way in which there can be a specific deficit in x is if there is a deficit in n.[5] Because of this, we would not want to say that the neural deficit *causes* the deficit in EXPRAIS. A deficit in one is *equivalent* to a deficit in the other. The two claims are identical and so cannot have a causal relationship, at least in the sense used elsewhere.[6]

We have shown above that, on our interpretation of the Leslie theory, EXPRAIS, as a cognitive primitive, will have a simple equivalent in neural circuitry, and there will be a one-to-one mapping between the two. The neural assembly would be equivalent to EXPRAIS because it would refer to the identical element. A further and stronger step which could be made in the argument is to suppose that such cognitive primitives are invariably instantiated in the brain in exactly the same way. A particular cognitive primitive is always instantiated by a particular, identifiable neuron, type of synapse, or assembly of neurons, N. Further, each occurrence of activity in N can be taken as evidence for the operation of this cognitive primitive. We suspect that theories which require what might be loosely called "innate" structures might require strong cognitive primitives of this form.[7]

In the previous section, our starting point was a cognitive primitive which, at the moment, lacks a neural instantiation. Next, we take as our starting point a structure identified on the biological side. One candidate as a biological primitive (in the sense we are developing) is "mirror neurons" (Rizzolatti et al, 1996; Gallese et al, 1996). These are neurons in the prefrontal lobe of the monkey cortex which respond selectively both when the animal sees another animal pick up an object in a particular way and when the animal picks up an object itself in the same way. The observed actions which most commonly activate mirror neurons are grasping, placing, and manipulating. Other mirror neurons are even more specific, responding to the use of a particular grip, such as a precision grip or a power grip, as well as to the animal itself picking up an object using the same kind of grip.

It is important to understand what these neurons are not for. To start with, they cannot be interpreted in terms of the preparation for an impending movement since the neurons still fire when the monkey sees a specific movement while it is engaged in an action unrelated to the observed movement. Nor can the response be interpreted in terms of the anticipation of a reward, since there is no response from the neuron when the experimenter picks up a piece of food with a pair of pliers prior to the monkey picking it up, whereas if the experimenter picks the food up with his fingers, the target neuron does fire (Rizzolatti and Fadiga, 1998). According to Rizzolatti and Fadiga (1998), the most likely interpretation of the mirror neurons is that their discharge generates an internal representation of the observed action. Since there are a number of neurons which specialize in the same action, we can assume that, in respect of "generating an internal representation," these collections can be considered as functional units. What we propose is that these collections can be seen as *equivalent to* cognitive elements. To understand this, let us take two collections of these mirror neurons, one specializing in grasping, *MNg*, and one specializing in placing, *MNp*.[8]

Let us now consider a possible *cognitive* theory about the recognition of actions which could be seen as equivalent to the biological theory outlined above. This model, illustrated in figure 15.4, is based on the early logogen model (Morton, 1969). It considers that there are units, equivalent to logogens, which fire when there is enough evidence in the input to conclude that a grasping action, *G*, or a placing action, *P*, and so on, is occurring. Such firing would have the effect of "generating an internal representation" of that action, which would be interpreted by cognitive processes. In addition, such units could be stimulated by inputs from the cognitive processes (an internal action goal), as a result of which instructions would be sent to the effector systems to use the specified action on an object of current attention. Under this, or a similar theory, the unit

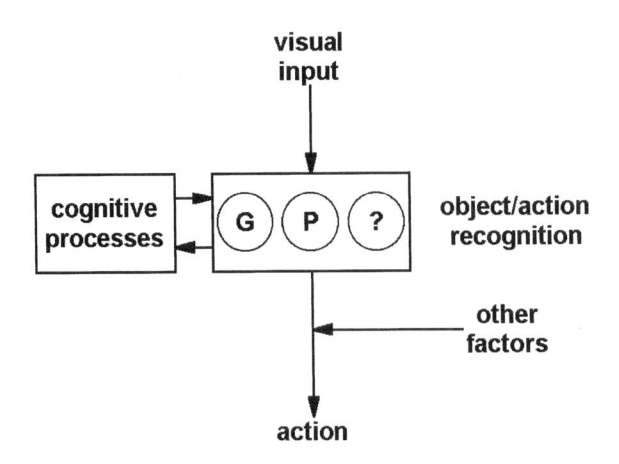

Figure 15.4
Cognitive model of grasping action meanings. Whether an action follows output of the units will depend on other factors.

labeled *G* would be equivalent to the neural collection *MNg*, and the unit labeled *P* would be equivalent to the neural collection *MNp*. In either a brain theory or a cognitive theory, the sets of terms could be used interchangeably without any change in meaning. That is, a statement about *MNg* would be identical to a statement about *G*, and vice versa. Equally, any deficit in *MNg* at the biological level would be equivalent to a deficit in *G* at the cognitive level.

To summarize, then, we are claiming that in some cases there will be equivalence between biological and cognitive descriptions of functional entities, and that the relation between the two is to be contrasted with the causal relationship which is the primary concern of causal modeling.[9]

Causal Influences from Cognition to Brain

In our previous writing on causal modeling, causal arrows have always gone from the biological or brain level to the cognitive level. This is because the developmental disorders that we wish to explain are known to have a biological origin before birth. There could be other disorders which may be best explained by postulating a *causal* influence from the cognitive to the biological level. We are unaware of any such disorders, but can illustrate the methodology in relation to acquired disorders.

Consider someone involved in a traumatic event such as a traffic accident. According to some theories (e.g., Metcalf and Jacobs, 1998), the fear (cognitively mediated) engendered by the event leads to the memory record being stored without the involvement of the hippocampus. The consequence of this is that the memory becomes intrusive and the person suffers from flashbacks in the condition known as post-traumatic stress disorder (PTSD). This sequence can be seen as a cognitive state having brain consequences which lead to further cognitive outcomes and is illustrated in figure 15.5.

A further example is in relation to pain. The traditional theory of pain, as proposed by Descartes, was that pain fibers relay information to a "pain center" in the brain. Activity in the pain center is then experienced as pain. This is the usual brain-to-cognition causal relationship. However, as Derbyshire (1997) pointed out, it has become clear that there is no

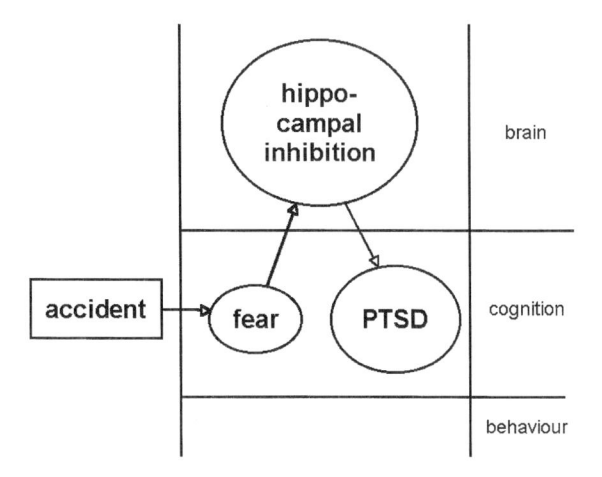

Figure 15.5
A model for PTSD in which a cognitive change leads to a change in brain state with further cognitive consequences.

direct relationship between the amount of stimulation of pain fibers and the experience of pain. Rather, with something like dental pain, psychological factors such as the circumstances of an operation, the patient's understanding of and attention to the trauma, and levels of experienced anxiety intervene to produce the final perception of pain. Even for the apparently simple case of dental pain, then, a model is required where the social context and psychological factors mediate at the biological level in the perception of pain. Derbyshire also summarized current work on the experience of pain by sufferers of rheumatoid arthritis. The theory specifies feedback from the psychological to the biological level, such that some aspects of a "negative coping strategy" can lead to an increase in inflammation.

Causal feedback from the cognitive to the brain level should be contrasted with a further kind of relation between the two, which philosophers term "token-token identity." This refers to the way in which cognitive elements (such as processes or beliefs) are embodied in the brain in cases other than those covered by the definition of equivalence. To illustrate, consider that any change at the cognitive level requires a change at the brain level. However, such changes are not usually part of a causal model. You change a belief from B to notB; there is a change in the brain.

However, the change is not systematic; the identity relation is not consistent because if the belief changes back to B there will be a further change in the brain—but it will not change back to where it was. Furthermore, for every individual, the brain state change which accompanies the shift from B to notB will be different. Thus, take the case that we come to believe something new: that this article has to be completed by the end of March, for example, rather than by the beginning of March. This welcome change in belief will, of course, be accompanied by some change in brain state. However, the causal consequences of the change of belief cannot be traced or predicted from the change in brain state. We have to consider the cognitive (meaning) representation in order to trace the causal consequences. Indeed, although the change in belief is identical for both of us and the causal consequences are similar, the accompanying changes in the brain are very unlikely to have any relationship. We have not been able to conceive of circumstances where such changes in brain states would enter into a causal model.

Why We Need Cognition

As Rugg (2000) has pointed out, without cognition, localization of function reverts to phrenology, with such human traits as "cautiousness," "conjugality," and "veneration" being assigned specific locations in the brain. The slight difference is that nowadays the labels correspond to redescription of patterns of data rather than terms drawn from intuitive theories of the mind. If a subject performs tasks in a scanner, how are we to interpret the patterns of activation? Our only realistic option is to do so in relation to the cognitive theory which drove the selection of tasks. The interpretation will only be as good as that cognitive theory.

In conclusion, let us add some history. *Cognition* has thrived in the years since 1972 and the reasons are clear to us. Initially it enabled people to develop the science of the mind without being crippled by the behaviorist empiricism of referees. Later, with the growth of cognitive neuropsychology, *Cognition* allowed biologists and those who study behavior to talk to one another. It is a role which suits *Cognition* and it is what cognition can uniquely accomplish.

Notes

1. The use of boxes is purely stylistic and anyone troubled by the possibility of surplus meaning in the convention can use ellipses or any other form or omit them completely.

2. While relevent genes are, to be sure, found in all parts of the body, their site of action in the example is in the brain. In other examples it might be more appropriate to label the first level as "biological" rather than "brain." This does not seem to change any of the properties of the notation.

3. Note that in theories of other deficits there may be more than one possible causal route specified. Thus, a child may be unable to perform phoneme deletion tasks either because of a deficit in phonological representation or because of a deficit in metarepresentational skills (Morton and Frith, 1993a,b). The notation must make it clear whether the presence of two causal arrows into a particular deficit is to be understood as a conjunction or a disjunction.

4. Note that you do not have to agree with the specific illustrative theory in order to appreciate the logic of the point being made. It could turn out that EXPRAIS was analyzable into two parts, Y and Z, each of which would either be a primitive or analyzable. And so on. Equally, if you don't like EXPRAIS, substitute your own theory—of equal specificity, of course. Then apply the same logic. Note that if you postulate a primitive that is the result of learning, then its absence could be caused through a problem in some learning mechanism.

5. There remains the tricky possibility that n is normal but disconnected from other neural assemblies which support the function x, for example, by transmitting the results of some computation. We will leave it to others to discuss whether, under these circumstances, we would want to talk about a deficit in x.

6. Barry Smith (personal communication, 2000) has suggested that the relationship between the deficit in n and the deficit in x is one where the one is a *causally necessary* condition of the other. Therefore, the lack of the first, or a deficit in the first, would be causally sufficient for a deficit in the latter.

7. It has been suggested to us that this strong relationship between a cognitive primitive and its neural equivalent can be characterized as what some philosophers call a "type-type" identity. However, Barry Smith (personal communication, 2000) has pointed out that type-type identity refers to such identities as the mean molecular energy and temperature or water and H_2O. Since it is clear that any well-defined cognitive function (computation) can be carried out in a variety of substrates, one cannot claim that there is a type-type identity with any one of those substrates, except (in the case of EXPRAIS) in the restricted context of the human brain where prosthetic devices are excluded.

8. Let us further note that while these collections of neurons could, in principle, play a role in imitation, their activation cannot be seen as being equivalent to imitation. The reason for this is that, while apes and humans imitate the actions of others, it seems that monkeys do not. According to Rizzolatti and Fadiga (1998),

"monkeys, although endowed of a mechanism that generates internal copies of actions made by others, are unable to use them for replicating those actions" (p. 91). We might hypothesize that the neural circuits involved in the imitation of action in the apes included collections of neurons such as *MNg* and *MNp*.

9. In considering the lack of constraints from biological facts onto cognitive theory, Mehler, Morton, and Jusczyk (1984) made an exception of cases of a one-to-one relationship between the levels—effectively what we mean by equivalence in this discussion.

References

American Psychiatric Association. (1987). *Diagnostic and Statistical Manual of Mental Disorders*, 3d ed., revised (DSM-III-R). Washington DC: American Psychiatric Association.

British Psychological Society (1999). *Dyslexia, literacy and psychological assessment: Report by a Working Party of the Division of Educational and Child Psychology*. Leicester, UK: British Psychological Society, [Commission on Professional and Hospital Activities. (1992). *International classification of diseases*, 10th ed. (ICD-10). Ann Arbor, MI: Commission on Professional and Hospital Activities.]

Derbyshire, S.W.G. (1997). Sources of variation in assessing male and female responses to pain. *New Ideas in Psychology, 15,* 83–95.

Frith, U., Morton, J., and Leslie, A. M. (1991). The cognitive basis of a biological disorder: Autism. *Trends in Neurosciences, 14,* 433–438.

Gallese, V. (in press). Actions, faces, objects and space: How to build a neurobiological account of the self. In *Proceedings of the Third Conference of the Association for the Scientific Study of Consciousness (ASSC)*.

Gallese, V., Fadiga, L., Fogassi, L., and Rizzolatti, G. (1996). Action recognition in the premotor cortex. *Brain, 119,* 593–609.

Leslie, A. M. (1987). Pretense and representation: The origins of "theory of mind." *Psychological Review, 94,* 412–426.

Lindgren, S. D., De Renzi, E., and Richman, L. C. (1985) Cross-national comparisons of developmental dyslexia in Italy and the United States. *Child Development, 56,* 1404–1417.

Mehler, J., Morton, J., and Jusczyk, P. W. (1984). On reducing language to biology. *Cognitive Neuropsychology, 1,* 83–116.

Metcalf, J., and Jacobs, W. J. (1998). Emotional memory: The effects of stress on "cool" and "hot" memory systems. *Psychology of Learning and Motivation, 38,* 187–222. Academic Press: NY.

Morton, J. (1969). Interaction of information in word recognition. *Psychological Review, 76,* 165–178.

Morton, J. (1986). Developmental contingency modelling. In Van Geert, P.L.C. (Ed.), *Theory building in developmental psychology* (pp. 141–165). Amsterdam: Elsevier North-Holland.

Morton, J., and Frith, U. (1993a). What lesson for dyslexia from Down's syndrome? Comments on Cossu, Rossini, and Marshall (1993). *Cognition, 48,* 289-296.

Morton, J., and Frith, U. (1993b). Approche de la dyslexie développementale par la modélisation causale. In J.-P. Jaffré (Ed.), *Les actes de la villette.* Paris: Nathan.

Morton, J., and Frith, U. (1995). Causal modelling: A structural approach to developmental psychopathology. In D. Cicchetti and D.J. Cohen, (Eds.), *Manual of developmental psychopathology* Vol. 1 (pp. 357–390). New York: Wiley.

Rizzolatti, J., and Fadiga, L. (1998). Grasping objects and grasping action meanings: The dual role of monkey rostroventral premotor cortex (area F5). In G. R. Bock and J. A. Goode (Eds.), *Novatis Foundation Symposium 218: Sensory guidance of movement* (pp. 81–103). Chichester, UK: Wiley.

Rizzolatti, G., Fadiga, L., Fogassi, L., and Gallese, V. (1996). Premotor cortex and the recognition of motor actions. *Cognition and Brain Research, 3,* 131–141.

Rugg, M. (2000). *Human memory: A perspective from cognitive neuroscience.* Inaugural professorial lecture, 8 March 2000. University College, London.

16

Counting in Animals and Humans

Rochel Gelman and Sara Cordes

On Phylogenetic Continuity and Possible Ontogenetic Implications

Throughout the century many anthropologists, psychologists, and historians of mathematics have assumed that there is a sharp discontinuity between the arithmetic abilities of animals, infants, young children, and early or "primitive" societies on the one hand, and older children and acculturated societies on the other hand. Again and again reports of birds, dogs, apes, bees, and so on keeping track of the number of items in small collections are ascribed to a "number sense faculty"[1] (e.g., see Dantzig, 1967; McCleish, 1991; Ifrah, 1985). Animals' seeming failure to work with larger set sizes was taken as evidence that they could not count. Indeed, the ability to count was considered too abstract a capacity for animals. "They never conceive absolute quantities because they lack the faculty of abstraction" (Ifrah, 1985, p. 4).

The contrast between a perceptual, non-quantitative mechanism for enumerating small sets of items (i.e., 1 to 5) and a "true" quantitative mechanism for representing larger numbers also is fundamental to ontogenetic and cultural discontinuity theories (e.g., see Baroody, 1992; Fuson, 1988; Fischer, 1992; Piaget, 1952; McCleish, 1991; Sophian, 1995). The Mehler and Bever (1967) demonstration that two-year-olds succeeded on a modified version of Piaget's famous number conservation task has been downplayed on the grounds that rather small set sizes (4–6) were used. A similar argument is used with respect to findings from Gelman's magic task, which was comprised of a two-phase procedure. During phase 1, expectancies for two Ns were established; phase 2 followed after a surreptitious transformation occurred (e.g., see Bullock and

Gelman, 1977; Gelman, 1972; Gelman and Gallistel, 1978). A combination of children's surprise levels, choices, and verbal explanations across the two phases of the experiment revealed early knowledge of a number-invariance scheme. These young children treated surreptitious changes in length, density, item kind, and color as irrelevant to the expected number(s). In contrast, they inferred that addition or subtraction had to have occurred when they encountered unexpected changes in number(s). Even 2½-year-olds used number-ordering relations, that is, they could pair the expectancy phase values of 1 vs. 2 with the unexpected values of 3 vs. 4. Even though the children in the various magic experiments often counted and explained their decisions, these results frequently are attributed to a perceptual number apprehension device, or "subitizing."

When subitizing is used as above, it typically is taken to mean that "twoness" and "threeness" are like "cowness" and "treeness," that is, that there are unique percepts for each set of small numbers. Another way of putting the claim is to predict that the reaction time function in the small-number range is flat; that is, it should take the same amount of time to identify the quality "twoness" as it does to identify "threeness" and "fourness." An alternative interpretation of subitizing allows that the items are individuated in the enumeration process. In fact, there are increases in reaction time as a function of set size, even within the small number range (Balakrishnan and Ashby, 1992; Folk, Egeth, and Kwak, 1988; Klahr and Wallace, 1975; Trick and Pylyshyn, 1993; see Dehaene, 1997; Gallistel and Gelman, 1992, for reviews).

Proposals that infants use object files (Carey, chapter 17; Simon, 1999) represent an alternative way to put the assumption that there is a limit on the number of items that can be processed at a given time, and that small sets of N are not processed in a truly numerical way. They are favored because they provide an account of how items are individuated within the small-number range. Since there is a limit to the number of object files that can be opened at one time (Trick and Pylyshyn, 1993), it is not surprising that authors who endorse the object-file account (e.g., see Carey, chapter 17) favor a discontinuity account of how small and larger numbers are processed.

Some discontinuity arguments are coupled with the idea that the ability to engage in numerical cognition depends on acculturation and the development of abstract reasoning structures. There are at least four reasons to challenge a strong phylogenetic discontinuity hypothesis. First, one need not assume that an abstract classification scheme mediates the ability to count. Second, there is an ever-growing body of findings that animals do count nonverbally, well beyond numbers in the range of 2 to 5 or 6. Third, there is evidence that adult humans share a nonverbal counting mechanism with animals, one that operates in both the small-number and large-number ranges (see below, as well as Cordes et al., submitted). Fourth, the set of principles governing the mechanism that generates nonverbal cardinal representations, or what Gelman and Gallistel (1978) dubbed "cardinal numerons," embeds an effective procedure for determining a successor.

What about the matter of ontogenetic continuity? There is evidence that adult humans can use the same nonverbal counting mechanism when working in either the small-number or large-number range. As we shall see, this is relevant to the interpretation of infant number data. Similarly, our findings that there are nonverbal generative processes for achieving representations of discrete (as opposed to continuous) quantities are relevant to an account of the differential ease with which humans come to learn verbal and other symbolic systems and why humans find it easier to re-represent natural (counting) numbers as opposed to rational and other kinds of real numbers.

On The Nature of Natural Numbers

There are extensive treatises on the nature of number and it is not our intent to provide an exhaustive review. Our goal is to bring to the fore three lines of theoretical reasoning about the psychological nature and generation of counting numbers. The first two are more "domain-general" than not. That is, they constitute efforts to build up the ability to generate and reason with natural numbers from non-numerical primitives. Of these two, one is rooted in the assumption that a given count or number represents an abstract class or set. The second, although not independent of the first, places considerable emphasis on the role of language. Neither of these accounts is applied to animal data. The third account is domain-specific and assumes that counting principles and its representations, in combination with arithmetic reasoning principles, constitute a foundational domain unto themselves. It is applied to accounts of nonverbal counting and arithmetic reasoning, be this in animals, preverbal children, or groups who use nonverbal counting systems.

The Classification and Set-Theoretic Approach

This view is very much related to traditional theories of concept acquisition and set-theoretic definitions of numbers. From the developmental perspective, the idea is that, at first, children cannot even classify like items with identical properties (e.g., Piaget, 1952; Vygotsky, 1962; Werner, 1948). Then, they move on to classifying together items that are perceptually identical, then items that differ in a property but share a shape, and so on, until items that differ considerably with respect to their surface properties are treated as members of an abstract class, "things." It is assumed that children have to develop classification structures in order to classify at this level of abstraction. In the case of numbers, the abstract capacity to count is related to the idea that a given number, say, 5, represents the set of all sets of that number, be these sets of five dogs, five ideas, five things in a room, or Lewis Carroll's miscellaneous collection of five kinds of things: ships, shoes, sealing wax, cabbages, and kings. Movement from the concrete to the abstract takes a very long time, even for man. Sir Bertrand Russell, who made fundamental contributions to the set-theoretical approach to number, put it thus: "It must have required many ages to discover that a brace of pheasants and a couple of days were both instances of the number two" (quoted in Dantzig, 1967, p. 6). Further discussion of the set-theoretical approach is presented in chapter 11 of Gelman and Gallistel (1978).

The Language Dependency Approach

There are at least two kinds of language-dependency arguments. One treats language as a necessary prerequisite. In this case there can be no such thing as preverbal or nonverbal counting. Further, understanding of the meaning of the count numbers, as well as their arithmetic use, emerges from inductions based on the use of counting words in the context of subitizing, and rote counting experiences. Fuson's (1988) treatment of verbal counting is a possible example of this account. So too is

McLeish's (1991), who wrote: "The reason for animals' inability to separate numbers from the concrete situation is that they are unable to think in the abstract at all—and even if they could, they have no language capable of communicating, or absorbing, such abstract ideas as 'six,' or a 'herd'" (p. 7).

A second kind of language-dependent argument allows that there is, or could be, a system for nonverbal counting and arithmetic that does not generate discrete values. However, this system is seen as both separate from and discontinuous with a verbal counting system. Carey (chapter 17) develops one form of such an account but there are others (e.g., Bloom, 2000; Butterworth, 1999; Dehaene, 1997). The verbal counting system usually is treated as being closely related to and emergent from the semantic-syntactic linguistic system of quantifiers. If the ability to count is indeed dependent on the development of language, then animals, who are nonlinguistic, and infants, who are prelinguistic, definitely cannot really count.

The Domain-Specific Approach

We favor a domain-specific approach to the count numbers and their arithmetic function and define a domain of knowledge in much the same way that formalists do, by appealing to the notion of a set of interrelated principles. A given set of principles, the rules of their application, and the entities to which they apply together constitute a domain. Since different structures are defined by different sets of principles, we can say that a body of knowledge constitutes a domain of knowledge to the extent that we can show that a set of interrelated principles organize the entities and related knowledge, as well as the rules of operation on these. Counting is part of a number-specific domain, because the representatives of numerosity (what we call numerons) generated by counting are operated on by mechanisms informed by, or obedient to, arithmetic principles. For counting to provide the input for arithmetic reasoning, the principles governing counting must complement the principles governing arithmetic reasoning. For example, the counting principles must be such that sets assigned the same numeron are in fact numerically equal and a set assigned a greater numeron is more numerous. The counting principles and

COUNTING PRINCIPLES
- HOW TO'S
1. One-one
2. Stable order
3. Cardinal
- PERMISSIONS
4. Order Irrelevance
5. Item Kind Irrelevance

Numerons that stand for the cardinal value (N) of counted collection of discrete entities.
If numerals mapped to counting list, **cardinal count word** re-represents cardinal quantity

ARITHMETIC REASONING PRINCIPLES
- ADDITION & SUBTRACTION

A/S applied to cardinal values
If A/S, relevant operation and N changes, cardinal value > or <
IF N-irrelevant operation, N is same (=) cardinal value

Figure 16.1
The counting principles and their relation to the arithmetic operations of addition, subtraction, and ordering. From Gelman and Gallistel (1978).

their relation to the arithmetic operations of addition, subtraction, and ordering are illustrated in figure 16.1.

Note that there is nothing in this formulation that yokes counting to the use of number words or other numerical symbols. Indeed, this is why we introduced terms to distinguish between nonverbal and verbal or symbolic representations: numerons and numerlogs (Gelman & Gallistel, 1978). As long as the generative process honors the one-one, stable ordering, and cardinal principles, and tolerates a wide range of entities, we can say that it is a candidate counting device. A pulse generator and a computer are examples of such devices.

Nor is it necessary to assume that a complex hierarchization scheme mediates the ability to come to understand that one can count a heterogeneous collection if one simply thinks of them as "things." Under these circumstances, the primary requirement is thingness or individuation of the members of the to-be-counted set. That is, the only requirement is that each thing can be treated as separable one from each other, either physically or mentally. Often it is the case that things being counted are separably movable items; processes that serve figure-ground abilities can help us sort things from non-things. Similarly, perceptual tendencies to separate auditory events (tones or speech units) can help one keep track of already-counted vs. to-be-counted inputs. Of course, there is much work to do on the question of what we will tolerate as a countable part, whether young children will count the Lewis Carroll set as five, and so on (Shipley & Shepperson, 1990). But the research agenda as to what counts as a separate countable is different from one that ties counting prowess to the development of the ability to use hierarchical classification systems.

In the next section we present our model of this counting device, namely, that it is just the machinery that generates the real numbers that represent countable, ordered, quantities, that is, the cardinal numerons (Gallistel and Gelman, 1992; Gallistel, Gelman, and Cordes, in press).

On Nonverbal Counting and the Accumulator Model

We favor a direct descendant of the Meck and Church (1983) accumulator model for animal counting data. As shown in figure 16.2, the model uses a discrete process to generate ordered quantitative representations for both animals and humans (Gallistel and Gelman, 1992; Whalen, Gallistel, and Gelman, 1999). In addition, it is assumed that humans build bidirectional mappings between the quantities generated by the counting processes and the various count words and symbols that stand for a given cardinal value. To expand on the nature of the model, we begin by reviewing some animal counting data. Then we turn to evidence that humans share the nonverbal counting mechanism. To end, we return to consider the idea that there are small-number mechanisms that are independent of counting, be these perceptual or object-file.

Evidence that Animals Count Nonverbally

As it turns out, there is a set of experiments showing a variety of animals possess some surprising numerical abilities. This includes the ability to reason about and count rather large set sizes. Of course, whatever the process by which animals do count, it surely is a nonverbal one. Brannon and Terrace (1998; 2000) showed that monkeys can order stimuli that display as many as nine items. Once trained to arrange sets of one, two, three, and four items in ascending order, they spontaneously generalized

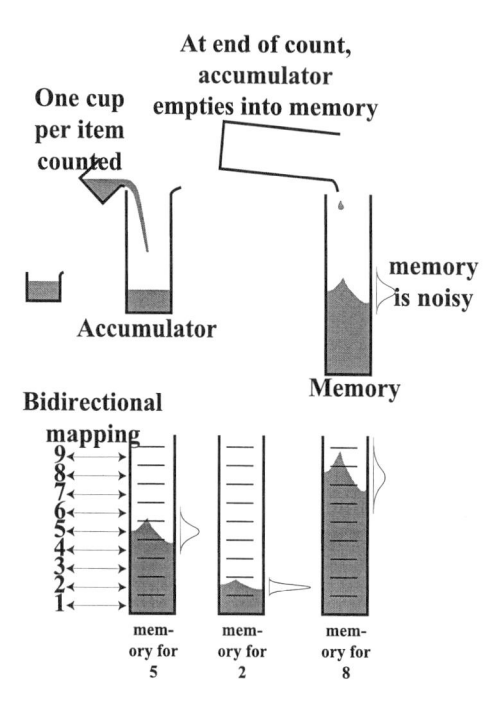

Figure 16.2
The accumulator model. A magnitude representing a numerosity is formed through accumulation of "cupfuls" of activation, whereby each "cup" represents one item, producing a total magnitude. Note that each discrete "next pour" in the process picks out the next magnitude. Accumulated magnitudes may be compared to a magnitude generated from memory or mapped to verbal or symbolic numerals for quantities. However, magnitudes read from memory have inherent scalar variability and so different retrievals will produce different magnitudes according to its probability density function. The greater the magnitude, the more likely an error.

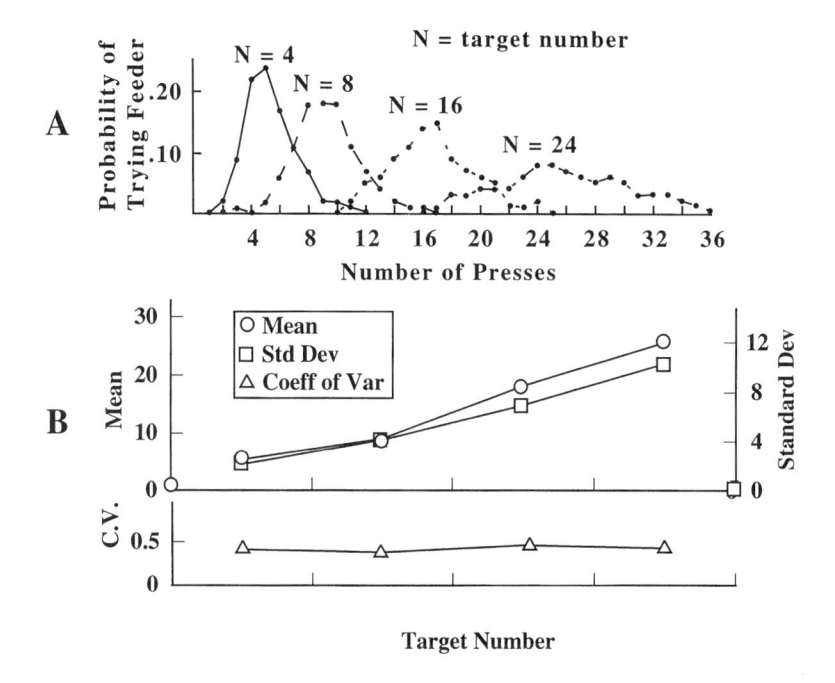

Figure 16.3
(A) The distribution in the number of presses rats make when attempting to approximate a given target number N (N = 4, 8, 16, 24). (B) Replot of the data to illustrate the signature of scalar variability. From Platt and Johnson (1971).

to novel sets of items containing five to nine items. Platt and Johnson's (1971) study of the ability of rats to count is an especially nice example of nonverbal counting in animals. Their data are shown in figure 16.3.

Platt and Johnson's rats had to press a lever for an experimenter-determined number of times in order to arm a feeder. The target values of N varied from 4 to 24. If a rat pressed too few times, the counter that was keeping track of the number of presses was reset, meaning that the rat had to start his count anew. The rats' responses were systematically controlled by the target value of the requisite number of presses in a given block of trials. As the target value increased for a block of trials, so did the modal number of presses. In addition, set size increases led to a systematic effect on the animal's accuracy, that is, the width of the distribution was proportional to the mean number of responses for a given experimental N. As a result, the outcome of dividing the standard deviation by the mean was a flat function. That is, the data from the experiment can be said to exhibit scalar variability. This characterization of the rat's counting behavior is common to a number of other studies with animals and is well modeled by the nonverbal counting mechanism shown in figure 16.2.

In our model of nonverbal counting (Gelman and Gallistel, 1978; Gallistel and Gelman, 2000; Gallistel, Gelman, and Cordes, in press) cardinal numerons are represented by magnitudes generated by an accumulator process. As shown in figure 16.2, each unit enumerated is represented by an additional increment in the accumulator, analogous to pouring a fixed amount of liquid into a beaker. The cardinal value of the counted set of items or events is represented by the final magnitude in the accumulator (analogous to the total amount of liquid in the beaker). As the magnitude in memory increases, there is a proportional increase in the variability of the magnitudes read from that memory—in the trial-to-trial variability of the magnitudes actually remembered. The proportional variability in the magnitudes read from memory leads to scalar variability in the behavioral data, because the processes that translate remembered numerical magnitudes into numerically governed behavior make ratio comparisons between remembered (target) and current estimates of numerosity. These and other findings imply that a system of simple arithmetic reasoning (addition, subtraction, ordering) mediates the translation of numerical magnitude into observable behavior (see Gallistel, Gelman, and Cordes, in press, for evidence that the reasoning processes also include multiplication and division).

Although the product of the accumulator processes is a magnitude, its processes are discrete and embody the counting principles. That is, the counting processes are governed by the how-to counting principles, these being (1) the one-one principle: there is one and only one discrete pour (or, if you prefer, pulse) for every item, and as many pours as there are items to count; (2) the ordering principle: the process is a sequential one, generating one and only one quantity at a time; and (3) the cardinal principle: the last magnitude generated by this discrete process has a special status—it is written to memory and represents the cardinal value of the counted set. Finally, the memory is variable—this is indicated by the sloshing line in the figure. Since it is a discrete process that renders successive quantities, there is a perfectly clear sense in which one gets the next magnitude. Given that there is a next relation that holds among the quantities, it follows that these represent a discrete quantity, that is, a countable quantity. This sets the stage for our idea that humans come to map nonverbal quantity representations generated by counting to verbal or other symbolic representations of the natural numbers (see the bottom of figure 16.2). Several lines of evidence are consistent with this proposal. In what follows we focus on the evidence that adults use a nonverbal counting mechanism that shares the key properties of the animal one presented above. The final section of the chapter returns to the mapping proposal.

Adults Also Count Nonverbally

It has been known, at least since the work of Moyer and Landauer (1967), that the mean reaction latency for adult human judgments of the numerical order of numerals increases with relative numerical proximity: the smaller the percent difference, the longer the mean latency (Dehaene, Dupoux, and Mehler, 1990; Holyoak, 1978; Holyoak and Mah, 1982; Moyer and Landauer, 1973). The increase in reaction latency suggests that adult human judgments of numerical order are based on an underlying magnitude representation of numerosity. Whalen et al. (1999) have presented evidence that adult humans can use a nonverbal counting mechanism to generate magnitudes. In a key press task, adults were first shown an Arabic numeral and asked to rapidly press a key, without counting, until they felt they had arrived at the represented number. The number of key presses increased linearly with target number and so did the trial-to-trial variability in the number of presses for a given target number. On a different task, subjects saw a rapidly, but arhythmically, flashing dot and indicated how many flashes they felt they had seen, without counting. Similar results to the key press task were obtained, providing further support for nonverbal numerical competence in adult humans similar to the one observed in nonverbal animals.

More recently, Cordes et al. (in press) asked adult subjects to talk while doing the nonverbal counting task. Depending on the condition, subjects either recited "Mary had a little lamb" or the word "the" over and over again, while pressing a key as fast as they could as many times to represent the value of the target Arabic numeral. We reasoned that these talk-aloud conditions would be an effective way of preventing verbal counting, subvocal or otherwise, because you have to say something just at the moment when you would say the count word if you were counting verbally.

Again, we obtained the signature pattern of scalar variability. Together, the above results support both of our proposals about human counting: we have available a nonverbal counting mechanism and engage in a bidirectional mapping process. The latter allows them to map either from a given numeric symbol to the nonverbal counting of taps or the nonverbal counting of flashes to numerical symbols.

At first blush, the idea that adults use any nonverbal counting mechanisms to deal with any arithmetic tasks, let alone one that has scalar variance, might seem decidedly odd. After all, the great advantage of the language of mathematics is that it makes things precise and exact. But ponder the fact that it takes longer for people to judge that $99 > 98$ than it does to judge that $4 > 3$, even though the differences are exactly equal. Similarly, it takes people longer to judge that $9 > 8$ than it does to judge that $9 > 4$. These size and distance effects are further evidence that adult humans can and do use a nonverbal counting mechanism that exhibits scalar variability. Still, the fact that we also know that, mathematically, $(99 - 98) = (4 - 3)$ makes it obvious that there is more than one way we can represent, define and reason about numbers (see also Dehaene et al., 1999; Droz, 1992).

Discontinuity Between Nonverbal and Verbal Counting?

As indicated at the start of this chapter, the notion of a non-numerical mechanism, be it a perceptual apprehension or an object-file one, is contrasted with the ability to count and reason about abstract representations of number and quantity. The distinction is grounded in one or both of two different kinds of accounts. First, there is the presumption that there is either a non-numerical, or at best, a quasi-numerical mechanism for processing displays that contain small set sizes, about three to five. This proposal is a common feature of histories of mathematics or the anthropology of "early man." It also has support from a number of developmental and cognitive psychologists. Many who study the abilities of infants, young children, and adults to discriminate between different set sizes also favor the hypothesis that a non-numerical mechanism accounts for the limited ability of infants to discriminate between two and three but not larger set sizes, young children's difficulties counting verbally with set sizes that go beyond three or four, and the fact that children and adults respond more rapidly to sets in the small number range as opposed to ones larger than six or seven. Whatever the details of these accounts, they all share a discontinuity assumption. This is that the nonverbal representation of the small numbers is fundamentally different from the nonverbal representation of the larger numbers.

If small numerosities were represented by something like sets of tally marks, say, object files, and larger numerosities by magnitudes (Carey, chapter 17; Uller et al., 1999), then these two modes of representation would seem to be like oil and water. How does one subtract a set of two tally marks from a magnitude representing, say, numerosity 8? By what mechanism does the brain carry out binary combinatorial operations that span the two ranges? If small numerosities were represented by something like a perceptual classification process, say, where twoness and threeness are like cowness and treeness and numerosities greater than four were represented by quantities, the combinatorial still exists. Indeed, it gets worse since there is nothing we know about perceptual classification processes that leads to an ordering rule such that cowness somehow represents more than treeness. Nor is it clear how one would perform operations of addition and subtraction. Although it is possible to order small sets of tally marks and even add or subtract them, there is the problem that the operation of addition would not be closed, that is, the addition of 3 and 3 would be impossible because the value 6 is not included in the small-number system. Further, unto themselves, object files do not generate a "next" entry. For this to happen, at least the operation of addition has to be added to the number-processing repertoire. We do not mean to deny that infants use object files to individuate items. We do deny that, on their own, they lack intrinsic quantitative properties (Simon, 1999).

Returning to figure 16.3, we draw attention to two characteristics of the data pattern. First, the animals made errors on some trials in the small number range. Second, there is no indication that there is a change in the statistical functions describing performance across the full range of the numbers used in the experiment. That is, there is no clear reason to hold that performance on the small N of 4 was governed by a different process than performance on the larger set size values. This consideration led Cordes et al. (in press) to include numerals representing values in the small, "number sense" range (2, 3, 4, and 5) as well as ones assumed to be in the nonverbal counting range (8, 13, 20, and 32). Each of four subjects were run twenty trials on every one of the set sizes. As already indicated above, these "talking" while nonverbal counting data are characterized by scalar variability. This means that all subjects made mistakes, even when given the target values of 2, 3, and 4. Note that the shapes of the relevant distributions shown in figure 16.4 are very much like those in figure 16.3, the latter being from data collected with rats. Moreover, as reported in Cordes et al., the coefficients of variation for targets 2, 3, 4, and 5 do not differ from those beyond that range. So, the variability in mapping the nonverbal counting system to a graphemically specified target number is scalar all the way down.

The probability that adults in the Cordes et al. study made an error when working with small Ns was small. Had we not used a large number of trials, we might have missed the systematic tendency to err even in this range. This does not mitigate the fact that the function describing the relationship between number of responses and errors is the same in the small-number and large-number range. Given that animals and adults err when generating a nonverbal representation of small values of number, there is a straightforward design implication for studies with young children that uses a variant of a discrimination paradigm. Given that the underlying variance of the memory for a given quantity representation increases systematically, as does set size, the probability of X number of children succeeding decreases as a function of N. Put differently, the probability of making a memory error when comparing the values in question will increase as does N. As a result, the fact that a given N of infants pass discrimination or habituation tasks with small, but not with somewhat larger, numbers could well be due to a statistical artifact as opposed to a common underlying competence mechanism that deals both with Ns outside and inside of the "subitizing range." In order to protect

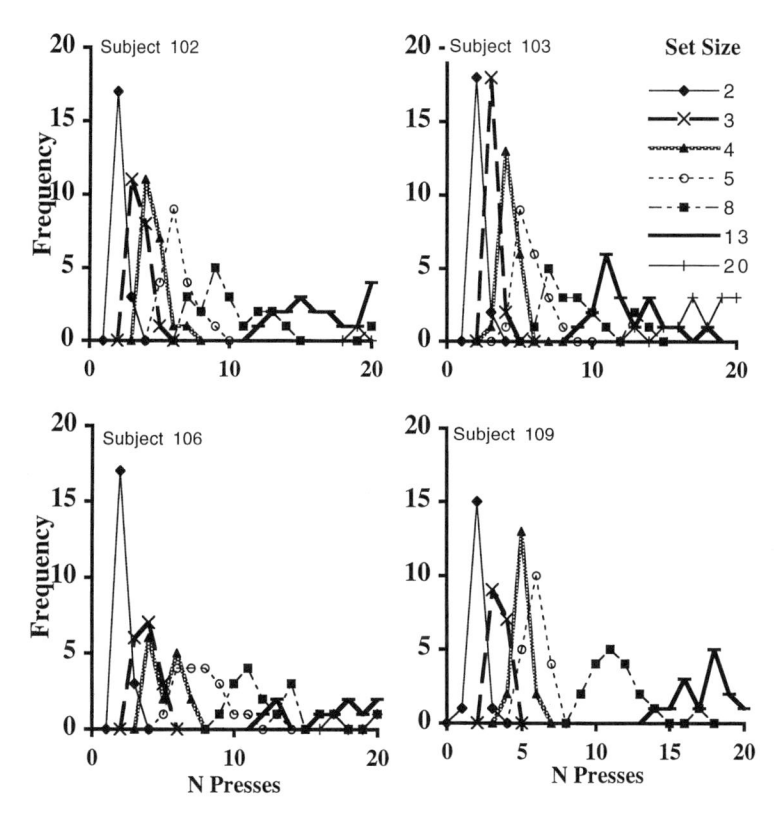

Figure 16.4
The distributions in the number of presses adult humans make when attempting to rapidly press a key a given target number of times ($N = 2, 3, 4, 5, 8, 13, 20, 32$), without counting, while stating "the" coincident with every key press. Data are from the four subjects in Cordes et al. (submitted) who were run on the set sizes of 2 to 5.)

against this possibility, it is important to find ways to run repeated trials with infants and young children. Since this is not an easy task to accomplish, investigators should at least increase the number of participants. Studies that incorporate these design features are needed in order to choose between a position like ours, which assumes there is a continuity between the ability of animals and humans to engage in nonverbal counting, and ones that favor a variant of a discontinuity hypothesis.

The Cordes et al. results bear on another issue raised by Carey (chapter 17). She states that "analog magnitude representational systems do not have the power to represent natural number" (p. x). This is not so. The natural numbers are a special subset of the real numbers. Since our model provides an effective procedure for determining a nonverbal successor, it encompasses a foundational feature of the natural numbers. This issue illustrates the importance of the question of how the implicit becomes explicit and whether or not the implicit system underlies learning about the re-representational tools, be these linguistic or alternative symbolic systems.

On Learning About Verbal and Other Symbolic Re-Representations of Number

The availability of a nonverbal arithmetic-counting structure facilitates the verbal counting system. This is because learning about novel data is aided whenever it can take advantage of an existing mental structure. The learner is more likely to select as relevant those pieces of cultural data that cohere. When the use-rules organizing the sequential sounds of a culture's particular count list honor the counting principles, then they can be mapped to the available, implicit list. In this case, young learners have a way to identify inputs that can be mapped to the counting principles and then the principles can begin to render the lists numerically meaningful. Preverbal counting principles provide a conceptual framework for helping beginning language learners identify and render intelligible the individual tags that are part of a list that is initially meaningless. They also provide a structure within which they build an understanding of the words (Gelman, 1993).

Of course, it is one thing to master a culture's language rules that makes it possible to generate successive next numerals forever. The issues of early variability are complex (Gelman and Greeno, 1989; Gelman, 1993). Here we focus on ones related to the topic of continuity and discontinuity. First, even adults, unlike computers, are less than enthusiastic about memorizing long lists of sounds that are not intrinsically organized. There is nothing about the sound "two" that predicts the next sound will be "three," or the sound "three" that predicts the next sound will be "four" and so on. This is a nonmathematical, information-processing reason to expect the task of committing even the first nine count words to memory to take a considerable amount of time, and it does—as much as two to four years (Miller et al., 1995). Since the nonverbal counting process allows for the generation of a successor for each representation, there is yet a further serial learning problem the child has to confront. This is the mastery of the structure of the base rules embedded in a particular language's counting system.

The English count list lacks a transparent decade rule for at least the first forty entries and, probably, first 130 count words. Learning this rule ends up taking a surprisingly long time. Hartnett (1991) found that a number of kindergarten and first grade students had yet to catch on to the procedure for generating the count words in the hundreds, most probably because many English-speaking children think *one hundred* is the next decade word after *ninety*. If so, in order for them to induce the 100s rules in English, they need to encounter relevant examples. Cross-cultural findings are consistent with this argument. Chinese has a more transparent base-10 count rule for generating subsequent count words, even for the tens and decades. Although the rate at which English- and Chinese-speaking children learn the first nine count words is comparable, Chinese-speaking children learn the subsequent entries at a much faster rate (Miller et al., 1995; Miller and Stigler, 1988). Cultural variables also influence the extent to which limited count lists are used. One of us (R. G.) recalls a conversation with a male member of the Ethiopian Falashal community who emigrated to Israel.[2] He was readily able to provide the count words in Aramaic, his native language, well into the thousands. However, when we asked him to name the word for one million, he used the Hebrew word. We pointed this out and he responded, "There is nothing we counted to one million when I lived there." Other examples of the rapid assimilation of well-formed generative count lists as soon as a culture encounters them are provided in Crump (1990), Gvozdanoviäc (1999), and Zaslavsky (1973).

A culture's rapid uptake of a count list that is better than one already in use is especially interesting. For us, these are examples of minds in search of the kind of list that best serves counting principles. It is as if the minds of people in these cultures are already prepared to adopt counting tools that will allow them to generate verbal counting successors. A related phenomenon was reported by Hartnett and Gelman (1998). They engaged children in kindergarten as well as grades 1 and 2 in a thought experiment designed to encourage them to reach an explicit induction of the successor principle. To do this they asked children to ponder a larger number, then add 1 to that, then again add 1, and so on. Periodically, a child was asked if the number he or she was thinking about was the largest there could be or whether there were more numbers. Then, depending on the child's answer, they were asked what the biggest number was and whether they could add to it or what was the next. A very large percentage of the children caught on to the successor principle; indeed, they went around telling their friends about their new knowledge.

The majority of children in the Hartnett and Gelman (1998) study listed numbers at least as great as 100 when first asked what they thought was a big number. Some, however, said numbers well below 100, for example, 29. These children resisted the idea that there is always a larger or next number. Still, it would be premature to conclude that these younger children lacked knowledge of the successor principle. They did know that there would be yet another, and another, and another, and . . . another dot—given a dot-generating machine and an endless supply of paper. It therefore is more likely that one needs to know the rules for making up the next verbal number. Indeed, one child said, "You can't do that, unless someone makes up a new count word." These are all considerations about the kinds of explicit knowledge that develop, and not whether the successor principle is implicitly available. This highlights why it is critical to keep separate the processes which generate the representation and the nature of that representation.

Differences in the structure of the linguistic or symbolic entities that notate different kinds of numbers are related to the question of whether learning about them is facilitated by existing nonverbal mathematical representations. Knowledge of the counting principles can facilitate learning how to interpret the words and marks for the integer. They are discrete and are used in a way that maps directly to the counting principles. Most important, they map to the *unique* mean values that the nonverbal system generates with the counting processs. Note that there is no corresponding word for the variance associated with the mean. Hence, the verbal instantiation of the counting procedure highlights what are the mathematical features of the system: it generates discrete, successively ordered values.

This is part of our account of how it can be that people can interpret a count word at two different levels: (1) as a map to its nonverbal representation and hence as a quantity; (2) as a mathematical entity that has a precise mathematical meaning.

Although count words map in a straightforward way to discrete nonverbal entities, that is, the means of the distributions generated by application of the nonverbal counting process, the same is not true for rational numbers or any other real numbers. Worse yet, the mathematical principles underlying the numberhood of fractions and decimals are *not* consistent with the verbally instantiated principles of counting. In fact, the verbal-notational system that children master for counting numbers is inconsistent with that for fractions and decimals. One cannot count things to generate a fraction. Formally, a fraction is defined as the division of one cardinal number by another; this definition solves the problem that there is a lack of closure of the integers under division. To complicate matters, some counting number principles do not apply to fractions. Rational numbers do not have unique successors; there is an infinite number of numbers between any two rational numbers. So, one cannot use counting-based algorithms for ordering fractions; for example, 1/4 is *not* more than 1/2. Neither the nonverbal nor the verbal counting principles map to a tripartite symbolic representation of fractions—two cardinal numbers X and Y separated by a line—whereas the formal definition of a fraction does.

Hartnett and Gelman (1998) reasoned that if children bring to their early school mathematics lessons the idea that the language of mathematics is grounded in counting principles and related rules of addition and subtraction, their constructivist tendencies could lead them to distort fraction inputs to fit their counting-based number theory. These authors reasoned that, therefore, early verbal knowledge of numbers might serve as a barrier to learning about the meaning and notational system for fractions. In contrast, they expected that children of the same age would have a relatively easy time engaging in a thought experiment designed to encourage the explicit induction of the successor principle, here because the structure of the to-be-learned inputs mapped to their existing arithmetic-counting structure. These predictions were supported. The children in the Hartnett and Gelman (1998) study were in kindergarten through grade

3, rather young in school age. However, even college students persist in treating rational numbers as if they were odd variants of natural numbers. For example, students in a chemistry class at UCLA said they could not graph a particular problem "because their graph paper did not have enough squares."

The Hartnett and Gelman findings highlight the need to pay attention to the extent to which the nonverbal representations of numbers map readily to the re-represented verbal and notational systems of number. The absence of an easy map between the nonverbal representations of rational and other real numbers and verbal or symbolic representations of these numbers is a serious impediment to learning the cultural creations for representing such numbers.

Acknowledgment

Work for this paper was partially supported by NSF grants DFS-9209741 and SRB-97209741 to R.G. We thank Emmanuel Dupoux, C. R. Gallistel, Beth Levin, Dana Thadani, and Osnat Zur for help comments on earlier versions.

Notes

1. The meaning of the phrase "number sense faculty" is used here as it was by the cited writers. In modern psychological writings, its meaning converges with the use of "subitizing" when it is meant to refer to a non-numerical perceptual processing mechanism. It should not be confused with Dehaene's (1997) book title *Number Sense*, which is about a wide range of numerical cognitions, their origins, and possible brain localizations.

2. R.G. is grateful to Professor Iris Levin of Tel Aviv University, her husband, and father for arranging and joining her at the interview.

References

Balakrishnan, J. D., and Ashby, F. G. (1992). Subitizing: Magical numbers of mere supersition? *Psychological Research, 54,* 80–90.

Baroody, A. J. (1992). The development of preschoolers' counting skills and principles. In J. Bideau, C. Mejac, and J-P. Fischer (Eds.), *Pathways to number: Children's developing numerical abilities* (pp. 99–126). Hillsdale, NJ: Erlbaum.

Bloom, P. (2000). *How children learn the meanings of words.* Cambridge, MA: MIT Press.

Brannon, E. M., and Terrace, H. S. (1998). Ordering of the numerosities 1 to 9 by monkeys. *Science, 282,* 246–249.

Brannon, E. M., and Terrace, H. S. (2000). Representation of the numerosities 1–9 by Rhesus macaques. *Journal of Experimental Psychology: Animal Behavior Processes, 26,* 31–49.

Bullock, M., and Gelman, R. (1977). Numerical reasoning in young children: The ordering principle. *Child Development, 48,* 427–434.

Butterworth, B. (1999). *The mathematical brain.* London: Macmillan.

Cordes, S., Gelman, R., Gallistel, C. R., and Whalen, J. (in press). Variability signatures distinguish verbal from nonverbal counting for both small and large numbers. *Psychonomic Bulletin and Review.*

Crump, T. (1990). *The anthropology of numbers.* Cambridge, UK: Cambridge Univ. Press.

Dantzig, T. (1967). *Number, the language of science.* New York: Free Press.

Dehaene, S., Dupoux, E., and Mehler, J. (1990). Is numerical comparison digital? Analogical and symbolic effects in two-digit number comparison. *Journal of Experimental Psychology: Human Perception and Performance, 16,* 626–641.

Dehaene, S. (1997). *The number sense.* Oxford: Oxford University Press.

Dehaene, S., Spelke, E. S., Pinel, P., Stanescu, R., and Tsivkin, S. (1999). Sources of mathematical thinking: Behavior and brain-imaging evidence. *Science, 284,* 970–974.

Droz, R. (1992). The multiple roots of natural numbers and their multiple interpretations. In J. Bideau, C. Mejac, and J-P. Fischer (Eds.), *Pathways to number: Children's developing numerical abilities* (pp. 229–244). Hillsdale, NJ: Erlbaum.

Fischer, J. P. (1992). Subitizing: The discontinuity after three. In J. Bideau, C. Mejac, and J-P. Fischer (Eds.). *Pathways to number: Children's developing numerical abilities* (pp. 191–208). Hillsdale, NJ: Erlbaum.

Folk, C. L., Egeth, H. E., and Kwak, H. (1988). Subitizing: Direct apprehension or serial processing? *Perception and Psychophysics, 44,* 313–320.

Fuson, K. C. (1988). *Children's counting and concepts of number.* New York: Springer-Verlag.

Gallistel, C. R., and Gelman, R. (1992). Preverbal and verbal counting and computation. *Cognition, 44,* 43–74.

Gallistel, C. R., and Gelman, R. (2000). Nonverbal numerical cognition: From reals to integers. *Trends in Cognitive Science, 4,* 59–65.

Gallistel, C. R., Gelman, R., and Cordes, S. (in press). The cultural and evolutionary history of the real numbers. In S. Levinson and P. Jaisson (Eds.), *Culture and Evolution.* Oxford: Oxford University Press.

Gelman, R. (1972). Logical capacity of very young children: Number invariance rules. *Child Development, 43*, 75–90.

Gelman, R. (1993). A rational-constructivist account of early learning about numbers and objects. In D. Medin (Ed.). *Learning and motivation*, Vol. 30 (pp. 61–96). New York: Academic Press.

Gelman, R., and Gallistel, C. R. (1978). *The child's understanding of number.* Cambridge, MA: Harvard University Press.

Gelman, R., and Greeno, J. G. (1989). On the nature of competence: Principles for understanding in a domain. In L. B. Resnick (Ed.), *Knowing and learning: Essays in honor of Robert Glaser* (pp. 125–186). Hillsdale, NJ: Erlbaum.

Gelman, R., and Meck, E. (1983). Preschoolers' counting: Principles before skill. *Cognition, 13*, 343–359.

Gvozdanoviäć, J. (1999). *Numeral types and changes worldwide.* Berlin: Mouton de Gruyter.

Hartnett, P. M. (1991). *The development of mathematical insight: From one, two, three to infinity.* Ph.D., University of Pennsylvania.

Hartnett, P. M., and Gelman, R. (1998). Early understandings of number: Paths or barriers to the construction of new understandings? *Learning and Instruction: The Journal of the European Association for Research in Learning and Instruction, 8*, 341–374.

Holyoak, K. J. (1978). Comparative judgments with numerical reference points. *Cognitive Psychology, 10*, 203–243.

Holyoak, K. J., and Mah, W. A. (1982). Cognitive reference points in judgments of symbolic magnitudes. *Cognitive Pscholology, 14*, 328–352.

Ifrah, G. (1985). *From one to zero.* New York: Viking.

Klahr, D., and Wallace, J. G. (1995). The role of quantification operators in the development of conservation of quantity. *Cognitive Psychology, 4*, 301–327.

McCleish, J. (1991). *Number; The history of numbers and how they shape our lives.* New York: Fawcett Columbine.

Meck, W. H., and Church, R. M. (1983). A mode control model of counting and timing processes. *Journal of Experimental Psychology: Animal Behavior Processes, 9*, 320–324.

Mehler, J., & Bever, T. G. (1967). Cognitive capacity of very young children. *Science, 158*, 141–142.

Miller, K. F., Smith, C. M., Zhu, J., and Zhang, H. (1995). Preschool origins of cross-national differences in mathematical competence: The role of number-naming systems. *Psychological Science, 6*, 56–60.

Miller, K. F., and Stigler, J. W. (1987). Counting in Chinese: Cultural variation in a basic cognitive skill. *Cognitive Development, 2*, 279–305.

Moyer, R. S., and Landauer, T. K. (1967). Time required for judgments of numerical inequality. *Nature, 215*, 1519–1520.

Moyer, R. S., and Landauer, T. K. (1973). Determinants of reaction time for digit inequality judgments. *Bulletin of the Psychonomics Society, 1,* 167–168.

Piaget, J. (1952). *The child's understanding of number.* New York: Norton.

Platt, J., and Johnson, D. (1971). Localization of position within a homogeneous behavior chain: Effects of error contingencies. *Learning and Motivation, 2,* 386–414.

Simon, T. (1999). Numerical thinking in a brain without numbers? *Trends in Cognitive Science, 3,* 363–364.

Sophian, C. (1995). *Children's numbers.* Madison, WI: Brown and Benchmark.

Trick, L. M., and Pylyshyn, Z. (1993). What enumeration studies can show us about spatial attention: Evidence for limited capacity preattentive processes. *Journal of Experimental Psychology: Human Perception and Performance, 19,* 331–351.

Uller, C., Carey, S., Huntley-Fenner, G., & Klatt, L. (1999). What representations might underlie infant numerical knowledge? *Cognitive Development, 13,* 1–43.

Vygotsky, L. S. (1962). *Thought and language.* Cambridge, MA: MIT Press.

Werner, H. (1948). *Comparative psychology of mental development.* Chicago: Follett.

Whalen, J., Gallistel, C. R., and Gelman, R. (1999). Non-verbal counting in humans: The psychophysics of number representation. *Psychological Science, 10,* 130–137.

Zaslavsky, C. (1973). *Africa counts.* Boston, Prindle, Wever & Schmidt.

17

On the Very Possibility of Discontinuities in Conceptual Development

Susan Carey

Continuity vs. Discontinuity in Conceptual Development

Jacques Mehler is one of many cognitive scientists who deny the very possibility of conceptual discontinuities in development (see also Fodor, 1975; Pinker, 1984; Macnamara, 1986), endorsing instead a strong version of the *continuity thesis*. The continuity thesis states that all the representational structures and inferential capacities that underlie adult belief systems either are present throughout development or arise through processes such as maturation or learning mechanisms that involve selecting among or concatenating existing representations.

Mehler's initial foray into the field of cognitive development was his celebrated exchange with Piaget in *Science* (Mehler and Bever, 1967, 1969). The conceptual domain in question was number representation: Mehler and Bever showed that 2½-year-olds succeeded on tasks that required the representation of numerical quantity, the operations of addition, and the capacity to compare two sets on the basis of numerical differences, in the face of conflicting perceptual information. Mehler and Bever showed a U-shaped developmental function, in which performance decreased over the next months, reaching the same level as in 2½-year-olds again at 4½. These findings were big news at the time, as most of the field of developmental psychology had assimilated Piaget's views that children did not create a representation of number until between ages four and six years. It was the success of the 2½-year-olds that challenged the Piagetian orthodoxy.

Voluminous research since that early exchange, some from Mehler's laboratory, has supported Mehler's position over Piaget's with respect to the ontogenetically early availability of representations of number. Infants habituate to sets of two or three individuals, dishabituating when presented a set of a different number (e.g., see Antell and Keating, 1983; Starkey and Cooper, 1980; Strauss and Curtis, 1981; Van Loosbroek and Smitsman, 1990; but see Clearfield and Mix, 1999, and Feigenson, Carey and Spelke, in press, for evidence that stimulus variables correlated with number may underlie infants' discrimination in these habituation studies involving small numbers of simultaneously presented visual objects). In most of the early studies, the individuals were pictured objects or dots. An important contribution from Mehler's laboratory was to extend the result to nonobject individuals (two- vs. three-syllable words: Bijeljac-Babic, Bertoncini and Mehler, 1991; see also Wynn, 1996, on infant discrimination of two jumps from three jumps). Xu and Spelke (2000) recently extended this finding to discriminations of 8 dots from 16 dots, in an experiment that successfully controlled for non-numerical bases of response.

Not only do infants discriminate sets of individuals on the basis of number, they also represent the results of additions and subtractions, as shown by Wynn's classic 1 + 1 = 2 or 1 violation of expectancy studies (Wynn, 1992a, 1995; see Simon, Hespos, and Rochat, 1995; Uller et al., 1999, for replications). Again, Mehler's laboratory made an important contribution to this literature, showing that infants cannot be anticipating the spatial location of the objects being placed behind the screen, their attention being drawn only when an object failed to appear where it had been represented to be or when an object appeared where none had been represented to be. Infants succeeded at a 1 + 1 = 2 or 1 and 2 − 1 = 2 or 1 experiment when the objects behind the screen were on a rotating plate, and thus the infant could not predict the layout of the resulting array (Koechlin, Dehaene, and Mehler, 1998).

Nevertheless, even in the face of this body of research supporting Mehler over Piaget concerning the early availability of number representation, I shall maintain that this domain, the representation of number, provides a parade case of *discontinuity* during development. To engage this issue, we must consider the nature of the representations that underlie infant success, and their relations to later-developing systems of number representation.

The Natural Numbers

My target here is not continuity within all of mathematical knowledge, or even all of arithmetic knowledge. I address the continuity assumption with respect to a modest, but important, part of arithmetic knowledge—the capacity to represent the positive integers, or natural numbers. The continuity assumption with respect to natural numbers is consistent with Leopold Kronecker's famous remark: "The integers were created by God; all else is man-made" (quoted in Weyl, 1949). Unlike Kronecker, I am concerned with conceptual primitives, not ontological ones, and if we would replace "God" with "evolution," we would be saying that evolution provided us with the capacity to represent the positive integers, the *natural* numbers, and that the capacity to represent the rest of arithmetic concepts, including the rest of the numbers (rational, negative, 0, real, imaginary, etc.) was culturally constructed by human beings. I assume that the rest of arithmetic is built upon a representation of the natural numbers; I shall not argue for this here. Rather, my goal is to convince you that God did not give man the positive integers either. Rather, the capacity to represent the positive integers is also a cultural construction that transcends evolutionarily given beginning points.

The historically and ontogenetically earliest explicit representational systems with the potential to represent natural numbers are integer lists. Most, but not all, cultures have explicit ordered lists of words for successive integers ("one, two, three, four, five, six . . ." in English; body parts in some languages; see Butterworth, 1999, and Dehaene, 1997, for examples of body-part integer lists). Integer lists are used in conjunction with counting routines to establish the number of individuals in any given set. In very important work, Gelman and Gallistel (1978) argued that if young toddlers understand what they are doing when they count (i.e., establishing the number of individuals there are in a given set), then, contra Piaget (1952) and supporting Mehler and Bever, they have the capacity to represent number. Gelman and Gallistel (1978) analyzed how integer list representations work: there must be a stably ordered list of symbols (the stable order principle). In counting, the symbols must be applied in order, in one-one correspondence to the individuals in the set being enumerated (one-one correspondence principle). The cardinal value of the set is determined by the ordinal position of the last symbol reached in the count (cardinality principle). These principles ensure another central feature of integer list representations, namely, that they embody the successor function: for any symbol in an integer list, if it represents cardinal value n, the next symbol on the list represents cardinal value $n + 1$. It is the successor function (together with some productive capacity to generate new symbols on the list) that makes the integer list a representation of natural number.

Gelman and Gallistel's Continuity Hypothesis

Gelman and Gallistel (1978) suggested that infants establish numerical representations through a nonverbal counting procedure: Babies represent a list of symbols, or "numerons," such as &, ^, #, $, @. Entities to be counted are put in one-one correspondence with items on this list, always proceeding in the same order through it. The number of items in the set being counted is represented by the last item on the list reached, and its numerical value is determined by the ordinal position of that item in the list. For example, in the list above, "^" represents 2, because "^" is the second item in the list.

Gelman and Gallistel's proposal for the nonlinguistic representation of number is a paradigm example of a continuity hypothesis, for this is exactly how languages with explicit integer lists represent the positive integers. On their hypothesis, the child learning "one, two, three, four, five . . ." need only solve a mapping problem: identify the list in their language that expresses the antecedently available numeron list. Originally learning to count should be no more difficult than learning to count in Russian once one knows how to count in English.

Wynn's Ease-of-Learning Argument against the Gelman-Gallistel Continuity Proposal

Children learn to count over the ages of two to four years, and contrary to the predictions of a numeron-list theory, learning to count is not a trivial matter (Fuson, 1988, Wynn, 1990, 1992b). Of course, this fact in itself does not defeat the continuity hypothesis, for we do not know in advance how difficult it is to learn an arbitrary list of words or to discover that one such list (e.g., "one, two, three . . .," " rather than "a, b, c . . ." or "Monday, Tuesday, Wednesday . . .") is the list in English that represents number. Nevertheless, Wynn's (1990, 1992b) studies show that children have difficulty discovering the meanings of specific number words even after they have solved these two problems. The sequence in which children learn the meanings of the number words therefore is at odds with that predicted by the continuity thesis.

Wynn began by confirming that young children do not know the numerical meanings of the words in the count sequence. First, she identified children who could count at least to six when asked "how many" objects there were in an array of toys. These two- to three-year-old children honored one-one correspondence in their counts, and used a consistently ordered list, although sometimes a nonstandard one such as "one, two, four, six, seven . . ." She then showed that if such a child were given a pile of objects and asked to give the adult "two" or "three" or any other number the child could use in the game of counting, most two- and three-year-old children failed. Instead, young children grabbed a random number of objects (always more than one) and handed them to the experimenter. Also, shown two cards depicting, for example, two vs. three balloons and asked to indicate which card had two balloons on it, young children responded at chance. Analyses of within-child consistency between the "give n" and "which card has n" tasks bolstered the conclusion that young children count for more than a year before they learn what the words in the count sequence mean; that is, before they learn how "one, two, three, four, five . . ." represents number.

There is one more observation of Wynn's that is important to the evaluation of the Gelman-Gallistel continuity hypothesis. Wynn (1990, 1992b) showed that from the beginning of learning to count, children know what "one" means. They can pick *one* object from a pile when asked, and they correctly distinguish a card with "one fish" from a card with "three fish." Further, they know that the other words in the count sequence contrast with one. They always grab a random number of objects greater than one, when asked to hand over "two, three, four . . ." objects, and they also successfully point to a card with three fish when it is contrasted with a card with one, even though their choices are random when three is contrasted with two. Thus, Wynn's studies provide evidence that toddlers learn the English count list and *identify the list as relevant to number* very early on (younger than age 2½): they know what "one" means, they use "two, three, four," and so on in contrast with "one," and they only use the number words above one when presented with sets greater than one. They are in this state of knowledge for a full year before they work out the principle that allows them to determine which number each numeral refers to. This state of affairs is impossible on the numeron list continuity hypothesis, whereby the English count list need only be identified and mapped onto the preexisting nonlinguistic numeron list that the infant already uses to represent number.

Thus, in spite of the evidence that prelinguistic infants represent number, it seems that the child is still in the process of constructing an integer list representation of number in the years two to four. How, then, do infants represent number, and how do their representations differ from the integer list representation?

Two Prelinguistic Representational Systems for Number

The argument that infants' representations of number are discontinuous with the integer list representation does not rest only upon Wynn's ease of learning argument. Rather, it depends upon the positive characterization of the nature of infant representations—their format and the computations defined over them. When we examine them in detail, we see precisely the nature of the discontinuities.

Two quite different nonlinguistic systems of number representation underlie infants' abilities to discriminate, compare, and compute over numbers of individuals in sets—analog magnitude representations and object-file representations.

Analog Magnitude Representations

Both human adults and animals deploy analog magnitude representations of number (see Gallistel, 1990; Dehaene, 1997, for reviews). Rather than being represented by a list of discrete symbols, in such systems number is represented by a physical magnitude that is proportional to the number of individuals in the set being enumerated. In analog magnitude number representations, each number is represented by a physical magnitude that is proportional to the number of individuals in the set being enumerated. An external analog magnitude representational system could represent 1 as "—", 2 as "——", 3 as "———", 4 as "————", 5 as "—————", ... 7 as "———————", 8 as "————————", and so on. In such systems, numerical comparisons are made by processes that operate over these analog magnitudes, in the same way that length or time comparisons are made by processes that operate over underlying analog magnitude representations of these continuous dimensions of experience. Importantly, there is a psychophysical Weber-fraction signature of analog magnitude representations: the discriminability of two numbers is a function of their ratio. Examining the external analogs above, it is easy to see that it is easier to discriminate 1 from 2 than 7 from 8 (what Dehaene, 1997, calls the magnitude effect), and it is easier to discriminate 1 from 3 than 2 from 3 (what Dehaene, 1997, calls the distance effect). This Weber-fraction signature applies to discrimination of continuous quantities as well, such as representations of lengths (as can be experienced directly by examining the above lengths), distances, time, and so forth, and is the primary evidence that number is being represented by a quantity that is linearly related to the number of individuals in the set. The Weber-fraction signature of infant number representations is shown in Xu and Spelke's (2000) finding that infants discriminated eight from sixteen dots, but failed to discriminate eight from twelve dots.

Why Analog Magnitude Representations Are Not Representations of Positive Integers

Analog magnitude representational systems do not have the power to represent natural number. This fact alone defeats the proposal that the analog magnitude system of numerical representation is continuous with the integer list representation. That is, learning an explicit integer list representation is not merely learning words for symbols already represented. To see this, let us consider Gallistel and Gelman's (1992) arguments for the strong proposal and the problems that arise.

There are many different ways analog magnitude representations of number might be constructed. The earliest proposal was Meck and

Church's (1983) accumulator model. The idea is simple—suppose the nervous system has the equivalent of a pulse generator that generates activity at a constant rate, and a gate that can open to allow energy through to an accumulator that registers how much has been let through. When the animal is in a counting mode, the gate is opened for a fixed amount of time (say 200 ms) for each individual to be counted. The total energy accumulated then serves as an analog representation of number. Meck and Church's model seems best suited for sequentially presented individuals, such as bar presses, tones, light flashes, or jumps of a puppet. Gallistel (1990) proposed, however, that this mechanism functions as well in the sequential enumeration of simultaneously present individuals.

Gallistel and Gelman (1992) argued that the accumulator model is formally identical to the integer list representational system of positive integers, with the successive states of the accumulator serving as the successive integer values, the mental symbols that represent numerosity. They point out that the accumulator model satisfies all the principles that support verbal counting: States of the accumulator are stably ordered, gate opening is in one-one correspondence with individuals in the set, the final state of the accumulator represents the number of items in the set, there are no constraints on individuals that can be enumerated, and individuals can be enumerated in any order. Thus, Gallistel and Gelman (1992) argue that the Meck and Church (1983) analog magnitude system is continuous with, and is likely to be the ontogenetic underpinnings of, an explicit integer list representational system and counting.

Unfortunately for this proposal, there is considerable evidence that suggests that the Church and Meck model is false, and that analog magnitude representations of number are not constructed by any iterative process. In particular, the time that subjects require to discriminate two numerosities depends on the ratio difference between the numerosities but not on their absolute value (Barth, Kanwisher, and Spelke, submitted). In contrast, time should increase monotonically with N for any iterative counting process. Moreover, subjects are able to discriminate visually presented numerosities under conditions of stimulus size and eccentricity in which they are not able to attend to individual elements in sequence (Intrilligator, 1997). Their numerosity discrimination therefore could not depend on a process of counting each entity in turn, even very rapidly.

Problems such as these led Church and Broadbent (1990) to propose that analog magnitude representations of number are constructed quite differently, through no iterative process. Focusing on the problem of representing the numerosity of a set of sequential events (e.g., the number of tones in a sequence), they proposed that animals perform a computation that depends on two timing mechanisms. First, animals time the temporal interval between the onsets of successive tones, maintaining in memory a single value that approximates a running average of these intervals. Second, animals time the overall duration of the tone sequence. The number of tones is then estimated by dividing the sequence duration by the average intertone interval. Although Church and Broadbent did not consider the case of simultaneously visible individuals, a similar noniterative mechanism could serve to compute numerosity in that case as well, by measuring the average density of neighboring individuals, measuring the total spatial extent occupied by the set of individuals, and dividing the latter by the former. Dehaene and Changeux (1989) described an analog magnitude model that could enumerate simultaneously presented visual individuals in a different manner, also through no iterative process.

The analog magnitude representational system of Church and Broadbent (1990) (as well as that of Dehaene and Changeux, 1989) differs from the original Meck and Church (1983) accumulator model in a number of important ways. Because the processes that construct these representations are not iterative, the analog magnitudes are not formed in sequence and therefore are less likely to be experienced as a list. Moreover, the process that establishes the analog magnitude representations does not require that each individual in the set to be enumerated be attended to in sequence, counted, and then ticked off (so that each individual is counted only once). These mechanisms do not implement any counting procedure.

Furthermore, none of the analog magnitude representational systems, even Church and Meck's accumulator system, has the power to represent natural number in the way an integer list representational system does. For one thing, analog magnitude systems have an upper limit, due to the capacity of the accumulator or the discriminability of the individuals in a set, or both, whereas base system integer list systems do not (subject to the coining of new words for new powers of the base). But the problem is much worse than that. Consider a finite integer list, like the body counting systems. Because it is finite, this system is also not a representation of the natural numbers, but it is still more powerful than analog magnitude representations, for it provides an *exact representation* of the integers in its domain.

Thus, all analog magnitude representations differ from any representation of the natural numbers, including integer list representations, in two crucial respects. Because analog magnitude representations are inexact and subject to Weber-fraction considerations, they fail to capture small numerical differences between large sets of objects. The distinction between 7 and 8, for example, cannot be captured by the analog magnitude representations found in adults. Also, noniterative processes for constructing analog magnitude representations, such as those proposed by Dehaene and Changeux (1989) and by Church and Broadbent (1990), include nothing that corresponds to the successor function, the operation of "adding 1." Rather, all analog magnitude systems positively obscure the successor function. Since numerical values are compared by computing a ratio, the difference between 1 and 2 is experienced as different from that between 2 and 3, which is again experienced as different from that between 3 and 4. And, of course, the difference between 7 and 8 is not experienced at all, since 7 and 8 or any higher successive numerical values cannot be discriminated.

In sum, analog magnitude representations are not powerful enough to represent the natural numbers and their key property of discrete infinity, do not provide exact representations of numbers larger than 4 or 5, and do not support any computations of addition or mutiplication that build on the successor function.

A Second Prelinguistic System of Number Representation: Parallel Individuation of Small Sets

I argued above that analog magnitude representations are not powerful enough to represent the natural numbers, even the finite subset of natural numbers within the range of numbers these systems handle. Here I describe a second system of representation that is implicated in many of the tasks that have shown sensitivity to number in infancy (Scholl and Leslie,

1999; Simon, 1997; Uller et al., 1999). In this alternative representational system, number is only implicitly encoded; there are no symbols for number at all, not even analog magnitude ones. Instead, the representations include a symbol for each individual in an attended set. Thus, a set containing one apple might be represented: "0" or "apple," and a set containing two apples might be represented "0 0" or "apple apple," and so forth. Because these representations consist of one symbol (file) for each individual (usually object) represented, they are called "object-file" representations.

Several lines of evidence suggest that object-file representations underlie most, if not all, of the infant successes in experiments that involve small sets of objects. Here I mention just two empirical arguments for this conclusion; see Uller et al. (1999) for a review of several others. First, success on many spontaneous number representation tasks do not show the Weber-fraction signature of analog magnitude representations; rather they show the set-size signature of object-file representations. That is, the number of individuals in small sets (one to three or four) can be represented, and numbers outside of that limit cannot, even when the sets to be contrasted have the same Weber fraction as those small sets where the infant succeeds. Second, in many experiments, the models of the small sets of objects are compared on the basis of overall size, rather than on the basis of number, abstracting away from size. Analog magnitude representations of number abstract away from the size of the individuals, of course.

The set-size signature of object-file representations is motivated by evidence that even for adults there are sharp limits on the number of object files open at any given moment, that is, the number of objects simultaneously attended to and tracked. The limit is around four in human adults. The simplest demonstration of this limit comes from Pylyshyn and Storm's (1988) multiple object tracking studies. Subjects see a large set of objects on a computer monitor (say, fifteen red circles). A subset is highlighted (e.g., three are turned green) and then become identical again with the rest. The whole lot is then put into motion and the observer's task is to track the set that has been highlighted. This task is easy if there are one, two, or three objects; performance falls apart beyond four. Trick and Pylyshyn (1994) demonstrated the relations between the limit on parallel tracking and the limit on subitizing—the capacity to directly enumerate small sets without explicit internal counting.

If object-file representations underlie infants' performance in number tasks, then infants should succeed only when the sets being encoded consist of small numbers of objects. Success at discriminating 1 vs. 2, and 2 vs. 3, in the face of failure with 3 vs. 4 or 4 vs. 5 is not enough, for Weber-fraction differences could equally well explain such a pattern of performance. Rather, what is needed is success at 1 vs. 2 and perhaps 2 vs. 3 in the face of failure at 3 vs. 6—failure at the higher numbers when the Weber fraction is the same or even more favorable than that within the range of small numbers at which success has been obtained.

This set-size signature of object-file representations is precisely what is found in some infant habituation studies—success at discriminating 2 vs. 3 in the face of failure at discriminating 4 vs. 6 (Starkey and Cooper, 1980). Although set-size limits in the infant addition and subtraction studies have not been systematically studied, there is indirect evidence that these too show the set-size signature of object-file representations. Robust success is found on $1 + 1 = 2$ or 1 and $2 - 1 = 2$ or 1 paradigms (Koechlin, et al, 1998; Simon, et al, 1995; Uller et al., 1999; Wynn, 1992a). In the face of success in these studies with Weber fraction of 1:2, Chiang and Wynn (2000) showed repeated failure in a $5 + 5 = 10$ or 5 task, also a Weber fraction of 1:2.

Two parallel studies (one with rhesus macaques; Hauser, Carey, and Hauser, 2000; one with ten- to twelve-month-old infants; Feigenson et al. (in press) provide a vivid illustration of the set-size signature of object-file representations. Both studies also address a question about spontaneous number representation left open by the addition-subtraction studies and by habituation studies, and both studies address two important questions about object-file models themselves. The question about spontaneous number representation is: Is it the case that nonverbal creatures merely discriminate between small sets on the basis of number, or do they also compare sets with respect to which one has more? The questions about object-file models themselves is: Is the limit on set sizes a limit on *each* set represented, or a limit on the total number of objects that can be indexed in a single task? And when two models are compared to establish more/less, is the basis of comparison number (one-one correspondence) or some continuous variable such as total volume or total surface area?

In these studies, a monkey or an infant watches as each of two opaque containers, previously shown to be empty, is baited with a different number of apple slices (monkeys) or graham crackers (babies). For example, the experimenter might put two apple slice/graham crackers in one container and three in the other. The pieces of food are placed one at a time, in this example: 1 + 1 in one container and then 1 + 1 + 1 in the other. Of course, whether the greater or lesser number is put in first, as well as whether the greater number is in the leftmost or rightmost container, is counterbalanced across babies/monkeys. Each participant gets only one trial. Thus, these studies tap spontaneous representations of number, for the monkey/baby does not know in advance that different numbers of pieces of food will be placed into each container, or even that they will be allowed to choose. After placement, the experimenter walks away (monkey) or the parent allows the infant to crawl toward the containers (infant). The dependent measure is which container the monkey/baby chooses.

Figures 17.1 and 17.2 show the results from adult free-ranging rhesus macaques and ten- to twelve-month-old human infants, respectively. What one sees is the set-size signature of object-file representations. Monkeys succeed when the comparisons are 1 vs. 2, 2 vs. 3, and 3 vs. 4, but they fail at 4 vs. 5, 4 vs. 8, and even 3 vs. 8. Infants succeed when the comparisons are 1 vs. 2 and 2 vs. 3, but fail at 3 vs. 4 and even 3 vs. 6. That is, they fail when one of the sets exceeds the limit on parallel individuation, even though the ratios (3:6, 3:8) are more favorable than some of those involving smaller sets with which they succeed.

These data show that rhesus macaques and human infants spontaneously represent number in small sets of objects, and can compare them with respect to which one has more. More important to us here, they show the set-size signature of object-file representations; monkeys succeed if both sets are within the set-size limits on parallel individuation (up to three in the case of ten- to twelve-month-old human infants, up to four in the case of adult rhesus macaques), and fall apart if one or both of the sets exceeds this limit. Also, it is of theoretical significance

Rhesus' ordinal choices

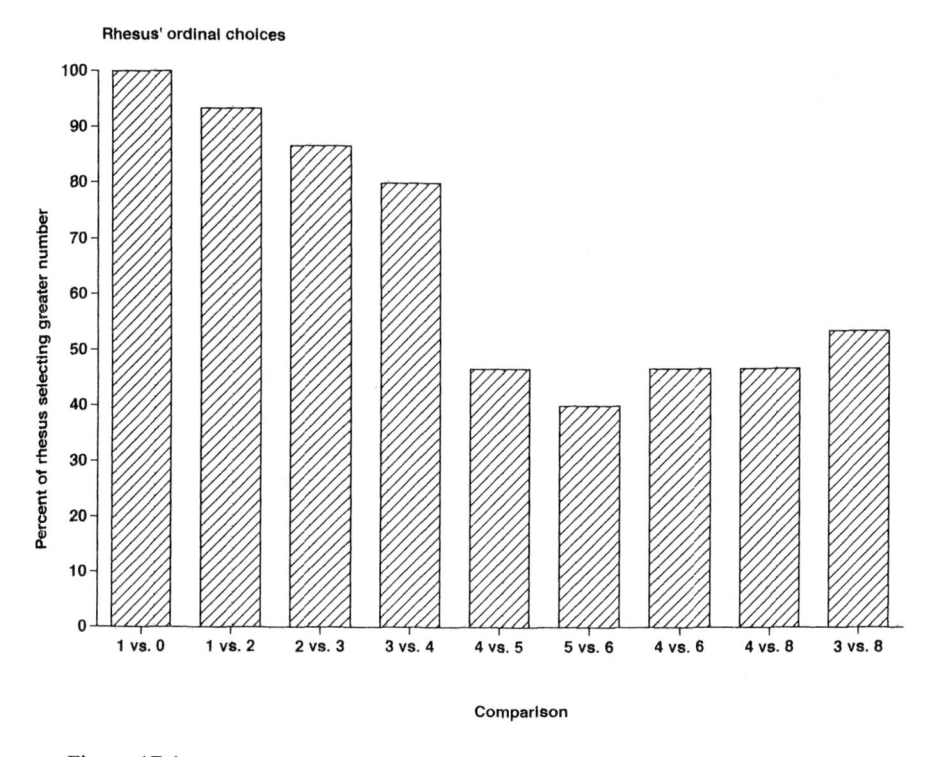

Figure 17.1
Adult rhesus macaques. Percent choice of the box with more apple slices.

to the object-file literature that infants and monkeys succeed in cases where the total number represented (infants, 5 in 2 vs. 3; monkeys 7 in 3 vs. 4) exceeds the limit on parallel individuation. Apparently, nonlinguistic primates can create two models, each subject to the limit, and then compare them in memory.

With respect to the third question—the basis of the ordinal comparison—the two populations gave different answers. A variety of controls ensured that both groups were creating models of the individual apples or crackers, and not responding to the total amount of time each container was handled or the total amount of attention drawn to each container. For instance, monkeys' performance is no different if the comparison is two apple slices and a rock into one container vs. three apple slices, even though now the total time placing entities into each con-

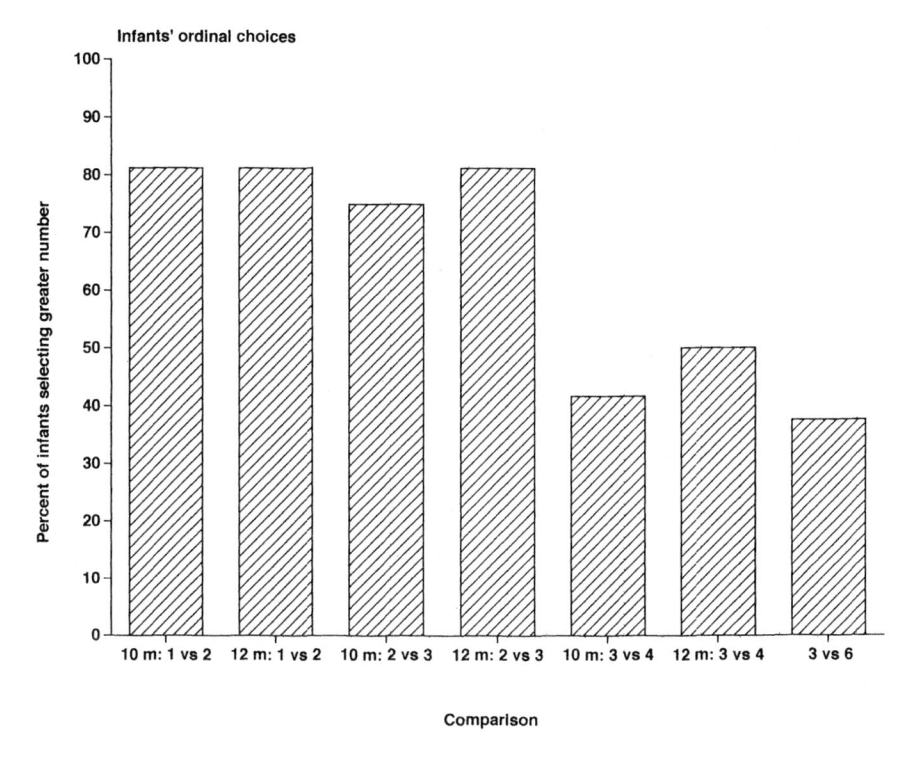

Infants' ordinal choices

[Bar chart showing percent of infants selecting greater number across comparisons:
- 10 m: 1 vs 2 — ~81%
- 12 m: 1 vs 2 — ~81%
- 10 m: 2 vs 3 — ~75%
- 12 m: 2 vs 3 — ~81%
- 10 m: 3 vs 4 — ~42%
- 12 m: 3 vs 4 — ~50%
- 3 vs 6 — ~38%]

Figure 17.2
Ten- and twelve-month-old infants. Percent choice of the box with more graham crackers.

tainer and the total amount of attention drawn to each container is equal. Similarly, infants' performance is no different if the comparison is two crackers into one container and one cracker plus a hand wave over the other container, even though now the total time that attention is drawn to each container is equated. Moreover, monkeys were responding to the number of apple slices placed in the containers rather than the total volume of apple placed into each container (even though that surely is what monkeys are attempting to maximize). Monkeys go to the container with three when the choice is one large piece ($1/2$ apple) vs. three small pieces (which sum to $1/2$ apple). We assume that although the monkeys are trying to maximize the total amount of apple stuff, they are making an equal-volume assumption and using number to estimate amount of stuff. (From ten feet away and with the slices shown briefly as they are placed into the container, apparently monkeys cannot encode the volume of each piece.)

Infants, in contrast, were responding on the basis of the total amount of cracker. Given the choice between a large cracker and two smaller ones, each ½ of the volume of the large one, infants were at chance, and given the choice between a large cracker and two smaller ones, each the volume of the large one, infants chose the single cracker with larger total volume. Infants' models of the contents of each container were subject to the limits on parallel individuation, but models were compared on the basis of total volume or surface area of the crackers. The same seems to be true of the classic habituation studies involving small sets of visual individuals. As pointed out above, these studies show the set-size signature of object-file representations. Clearfield and Mix (1999) and Feigenson, Carey, and Spelke (in press) point out that none of the habituation studies with small sets of objects successfully control for a variety of continuous variables correlated with number. When number is pitted against such variables, infants dishabituate to changes in total surface area, total contour length, or other continuous variables correlated with these, and do not respond to changes in number. Further, when continuous variables are removed as a basis for dishabituation, infants show no sensitivity to changes in number (Feigenson et al., in press). Apparently, in these habituation studies, total spatial extent of the individuals is a more salient feature of the array than is number.

This fact further elucidates the nature of object-file representations. As is known from the adult literature, properties may be bound to object files. In the infant case, size is apparently one such property, and the total size of the objects in small sets is a salient property of the whole collection.

It is not the case, however, that prelinguistic infants *never* operate on object-file representations on the basis of numerical equivalence (one-one correspondence). Van de Walle, Carey, and Prevor (2000) developed a paradigm combining the violation-of-expectancy methodology with reaching as a dependent measure. Infants reach into a box into which they cannot see. Sometimes an expected object has been surreptitiously removed, and persistence of search reveals when the infant represents unretrieved entities still in the box. Feigenson and Carey (in preparation) have recently shown that it is number of objects (one vs. two), not total surface area or volume of the sets, that determines the pattern of reaching of twelve-month-old infants.

Object-file representations are numerical in five senses. First, the opening of new object files requires principles of individuation and numerical identity; models must keep track of whether this object, seen now, is the same *one* as that object seen before. Spatiotemporal information must be recruited for this purpose, because the objects in many experiments are physically indistinguishable from each other, and because, in many cases, property or kind changes within an object are not sufficient to cause the opening of a new object file (Kahneman, Triesman, and Gibbs, 1992; Pylyshyn, in press; Xu and Carey, 1996). Second, the opening of a new object file in the presence of other active files provides a natural representation for the process of adding one to an array of objects. Third, object-file representations provide implicit representations of sets of objects; the object files that are active at any given time as a perceiver explores an array determine a set of attended objects. Fourth, if object-file models are compared on the basis of one-one correspondence, the computations over object-file representations provide a process for establishing numerical equivalence and more/less. Fifth, object files represent numerosity exactly for set sizes up to about four and are not subject to Weber's law.

Unlike the analog magnitude system of number representations, the object-file system is not dedicated to number representations. Number is only implicitly represented in it, as it contains no symbols for numbers. It does not remotely have the power to represent natural number, for two reasons. Most important, object-file models contain no symbols for cardinal values. The only symbols in such models represent the individual objects themselves. Second, object-file models have an upper bound at very low set sizes indeed.

Object-file representations cannot account for all the evidence from studies of number representations in infants. In particular, such representations cannot account for infants' successful discrimination of eight from sixteen dots (Xu and Spelke, 2000). These numbers are out of the range of object-file representations, and as mentioned above, infant discrimination of large numbers is subject to Weber-fraction constraints; the infants in Xu and Spelke's studies failed to distinguish eight from twelve dots.

Also, object-file representations cannot account for infants' success at discriminating sets of events (e.g., jumps of a puppet: Wynn, 1996) or sounds (e.g., syllables in words: Bijeljac-Babic et al., 1991) on the basis of number, although it is not yet known whether such stimuli also are subject to the set-size limitations of parallel individuation. Like Xu and Spelke, I conclude that infants have two systems of number-relevant representations: the object-file representations that are often deployed with small sets of individual objects and analog magnitude representations that are deployed with large sets of objects, and perhaps with sequences of events or sounds. The questions of the conditions under which each type of representation is activated, and whether analog magnitude representations are ever deployed with small sets of individuals, are still very much open.

On the Use of the Term *Number* in Claims of Prelinguistic Representation Thereof

When we say that infants or nonverbal animals represent *number*, it is very important to be clear about what we are claiming. It is necessary to specify the precise nature of the symbol systems that underlie the number-sensitive behavior, and ask in what senses they are representations of number—what numbers do they have the capacity to represent and what number-relevant computations do they support? I have argued above that neither of the candidate representational systems that underlie the behavior on nonlinguistic number tasks represents *number* in the sense of *natural number* or *positive integer*. Nonetheless, both systems support number-relevant computations, and the analog magnitude system contains symbols for approximate number, so they deserve to be called representations of number, as long as one does read too much into the term "representation of number."

In sum, even if the infant is endowed with both analog magnitude and object file systems of representation, the infant's capacity to represent number will be markedly weaker than that of the child who commands the integer list representation. Evolution did not make the positive integers. Neither object files nor analog magnitudes can serve to represent large exact numerosities: object files fail to capture number concepts such as "seven" because they exceed the capacity limit of four, and analog magnitude representations fail to capture such concepts because they exceed the limits on their precision (for infants, a 1:2 ratio; Xu & Spelke, 2000). Thus, the integer list system mastered during ages 2½ to 4 is a genuinely new representational resource, more powerful than, and discontinuous with, the number-relevant representational capacities of infants.

Questions That Remain

If the capacity to represent natural numbers is a cultural construction, transcending evolutionarily given representational capacities, how is it built? What are the processes that create a representational resource with more power than those that precede it? Is the integer list representation built from object-file representations, analog magnitude representations, both, or neither? Answering these questions is beyond the scope of this chapter, but see Carey and Spelke (in press) for a stab at it.

Acknowledgments

The research reported here was supported by NSF grants to Susan Carey, and to Susan Carey and Marc Hauser. The chapter is largely drawn from Carey, 2001.

References

Antell, S., and Keating, D. (1983). Perception of numerical invariance in neonates. *Child Development, 54,* 695–701.

Barth, H., Kanwisher, N., and Spelke, E. (submitted). Construction of large number representations in adults.

Bijeljac-Babic, R., Bertoncini, J., and Mehler, J. (1991). How do four-day-old infants categorize multisyllabic utterances? *Developmental Psychology, 29,* 711–721.

Butterworth, B. (1999) *What counts: How every brain is hardwired for math.* New York: Free Press.

Carey, S. (2001). Cognitive foundations of arithmetic: Evolution and ontogenesis. *Mind and Language, 16,* 37–55.

Carey, S. and Spelke, E. (in press). Bootstrapping the integer list: Representations of number. In J. Mehler and L. Bonati (Eds.), *Developmental cognitive science*. Cambridge, MA: MIT Press.

Chiang, W. C., and Wynn, K. (2000). Infants' representations and tracking of objects: Implications from collections. *Cognition, 77,* 169–195.

Church, R. M., and Broadbent, H. A. (1990). Alternative representations of time, number, and rate. *Cognition, 37,* 55–81.

Clearfield, M. W., and Mix, K. S. (1999). Number versus contour length in infants' discrimination of small visual sets. *Psychological Science, 10,* 408–411.

Dehaene, S. (1997). *The number sense: How the mind creates mathematics*. Oxford: Oxford University Press.

Dehaene, S., and Changeux, J. P. (1989). Neuronal models of cognitive function. *Cognition, 33,* 63–109.

Feigenson, L., and Carey, S. (in preparation). Reaching for one or two: A response to small numbers.

Feigenson, L., Carey, S., and Hauser, M. (in press). More crackers: Infants' spontaneous ordinal choices. *Psychological Science*.

Feigenson, L., Carey, S., and Spelke, E. (in press). Infants' discrimination of number vs. continuous spatial extent. *Cognitive Psychology*.

Fodor, J. (1975). *The language of thought*. New York: Crowell.

Fuson, K. C. (1988). *Children's counting and concepts of number*. New York: Springer-Verlag.

Gallistel, C. R. (1990). *The organization of learning*. Cambridge, MA: MIT Press.

Gallistel, C. R., and Gelman, R. (1992). Preverbal and verbal counting and computation. *Cognition, 44,* 43–74.

Gelman, R., and Gallistel, C. R. (1978). *The child's understanding of number*. Cambridge, MA: Harvard University Press.

Hauser, M., Carey, S., and Hauser, L. (2000). Spontaneous number representation in semi–free-ranging rhesus monkeys. *Proceedings of the Royal Society of London. Series B: Biological Sciences*.

Intrilligator, T. M. (1997). *The spatial resolution of attention*. Unpublished PhD dissertation, Harvard University, Cambridge, MA.

Kahneman D., Treisman, A., and Gibbs, B. (1992). The reviewing of object files: Object specific integration of information. *Cognitive Psychology, 24,* 175–219.

Koechlin, E., Dehaene, S., and Mehler, J. (1998). Numerical transformations in five-month-old human infants. *Mathematical Cognition, 3,* 89–104.

Macnamara, J. (1986). *A border dispute: The place of logic in psychology*. Cambridge, MA: MIT Press.

Meck, W. H., and Church, R. M. (1983). A mode control model of counting and timing processes. *Journal of Experimental Psychology: Animal Behavior Processes, 9,* 320–334.

Mehler, J., and Bever, T. G. (1967). Cognitive capacity in very young children. *Science, 158*, 141–142.

Mehler, J., and Bever, T. G. (1969). Reply by J. Mehler and T. G Bever. *Science, 162*, 979–980.

Piaget, J. (1952). *The child's conception of number.* New York: Humanities Press.

Pinker, S. (1984). *Language learnability and language development.* Cambridge, MA: Harvard University Press.

Pylyshyn, Z. W. (in press). Visual indexes, preconceptual objects, and situated vision. *Cognition.*

Pylyshyn, Z. W., and Storm, R. W. (1988). Tracking multiple independent targets: Evidence for a parallel tracking mechanism. *Spatial Vision, 3*, 179–197.

Scholl, B. J., and Leslie, A. M. (1999). Explaining the infant's object concept: Beyond the perception/cognition dichotomy. In E. Lepore and Z. Pylyshyn (Eds.), *What is cognitive science?* (pp. 26–73). Oxford: Blackwell.

Simon, T. J. (1997). Reconceptualizing the origins of number knowledge: A "non-numerical" account. *Cognitive Development, 12*, 349–372.

Simon, T., Hespos, S. and Rochat, P. (1995). Do infants understand simple arithmetic? A replication of Wynn (1992). *Cognitive Development, 10*, 253–269.

Starkey, P., and Cooper, R. (1980). Perception of numbers by human infants. *Science, 210*, 1033–1035.

Strauss, M., and Curtis, L. (1981). Infant perception of numerosity. *Child Development, 52*, 1146–1152.

Trick, L., and Pylyshyn, Z. (1994). Why are small and large numbers enumerated differently? A limited capacity preattentive stage in vision. *Psychological Review, 101*, 80–102.

Uller, C., Huntley-Fenner, G., Carey, S., and Klatt, L. (1999). What representations might underlie infant numerical knowledge? *Cognitive Development, 14*, 1–36.

Van de Walle, G., Carey, S., and Prevor, M. (2000). The use of kind distinctions for object individuation: Evidence from manual search. *Journal of Cognition and Development, 1*, 249–280.

Van Loosbroek, E., and Smitsman, A. (1990). Visual perception of numerosity in infancy. *Developmental Psychology, 26*, 916–922.

Weyl, H. (1949). *Philosophy of mathematics and natural science.* Princeton, NJ: Princeton University Press.

Wynn, K. (1990). Children's understanding of counting. *Cognition, 36*, 155–193.

Wynn, K. (1992a). Addition and subtraction by human infants. *Nature, 358*, 749–750.

Wynn, K. (1992b). Children's acquisition of the number words and the counting system. *Cognitive Psychology, 24*, 220–251.

Wynn, K. (1995). Origin of numerical knowledge. *Mathematical Cognition, 1*, 36–60.

Wynn, K. (1996). Infants' individuation and enumeration of physical actions. *Psychological Science, 7*, 164–169.

Xu, F., and Carey, S. (1996). Infants' metaphysics: The case of numerical identity. *Cognitive Psychology, 30*, 111–153.

Xu, F., and Spelke, E. S. (2000). Large number discrimination in 6-month-old infants. *Cognition, 74*, B1–B11.

18

Continuity, Competence, and the Object Concept

Elizabeth Spelke and Susan Hespos

In 1967, Mehler and Bever reported a groundbreaking study of number concepts and logical abilities in young children. Their studies focused on Piaget's famous "number conservation" task, which had revealed a striking discontinuity in children's cognitive performance. When children about five years of age are asked to judge the relative numerosities of two rows of objects that differ in length and spacing, they often respond on the basis of length rather than number; when older children are given the same task, in contrast, they focus consistently on number. This developmental change was taken by Piaget to reflect the emergence of a new system of logical operations and a new set of concepts, including the first true number concepts. To test this interpretation, Mehler and Bever presented two- to five-year-old children with a child-friendlier task that was logically identical to Piaget's. After judging that two rows had the same number of candies, children watched as the arrays were shortened, lengthened, or changed in number and then they were invited to take one row of candies. For comparison, the same children also were tested with clay pellets and were asked, as in Piaget's studies, which of two rows had more pellets. With the candies test, children chose the more numerous row at all ages; with the pellets, they succeeded at the youngest and oldest ages but failed at about four years.

To account for the task-dependent performance that they discovered, Mehler and Bever invoked the distinction between competence and performance and argued for continuity in children's competence. Children have an understanding of number that emerges early in the third year, if not before, that is continuously present thereafter, and that reveals itself in simple and motivating situations such as the candies task. They suggested that four-year-old children's failure in the pellets task reflected their discovery of a perceptual strategy—using length to judge number—that temporarily masked their underlying competence in the less motivating pellets task.

Many cognitive scientists now criticize developmental explanations like Mehler and Bever's that invoke the competence-performance distinction and the continuity thesis (e.g., see Thelen and Smith, 1993; Rumelhart and McClelland, 1986; Munakata et al., 1997). The pattern of task-dependent behavior discovered by Mehler and Bever nevertheless is one of many that continue to challenge developmental psychologists, who must explain why children of a given age appear to command a given concept or reasoning process in one context but not in another. Today, task-dependent performance patterns are puzzling investigators in domains ranging from earlier-developing number concepts (infant success: Wynn, 1998; preschool failure: Mix, Huttenlocher, and Levine, 1996), to theory of mind (success before three years: Clements and Perner, 1994; failure until four years: Perner and Wimmer, 1988), to naive physics (success in infancy: Baillargeon, 1995; failure in preschoolers: Hood, 1995; Berthier et al., 2000). But perhaps the most famous and vexing example of task-dependence has arisen in studies of infants' ability to represent hidden objects: an ability that appears to be present as early as two months when infants are tested by preferential looking methods, but that appears to be absent as late as 18 months when children are tested by manual search methods. This last example, we believe, provides a good test case for contrasting approaches to cognitive development.

Evidence for Task-Dependence in Object Concept Development

Young infants' patterns of preferential looking provide evidence that they represent the continued existence of objects that move from view (see Baillargeon, 1993, and Spelke, 1998, for reviews). For example, when five-month-old infants observe first a single object that is occluded by a moving screen and then a second object that joins the first behind the screen, they subsequently look longer when the screen is lowered to reveal one object (an unexpected outcome to adults, but one that is superficially familiar since only one object ever was visible in the scene at any given time) than when it is lowered to reveal two objects (Wynn, 1992). This extensively replicated finding (Baillargeon, 1994; Koechlin, Dehaene, and Mehler, 1998; Simon, Hespos, and Rochat, 1995; Uller et al., 1999; see also Spelke et al., 1995; Xu and Carey, 1996) provides evidence that infants continued to represent the existence of each object within the scene after it was occluded. As a second example, when 2½-month-old infants observe an object hidden either inside or behind a container and then see the object again when the container is moved to the side, they react differently depending on how the object is hidden (Hespos and Baillargeon, 2001a). Although both scenes look identical at the time the container is moved, infants look longer if the object had originally been placed inside it. This finding is one of many providing evidence that infants reason about the behavior of the hidden object in accord with a solidity principle, inferring that the hidden object could not pass directly through the side of the container (see Baillargeon, 1995).

In contrast to these findings, infants fail to retrieve hidden objects until about nine months of age in the simplest situations (Piaget, 1954), and their search for hidden objects sometimes fails to be guided by inferences about solidity until they are more than two years old (Berthier et al., 2000; Hood, 1995; Hood et al., 2000). For example, when 2½-year-old children view a ball rolling down a ramp toward a partly hidden barrier and are encouraged to retrieve the ball by reaching through one of four doors in a screen that covers the ramp and most of the barrier, they choose among the doors at random, apparently oblivious to the relation between the ball, ramp, and barrier (Berthier et al., 2000). Although many of the search tasks that infants fail are more complex than the preferential looking tasks at which they succeed, this is not always the case. Preferential looking experiments and reaching experiments produced contrasting findings, for example, when infants were tested on the same object permanence task at the same ages with the two measures (e.g., see Ahmed and Ruffman, 1998). What accounts for these discrepancies?

Hypothesis 1: Preferential Looking Reflects Purely Perceptual Representations

Some investigators have proposed that preferential looking methods reveal relatively shallow representations and processes. For example, such methods may assess only infants' sensory abilities (e.g., see Haith and Benson, 1997; Bogartz, Shinsky, and Speaker, 1997), midlevel visual abilities (e.g., see Scholl and Leslie, 1999), implicit knowledge (e.g., see Moscovitch, 1992), or representations divorced from action (Bertenthal, 1996; Spelke, Vishton, and von Hofsten, 1995). Further experiments, however, cast doubt on all these proposals. First, young infants have been found to reach for objects in the dark: a pattern that provides evidence that they represent the permanence of objects that lose their visibility due to darkness, and that this representation guides reaching actions (Hood and Willatts, 1986; Clifton et al., 1991). Moreover, young infants sometimes show very similar looking and reaching patterns when the task of reaching for an occluded object is simplified.

In particular, Hespos and Baillargeon (submitted) investigated whether infants would demonstrate, in a simple action on an occluded object, the same representational capacity that was previously revealed in preferential looking tasks. One set of experiments focused on infants' knowledge about occlusion and containment events, and more specifically infants' understanding that the height of an object relative to an occluder or container determined how much of the object could be hidden behind or inside it. Previous research (Hespos and Baillargeon, 2001b) indicated that at four months, infants look significantly longer at an event where an object is completely hidden behind an occluding screen that is only half as tall as the object than at an event in which the object was hidden behind a screen of equal height. In contrast, infants fail to look longer when the same object is hidden inside a short occluding container until 7^1/$_2$ months. To test whether infants would exhibit the same developmental pattern in an object-retrieval task, infants of 5^1/$_2$ and 7^1/$_2$ months of age were presented with a tall frog and encouraged to play with it. After a few seconds, the frog was removed and the infants were presented with two occluding screens or containers that had frog legs wrapped around the sides of them so that the feet were sticking out in front and could be grasped directly. The screens and containers were identical except for their height: one was tall enough to conceal the entire frog behind it, whereas the other was one-third the needed height. After the infants' attention was drawn to each occluding object, the apparatus was moved toward the infant, whose reaching was observed.

Infants of both ages reached significantly more often for the frog behind the tall screen, even though, in a control condition, they showed no intrinsic preference for the tall screen. These reaching patterns provide evidence that the infants appreciated that only the tall screen could conceal the frog. In contrast, infants reached more often for the frog inside the tall container at $7\frac{1}{2}$ but not $5\frac{1}{2}$ months of age. These experiments, and additional experiments on knowledge about support events, suggest that the representations of object occlusion, containment, and support emerge on the same developmental time course in preferential looking and reaching tasks. This knowledge evidently does not reside in representations that are purely sensory, implicit, or otherwise divorced from goal-directed action.

Hypothesis 2: Piagetian Search Tasks Reflect Developing Capacities for Means-Ends Coordination

A second family of explanations for infants' task-dependent performance with respect to occluded objects appeals to the demands that Piaget's search tasks place on the infant's developing actions. In particular, infants may reach for visible objects before they reach for occluded objects because the former can be retrieved by a simple, direct reach, whereas the latter can be retrieved only if the infant reaches around the occluder or removes it: actions that may overtax infants' capacities for means-ends coordination (e.g., see Baillargeon, 1993; Spelke et al., 1992). Young infants' successful reaching for objects in the dark and for objects with a visible, graspable part (the occluded frogs with protruding legs) is consistent with these explanations, because the nonvisible objects could be retrieved in these cases by a simple, direct reach. Motor limitations also could explain infants' success in Ahmed and Ruffman's (1998) preferential looking tasks and failure in their parallel reaching tasks, because the tasks differ in the motor demands they place on infants. Further experiments cast doubt on these explanations, however, because young infants sometimes fail to use object representations to guide their search for hidden objects even in tasks that put no undue demands on their action systems. These studies focused on six-month-old infants' predictive reaching for moving objects.

When a continuously visible, out-of-reach object begins to move smoothly toward them, infants of four months and beyond typically will attempt to grab it, initiating their reach before the object enters reaching space and aiming ahead of the object's current position so as to intercept it when it comes within their range (von Hofsten, 1980; von Hofsten et al., 1998). Accordingly, Spelke and von Hofsten (in press) placed an occluder over a portion of the path of the object on some trials, such that the object moved briefly out of view before it entered infants' reaching space. Because the occluder was out of reach, it did not serve as a barrier to infants' reaching: infants could catch the object by carrying out the same direct, predictive reach as in the trials on which the object was fully visible. The occluder therefore prevented infants from seeing the object but did not affect the motor actions needed to retrieve the object. Based on the hypothesis that limits to infants' object search reflected limits to motor control, we initially predicted that infants would succeed in reaching predictively for the temporarily occluded object.

Infants' head tracking in this situation suggested that they represented the hidden object and quickly learned to predict where it would reappear (von Hofsten, Feng, and Spelke, 2000). Contrary to our predictions, however, infants almost never reached for the object on trials with the occluder (Spelke and von Hofsten, in press). Although infants could have retrieved the object by executing the same direct reach that they used on trials without the occluder, they failed to do this when the object was hidden during the time that the reach would have begun. We conclude that limits to infants' means-ends coordination cannot fully account for infants' failure to search for hidden objects.

Hypothesis 3: Representations of Objects Are More Precise When the Objects Are Visible; the Loss of Visibility Therefore Causes a Performance Decrement on Any Task for Which Precision Is Required

Why do looking and reaching tasks sometimes yield different patterns of performance (e.g., see Ahmed and Ruffman, 1998; Spelke and von

Hofsten, in press) and at other times yield the same patterns of performance (e.g., see Hespos and Baillargeon, submitted)? We propose that a critical variable concerns the precision of the representation that a task requires. When reaching or preferential looking tasks require a precise representation, then infants' performance will be impaired by occlusion; when reaching or preferential looking tasks demand only an imprecise representation, then occlusion may have little effect on performance.

Consider, for example, the Ahmed and Ruffman (1998) experiments. In their preferential looking experiment, infants viewed an object that moved successively behind two screens and then was revealed either behind the last screen where it was hidden (consistent with continuity) or behind the other screen (inconsistent). To detect the inconsistency, infants had to represent that the object existed behind a given screen but did not need to represent the object's shape, height, or exact location. In their reaching experiment, in contrast, infants viewed an object that moved in the same sequence behind the two screens, and then they reached for it. To guide an appropriate reach for the object, infants had to represent not only that it existed behind the screen but also its shape, size, and exact location: without this information, appropriate reaching is impossible. If infants' representations of occluded objects are imprecise, therefore, they may succeed at the first task but fail the second, resorting instead to an action strategy ("do what worked before") not unlike the perceptual strategies described by Mehler and Bever (1967).

In contrast, consider the Hespos and Baillargeon (2001b, submitted) studies described above. In both the preferential looking and the reaching experiments, infants must represent the crude height of the frog in relation to the heights of the two occluders. Because the infant can obtain the frog by reaching for its visible foot, however, neither the reaching task nor the preferential looking task requires a precise representation of the hidden frog's location or shape. The two tasks therefore put equivalent demands on the precision of the infant's object representations and give equivalent findings.

Evidence that object representations are less precise when objects are occluded, and that this loss in precision affects infants' performance in preferential looking as well as reaching tasks, comes from studies by Baillargeon. In one well-known series of experiments, Baillargeon used a rotating screen task to test infant's representation of hidden objects (see

Baillargeon, 1993, for review). Infants were habituated to a screen that rotated through a 180-degree arc. In the test trials, a box was placed in the path of the rotating screen. In the consistent event, infants saw the screen come into contact with the box and reverse direction. In a set of inconsistent events, infants saw the screen rotate through part or all of the space that was occupied by the box. Infants as young at 3½ months old looked significantly longer when the screen rotated through all of the space occupied by the box, providing evidence that they represented the hidden box. In contrast, 4½-month-old infants failed to look longer at the inconsistent event when the screen rotated through 80 percent of the space occupied by the box, and 6½-month-old infants failed to look longer when it rotated through 50 percent of that space, although they did look longer at the 80 percent violation. Taken together, these studies suggest that there is a developmental increase in the precision of infants' representations of occluded objects. Experiments with visible reminders (a visible box next to an identical occluded box) further support this interpretation (Baillargeon, 1993).

The hypothesis that infants' object representations are more precise when objects are visible than when they are hidden could explain why infants reach predictively for moving, visible objects and look predictively at moving, temporarily occluded objects, but fail to reach predictively for moving, temporarily occluded objects. To catch a moving object, one must represent considerable information about the object, including its size, shape, path, and speed of motion. When the object is continuously visible, infants' representations evidently are adequate to guide appropriate reaching. When the object is hidden, however, their representation of its properties may become too imprecise to guide effective attempts to intercept it. Even an imprecise representation of an occluded object may suffice, however, to guide a look in the correct direction.

The visibility hypothesis likely applies not only to the object representations formed by infants but also to those formed by adults. To test whether loss of visibility impairs the precision of object representations throughout development, we ran an experiment on adults, modeled on the studies of predictive reaching in infants. The adults stood on the right side of a large board containing a target (the same display used in the reaching studies with infants) that moved by means of a hidden, motor-controlled magnet from the left side of the board in a linear path halfway across the board. The task for the adults was to extend the linear trajectory of the target to the right side of the board and to mark the point at the end of the board where the object could be intercepted. There were six different linear trajectories similar to those used with infants. In the first part of the experiment, the object was continuously visible. In the second part of the experiment, we introduced a large screen that occluded the second half of the object's trajectory—roughly the same part of the trajectory that was occluded for infants. The adults viewed the first part of the trajectory, saw the target pass under the occluder, and again made predictions about the endpoint of the trajectory. The adults were significantly more accurate in the visible condition than in the occluder condition. For adults, as for infants, object representations are more precise when objects are visible.

Despite its advantages, the visibility hypothesis cannot explain why babies reach more for objects in the dark than for occluded objects. This finding motivates the next hypothesis.

Hypothesis 4: Representations of Objects Are More Precise When No Other Objects Compete for Attention

Munakata and Stedron (in press) propose that infants' object representations, like those of adults, are competitive: the more an infant attends to one object, the less precise will be his or her representations of other objects. When an object is hidden behind an occluder, therefore, it may suffer a double loss of precision due both to its lack of visibility and to competition from its visible occluder. In contrast, an object that vanishes into darkness suffers a loss of precision only because of its lack of visibility and so should be represented more precisely than an occluded object, though less precisely than a visible one.

Jansson and von Hofsten (submitted) conducted an initial test of Munakata and Stedron's competition hypothesis by comparing infants' predictive reaching for objects that were obscured by darkness vs. occlusion. Six-month-old infants reached for a moving object on a series of trials in which the object was fully visible and then on a series of trials in which it was made invisible for one of three durations by one of two means: occlusion vs. darkness. The experiment revealed three effects. First, infants reached most frequently and accurately when the object was continuously visible, consistent with the visibility hypothesis. Second, reaching was more impaired by longer than by shorter periods of invisibility. Third, reaching was more impaired by occlusion than by darkness. All these findings are consistent with the thesis that reaching requires a precise representation of the goal object and that both loss of visibility and competition from a visible occluder reduce the precision of infants' representations.

Like the visibility hypothesis, the competition hypothesis likely applies to adults as well as infants. In our study of adults' predictive reaching, we investigated this possibility by presenting the same subjects who participated in the visible and occluder conditions with a third condition, in which the lights in the room were extinguished during the object's motion. The period of darkness was timed such that the object was visible and invisible at the same times and places in the darkness condition as in the occluder condition. Performance during the darkness condition, however, was significantly better than performance during the occluder condition. These findings support Munakata and Stedron's proposal that object representations are competitive both for adults and for infants.

Thus far, we have attempted to explain infants' performance on object permanence tasks by proposing that infants represent objects from very early ages, and that the precision of their representations is affected by two factors that also affect the precision of object representations in adults: visibility and competition. Together, the visibility and competition hypotheses support a continuity theory of object concept development, whereby infants and adults show common representational capacities and mechanisms. The continuity view nevertheless continues to face one problem: How can it account for developmental changes in infants' performance on object permanence tasks? To address this question, we propose one developmental change in object representations.

Hypothesis 5: Object Representations Show Qualitative Continuity over Human Development but Become Increasingly Precise as Infants Grow

Although adults reach best for fully visible objects, next best for objects obscured by darkness, and least well for objects obscured by occluders, they manage to reach for objects in all these cases. Why, in contrast, do young infants fail to reach at all when objects are occluded? We suggest that infants' basic capacities for representing objects are constant over development, but that their object representations increase gradually in precision as they grow. Young infants may fail to reach for occluded objects under many circumstances because the precision of their representations is too low to guide the hand effectively. At older ages, reaching still is less accurate when objects are occluded (as, indeed, it is for adults), but the precision of object representations rises to the point where reaching can be attempted. Gradual, quantitative changes in the precision of object representations therefore give rise to a qualitative change in infants' performance on Piaget's search tasks.

The thesis that object representations become more precise with growth and development is supported by Baillargeon's (1991, 1993) developmental studies of infants' reactions to inconsistent partial rotations in the rotating screen task. It gains further plausibility from studies of perceptual development, which have documented developmental increases in the precision of perceptual representations for a variety of tasks, including grating acuity, contrast sensitivity, motion detection, and auditory localization (see Kellman and Arterberry, 1998, for review). Research in progress is testing this hypothesis further by investigating developmental changes, from six to twelve months, in infants' predictive reaching for visible and temporarily invisible objects. As both the age of the infant and the duration of interrupted visibility vary, we expect to see variations in performance suggestive of a continuous process of development, rather than a qualitative, age-related shift caused by the emergence of a new representational capacity. Initial findings are consistent with this prediction.

Conclusions

Both Piaget's number conservation tasks and his object permanence tasks reveal striking developmental changes in children's performance. In both cases, however, we believe that children's performance depends on cognitive capacities that are continuous over human development. In particular, human infants have a capacity to represent occluded objects as early as two months of age, and that capacity undergoes no qualitative reorganization as children grow.

The evidence for ontogenetic continuity discussed here complements evidence for phylogenetic continuity in the capacity to represent objects. In particular, monkeys represent objects similarly to human infants both in preferential looking and in object search tasks (Hauser, MacNeilage, and Ware, 1996; Antinucci, 1990), and they progress through Piaget's stage sequence more rapidly than human infants do: a pattern that undermines views likening infants to scientists who form and change their theories (see Carey and Spelke, 1996). Most dramatically, newly hatched chicks have now been shown to pass the object search tasks that human infants fail until nine months of age (Regolin, Vallortigara, and Zanforlin, 1995). These findings make little sense if one thinks that search for hidden objects depends on the construction of a new conception of the world. They mesh well, however, with the view that basic mechanisms of object representation are constant over much of evolution and ontogeny, that their expression depends in part on the developing precision of representations, and that this development occurs at different rates for different species.

Because infants' actions on objects do change over development, the thesis that infants have a constant capacity for object representation requires that one distinguish competence from performance and that one analyze the factors that limit infants' performance at young ages, as did Mehler and Bever (1967). Our attempt to provide this analysis here suggests that two factors—visibility and competition—limit actions on objects in the same ways for infants and for adults. Infants and adults appear to form more precise representations of visible than of hidden objects, and they appear to maintain more precise representations when an object disappears into darkness than when it disappears behind a visible occluder. This analysis suggests there is developmental continuity not only in representational capacities but also in some of the factors that influence the expression of those capacities.

Jacques Mehler has spent a good part of his career conducting research based on the thesis that one can observe in infants the core of capacities we continue to exercise as adults. Guided by this thesis, he has crafted studies of infants that shed light on central aspects of speech and language in adults. Object concept development, however, has long seemed to provide a major challenge to any continuity theory. In Piaget's day, infants were thought to progress from an egocentric universe centered on their own actions to an objective world in which they viewed themselves as one entity among many. In more recent times, developmental changes in actions on objects were proposed to reflect other qualitative changes, including an emerging coordination in the functioning of multiple visual pathways (Bertenthal; 1996; Spelke et al., 1995), a progression from qualitative to quantitative reasoning (Baillargeon, 1995), an emerging capacity for recall (Mandler, 1992), or an emerging propensity for generating explanations (Baillargeon, 1998). In contrast to all these possibilities, we suggest that the nature and limits to human representations of objects are the same for infants and adults, and only the precision of those representations gradually changes. If we are right, then Mehler and Bever's approach to task-dependence, emphasizing developmental continuity and the competence-performance distinction, is as fitting today as it was in the 1960s.

References

Ahmed, A., and Ruffman, T. (1998). Why do infants make A not B errors in a search task, yet show memory for the location of hidden objects in a non-search task? *Developmental Psychology, 34,* 441–453.

Antinucci, F. (1990). The comparative study of cognitive ontogeny in four primate species. In P. S. Taylor and G. K. Rita (Eds.), *"Language" and intelligence in monkeys and apes: Comparative developmental perspectives* (pp. 157–171). New York: Cambridge University Press.

Baillargeon, R. (1991). Reasoning about the height and location of a hidden object in 4.5- and 6.5-month-old infants. *Cognition, 38,* 13–42.

Baillargeon, R. (1993). The object concept revisited: New directions in the investigation of infants' physical knowledge. In C. E. Granrud (Ed.), *Visual perception and cognition in infancy. Carnegie-Mellon Symposia on Cognition.* Vol. 23. Hillsdale, NJ: Erlbaum.

Baillargeon, R. (1994). Physical reasoning in young infants: Seeking explanations for unexpected events. *British Journal of Developmental Psychology, 12,* 9–33.

Baillargeon, R. (1995). Physical reasoning in infancy. In M. S. Gazzaniga (Ed.), *The cognitive neurosciences.* Cambridge, MA: MIT Press.

Baillargeon, R. (1998). Infants' understanding of the physical world. In M. Sabourin, F. Craik, and M. Robert (Eds.), *Advances in psychological science.* Vol. 2 (pp. 503–529). London: Psychology Press.

Bertenthal, B. I. (1996). Origins and early development of perception, action, and representation. *Annual Review of Psychology, 47,* 431–459.

Berthier, N. E., DeBlois, S., Poirier, C. R., Novak, M. A., and Clifton, R. K. (2000). Where's the ball? Two- and three-year-olds reason about unseen events. *Developmental Science, 36,* 394–401.

Bogartz, R., Shinsky, J., and Speaker, C. (1997). Interpreting infant looking: The event set by event set design. *Developmental Psychology, 33,* 408–422.

Carey, S., and Spelke, E. S. (1996). Science and core knowledge. *Philosophy of Science, 63,* 515–533.

Clements, W. A., and Perner, J. (1994). Implicit understanding of belief. *Cognitive Development, 9,* 377–395.

Clifton, R. K., Rochat, P., Litovsky, R., and Perris, E. (1991). Object representation guides infants' reaching in the dark. *Journal of Experimental Psychology: Human Perception and Performance, 17,* 323–329.

Haith, M. M., and Benson, J. (1997). Infant cognition. In D. Kuhn and R. Siegler (Eds.), *Handbook of Child Psychology,* 5th ed. Vol. 2: *Cognition, perception and language development.* New York: Wiley.

Hauser, M. D., MacNeilage, P., and Ware, M. (1996). Numerical representations in primates. *Proceedings of the National Academy of Sciences of the United States of America, 93,* 1514–1517.

Hespos, S. J., and Baillargeon, R. (2001a). Knowledge about containment events in very young infants. *Cognition, 78,* 204–245.

Hespos, S. J., and Baillargeon, R. (2001b). Infants' knowledge about occlusion and containment: A surprising decalage. *Psychological Science, 12,* 140–147.

Hespos, S. J., and Baillargeon, R. (submitted). Reasoning about support, occlusion, and containment events: Evidence from object-retrieval tasks.

Hespos, S. J., and Rochat, P. (1995). Dynamic representation in infancy. *Cognition, 64,* 153–189.

Hood, B. M. (1995). Gravity rules for 2- to 4-year-olds? *Cognitive Development, 10,* 577–598.

Hood, B. M., Carey, S., and Prasada, S. (2000). Predicting the outcomes of physical events: Two-year-olds fail to reveal knowledge of solidity and support. *Child Development, 71,* 1540–1554.

Hood, B., and Willatts, P. (1986). Reaching in the dark to an object's remembered position: Evidence for object permanence in 5-month-old infants. *British Journal of Developmental Psychology, 4,* 57–65.

Jansson, B., and von Hofsten, C. (submitted). Infants' ability to track and reach for temporarily occluded objects.

Kellman, P. J., and Arterberry, M. E. (1998). *The cradle of knowledge: The development of perception in infancy.* Cambridge, MA: MIT Press.

Koechlin, E., Dehaene, S., and Mehler, J. (1998). Numerical transformations in five-month-old human infants. *Mathematical Cognition, 3,* 89–104.

Mandler, J. M. (1992). The foundations of conceptual thought in infancy. *Cognitive Development, 7,* 273–285.

Mehler, J., and Bever, T. G (1967). Cognitive capacity of very young children. *Science, 158*, 141–142.

Mix, K. S., Huttenlocher, J., and Levine S. C. (1996). Do preschool children recognize auditory-visual numerical correspondences? *Child Development, 67*, 1592–1608.

Moscovitch, M. (1992). Memory and working with memory: A component process model based on modules and central systems. *Journal of Cognitive Neuroscience, 4*, 257–267.

Munakata, Y., and Stedron, J. (in press). Memory for hidden objects in early infancy: Behavior, theory, and neural network simulation. In J. W. Fagen and H. Hayne (Eds.), *Advances in infancy research*. Vol. 14. Norwood, NJ: Ablex.

Munakata, Y., McClelland, J. L., Johnson, M. H., and Siegler, R. S. (1997). Principles, processes, and infant knowledge: Rethinking successes and failures in object permanence tasks. *Psychological Review, 104*, 686–713.

Perner, J., and Wimmer, H. (1988). Misinformation and unexpected change: Testing the development of epistemic-state attribution. *Psychological Research, 50*, 191–197.

Piaget, J. (1954). *The construction of reality in the child*. New York: Basic Books.

Regolin, L., Vallortigara, G., and Zanforlin, M. (1995). Object and spatial representations in detour problems by chicks. *Animal Behavior, 49*, 195–199.

Rumelhart D. E., and McClelland, J. L. (1986). PDP models and general issues in cognitive science. In D. E. Rumelhart, J. L. McClelland, and PDP Research Group, (Eds.), *Parallel distributed processing*. Vol. 1: *Foundations*. Cambridge, MA: MIT Press.

Scholl, B. J., and Leslie, A. M. (1999). Explaining the infants' object concept: Beyond the perception/cognition dichotomy. In E. Lepore and Z. Pylyshyn (Eds.), *What is cognitive science?* Oxford: Blackwell.

Simon, T., Hespos, S., and Rochat, P. (1995). Do infants understand simple arithmetic? A replication of Wynn (1992). *Cognitive Development, 10*, 253–269.

Spelke, E. S. (1988). Where knowledge begins. Physical conceptions in infancy. In H. Azuma (Ed.), *Ninth Biennial Meeting of International Society for the Study of Behavioural Development*. Tokyo: The Center of Developmental Education and Research.

Spelke, E. S. (1998). Nativism, empiricism, and the origins of knowledge. *Infant Behavior and Development, 21*, 183–202.

Spelke, E. S., and von Hofsten, C. (in press). Predictive reaching for occluded objects by six-month-old infants. *Journal of Cognition and Development*.

Spelke, E. S., Breinlinger, K., Macomber, J., and Jacobson, K. (1992). Origins of knowledge. *Psychological Review, 99*, 605–632.

Spelke, E. S., Kestenbaum, R., Simons, D., and Wein, D. (1995). Spatiotemporal continuity, smoothness of motion and object identity in infancy. *British Journal of Developmental Psychology, 13*, 113–142.

Spelke, E. S., Vishton, P., and von Hofsten, C. (1995). Object perception, object-directed action, and physical knowledge in infancy. In M. Gazzaniga (Ed.), *The cognitive neurosciences*. Cambridge, MA: MIT Press.

Thelen, E., and Smith L. B. (1993). *A dynamical systems approach to the development of cognition and action*. Cambridge, MA: MIT Press.

Uller, C., Carey, S., Huntley-Fenner, G., and Klatt, L. (1999). What representations might underlie infant numerical knowledge? *Cognitive Development, 14,* 1–36.

von Hofsten, C. (1980). Predictive reaching for moving objects by human infants. *Journal of Experimental Child Psychology, 30,* 369–382.

von Hofsten, C., Feng, Q., and Spelke, E. S. (2000). Object representation and predictive action in infancy. *Developmental Science, 3,* 193–205.

von Hofsten, C., Vishton, P., Spelke, E. S., Rosander, K., and Feng, Q. (1998). Principles of predictive action in infancy: Tracking and reaching for moving objects. *Cognition, 67,* 255–285.

Wynn, K. (1992). Addition and subtraction in infants. *Nature, 358,* 749–750.

Wynn, K. (1998). Psychological foundations of number: Numerical competence in human infants. *Trends in Cognitive Sciences, 2,* 296–303.

Xu, F., and Carey, S. (1996). Infants' metaphysics: The case of numerical identity. *Cognitive Psychology, 30,* 111–153.

19

Infants' Physical Knowledge: Of Acquired Expectations and Core Principles

Renée Baillargeon

Over the past ten years, there have been at least two distinct trends in the research on infants' understanding of the physical world. Spelke and her colleagues (e.g., see Spelke, 1994; Spelke et al., 1992; Spelke, Phillips, and Woodward, 1995) have sought to ascertain whether core principles constrain, from a very early age, infants' interpretations of physical events. Two of the core principles proposed by Spelke are those of *continuity* (objects exist and move continuously in time and space) and *solidity* (two objects cannot exist at the same time in the same space).

Other investigators, myself included (e.g., see Aguiar and Baillargeon, 1999; Baillargeon, 1991; Hespos and Baillargeon, 2001b; Kotovsky and Baillargeon, 1998; Needham and Baillargeon, 1993; Sitskoorn and Smitsman, 1995; Wilcox, 1999), have attempted to uncover how infants' physical knowledge develops—what expectations are acquired at what ages, and what learning processes make possible these acquisitions.

Although until recently these two lines of investigation have coexisted largely independently, these carefree days are now over. The more we find out about how infants acquire their physical knowledge, the more absorbed we become by questions concerning the potential role of core principles in infants' interpretations of physical events.

This chapter is organized into two main sections. In the first, I review recent findings from my laboratory and elsewhere on infants' acquisition of physical knowledge. In the second section, I consider the implications of some of these findings for Spelke's (e.g., see Spelke, 1994; Spelke et al., 1992, 1995) claim that, from a very early age, continuity and solidity principles guide infants' reasoning about physical events. In particular, I point out that these findings indicate that infants *fail* to detect many salient continuity and solidity violations. I then propose a way in which these failures might be reconciled with Spelke's claim, and suggest a possible experimental test of this approach.

It seemed fitting to offer these speculations in the present context because Jacques Mehler, as we all know, has always been extremely supportive of new ideas in infant cognition. Unlike most journal editors, who seem inclined to tie their authors' hands and feet, Jacques Mehler, as editor of *Cognition*, has always allowed his authors sufficient rope to leap to new and provocative conclusions (or, of course, to hang themselves, depending on one's point of view). I am very grateful to Jacques for his openness and support over the years, and humbly dedicate the following pages to him.

How Do Infants Learn about the Physical World?

Infants' Identification of Initial Concepts and Variables

For many years, my collaborators and I have been exploring infants' acquisition of physical knowledge (for reviews, see Baillargeon, 1995, 1998, and Baillargeon, Kotovsky, and Needham, 1995). We have found that, when learning about support, occlusion, collision, and other physical events, infants first form an *initial concept* centered on a primitive, all-or-none distinction. With further experience, infants identify a sequence of discrete and continuous *variables* that refine and elaborate this initial concept, resulting in increasingly accurate predictions and interpretations over time. To illustrate this general pattern, I briefly describe the results of experiments on infants' expectations about support and occlusion events.

Support Events In our experiments on the development of infants' knowledge about support events (e.g., see Baillargeon, Needham, and DeVos, 1992; Needham and Baillargeon, 1993; for reviews, see Baillargeon, 1995, 1998, and Baillargeon et al., 1995), infants aged 3 to 12½ months were presented with support problems involving a box and a platform; the box was held in one of several positions relative to the platform, and the infants judged whether the box should remain stable when released. The results indicated that, by 3 months of age, infants have formed an initial concept of support centered on a simple *contact/*

no-contact distinction: they expect the box to remain stable if released in contact with the platform and to fall otherwise. At this stage, any contact with the platform is deemed sufficient to ensure the box's stability. In the months that follow, infants identify a sequence of variables that progressively revise and elaborate their initial concept. At about 4½ to 5½ months of age, infants begin to take into account the *type of contact* between the box and the platform. Infants now expect the box to remain stable when released on but not against the platform. At about 6½ months of age, infants begin to consider the *amount of contact* between the box and the platform. Infants now expect the box to remain stable only if over half of its bottom surface rests on the platform.[1] At about 8 months of age, infants begin to distinguish between situations in which the *side or middle portion* of the box's bottom surface rests on a platform; they recognize that, in the latter case, the box can be stable even if less than half of its bottom surface is supported.[2] Finally, at about 12½ months of age, infants begin to attend to the *proportional distribution* of the box; they realize that an asymmetrical box can be stable only if the proportion of the box that rests on the platform is greater than that off the platform.

Occlusion Events In our experiments on the development of young infants' expectations about occlusion events (e.g., see Aguiar and Baillargeon, 1999, in press; Baillargeon and DeVos, 1991; Luo, 2000; for reviews, see Baillargeon, 1998, 1999), infants aged 2½ to 3½ months watched a toy move back and forth behind a large screen; next, a portion of the screen was removed, and the infants judged whether the toy should remain hidden or become (at least partly) visible when passing behind the screen. The results indicated that, by 2½ months of age, infants have formed an initial concept of occlusion centered on a simple *behind/not-behind* distinction. When the entire midsection of the screen is removed to form two separate screens, infants expect the toy to become visible in the gap between them. However, if the screens remain connected at the top or at the bottom by a narrow strip, infants no longer expect the toy to become visible: they view the connected screens as a single screen, and they expect the toy to be hidden when behind it. Over the course of the next month, infants rapidly progress beyond their initial concept. At about 3 months of age, infants begin to consider the presence of a *discontinuity in the lower edge* of the screen. Although infants still expect the toy to remain hidden when passing behind two screens that are connected at the bottom by a narrow strip, they now expect the toy to become visible when passing behind two screens that are connected at the top by a narrow strip. Finally, at about 3½ months of age, infants begin to consider the relative *heights* of the toy and screen. When the toy passes behind two screens that are connected at the bottom by a narrow or wide strip, infants expect the toy to become partly visible if it is taller but not shorter than the strip.

Infants' Formation of Event Categories

How general or specific are the expectations that infants acquire about physical events? Do infants acquire *general* expectations that are applied broadly to all relevant events, or *specific* expectations that remain tied to the events where they are first acquired? Our initial investigations of infants' physical knowledge could not provide an answer to this question, because they focused on events such as support and occlusion events that implicated very different expectations. In recent experiments, my collaborators and I have begun comparing infants' acquisition of similar expectations across events (e.g., see Hespos, 1998; Hespos and Baillargeon, 2001a; Wang and Paterson, 2000). The experiments test whether an expectation revealed in the context of one event (e.g., height in occlusion events) is typically also revealed in the context of other relevant events (e.g., height in containment events).

The results we have obtained to date do not support the notion that infants acquire general expectations that are applied broadly to all relevant events. Rather, our results suggest that infants' expectations are *event-specific*: infants appear to "sort" physical events into narrow event categories, and to learn separately how each category operates. A variable acquired in the context of one event category is not generalized to other relevant categories; it is kept tied to the specific category where it is first identified. As a result, infants must sometimes "relearn" in one event category a variable they have already acquired in another category. When weeks or months separate the acquisition of the variable in the two categories, striking lags (or, to borrow a Piagetian term, décalages) can be observed in infants' responses to events from the two categories. To illustrate this pattern, I briefly describe the results of recent experiments on infants' responses to height and transparency information in occlusion, containment, and other events.

Height Information In a first series of experiments (Hespos and Baillargeon, 2001a), 4½- to 7½-month-old infants saw an object being lowered behind an occluder or inside a container; the heights of the object and occluder or container were varied, and the infants judged whether the object could be fully or only partly hidden. The occlusion and containment events were made as perceptually similar as possible (e.g., in some of the experiments, the occluders were identical to the containers with their backs and bottoms removed; at the start of the experiment, the occluders and containers were rotated forward so that the infants could inspect them). The results indicated that, at 4½ months of age, infants are surprised to see a tall object become fully hidden behind a short occluder. In marked contrast, 4½-, 5½, and 6½-month-old infants are *not* surprised to see a tall object become fully hidden inside a short container; only 7½-month-old infants reliably detect this violation. These results, together with those discussed in the last section, suggest that although infants realize at about 3½ months of age that the height of an object relative to that of an occluder determines whether the object can be fully or only partly hidden when behind the occluder (Baillargeon and DeVos, 1991), it is not until four months later, at about 7½ months of age, that infants realize that the height of an object relative to that of a container determines whether the object can be fully or only partly hidden when inside the container.[3]

In a second series of experiments (Wang and Paterson, 2000), 9-month-old infants saw an object either being lowered inside a container, being lowered inside a tube, or being covered with a rigid cover; the height of the container, tube, or cover was varied, and the infants judged whether the object could be fully or only partly hidden. As before, efforts were made to render the events as perceptually similar as possible (e.g., the tubes were identical to the containers with their bottoms removed, and the covers were identical to the containers turned upside down; prior to the experiment, the infants were allowed to inspect the containers, tubes, or covers). As expected, given the results of the previous experiments, the data showed that 9-month-old infants are surprised to see a tall object become fully hidden inside a short container. However, infants this age are not surprised to see a tall object become fully hidden inside a short tube or under a short cover. We are currently testing older infants to find out at what age infants begin to realize that the height of an object relative to that of a tube or cover determines whether the object can be fully or only partly hidden when inside the tube or under the cover.

Together, the results of these experiments suggest that infants view events involving occluders, containers, tubes, and covers as belonging to separate categories, and do not generalize information acquired in one category to the others. Infants begin to consider height information in occlusion events at about 3½ months of age, in containment events at about 7½ months of age, and in events involving tubes and covers at some point beyond 9 months of age.

Transparency Information In an ongoing series of experiments (Luo and Baillargeon, in preparation), 8½- and 10-month-old infants see an object being lowered behind a transparent occluder or inside a transparent container (the occluder and container are made of Plexiglas and their edges are outlined with red tape; the infants are allowed to inspect the occluder or container prior to being tested). The experiments examine whether the infants realize that the object should be visible through the occluder when placed behind it, or through the front of the container when placed inside it. The occluder and container events are highly similar perceptually (e.g., the occluder is identical to the front of the container). Our results to date indicate that, at 8½ months of age, infants expect an object to be visible when lowered behind a transparent occluder, but not when lowered inside a transparent container. It is not until infants are about 10 months of age that they are surprised when an object is lowered inside a transparent container which is then revealed to be empty. We are now conducting experiments with younger infants to find out at what age infants first succeed at reasoning about transparency information in occlusion events.

These transparency experiments provide further evidence that infants view containment and occlusion events as belonging to distinct categories, and learn separately about each category. Infants identify the variable transparency first in the context of occlusion events, and only after some time in the context of containment events.

Additional Remarks On reflection, it is not very surprising that infants should use a learning strategy of forming narrow event categories and identifying variables separately for each category. Overall, this strategy must greatly facilitate infants' acquisition of physical knowledge; after all, breaking down the task of learning into smaller, more manageable components is a time-honored solution to the difficulties of knowledge acquisition.

Future research will need to address many questions about the nature and formation of infants' event categories. For example, on what basis are these categories generated? Why are occlusion and containment, in particular, regarded as distinct categories? In many cases (and contrary to those examined in this section), occlusion and containment outcomes differ: for example, an object that has been lowered inside a container typically moves with it when displaced, whereas an object that has been lowered behind an occluder does not. Could such causal regularities (which even 2½-month-old infants can detect; Hespos and Baillargeon, 2001b) provide the basis for infants' event categories (e.g., see Keil, 1995; Leslie, 1994; Pauen, 1999)?

What of other distinctions infants appear to draw, such as that between events including containers and tubes? Do infants recognize that in some cases tube outcomes differ from containment outcomes (e.g., an object that has been lowered inside a tube typically moves with it when slid to the side but not when lifted)? Or do infants possess a notion of a prototypical container, and do not categorize as containment events involving tubes or other nonprototypical containers (e.g., a box with a back much taller than its other three sides)?

Finally, at what point in development do infants begin to weave together their knowledge of different event categories? And what role do language and other cognitive processes play in this unification or redescription process (e.g., see Karmiloff-Smith, 1992)?

How Do Infants Identify Variables?

The results presented in the previous sections suggest that infants form narrow event categories and identify variables separately for each category. How do infants go about identifying these variables? My colleagues and I (e.g., see Aguiar and Baillargeon, 1999; Baillargeon, 1999; Hespos and Baillargeon, 2001b) have proposed that what typically triggers the identification of a variable in an event category is exposure to contrastive outcomes that cannot be explained or predicted by infants' current knowledge of the category. When infants register these contrastive outcomes, they seek out the conditions that map onto the outcomes.[4] Identification of these condition-outcome relations signals the identification of a new variable.[5]

This brief description leaves many questions unanswered about the process responsible for infants' identification of variables. Clearly, a great deal of research will be needed to fully specify the nature of this process. Nevertheless, it is possible to offer educated guesses about some of the factors likely to affect the ages at which specific variables are identified. Two such factors are briefly discussed below.

Exposure to Relevant Outcomes One factor likely to affect the age at which infants identify a variable is age of exposure to contrastive outcomes for the variable. Obviously, if infants are not exposed to contrastive outcomes for a variable, they will not begin the process of seeking out the conditions responsible for the outcomes. To illustrate, consider the finding, discussed earlier, that infants do not identify amount of contact as a support variable until about 6½ months of age (e.g., see Baillargeon et al., 1992). We have suggested that infants do not acquire this variable sooner in part because they are not exposed to appropriate contrastive outcomes sooner. In their daily lives, infants often see their caretakers place objects on supports (e.g., plates on tables or bottles on counters). However, in most instances, the objects remain stable when released; only in rare accidental cases do the objects fall. Hence, it is typically not until infants themselves begin to deposit objects on supports (presumably after 6 months of age, when they begin to sit independently; e.g., see Rochat, 1992) that they finally have the opportunity to notice that objects placed on supports sometimes remain stable and sometimes do not. At that point, infants begin to seek out the conditions that are responsible for these different outcomes, and eventually come to the conclusion that an object on a support can be stable when a large but not a small portion of its bottom surface rests on the support.[6]

Availability of Data on Relevant Conditions Another factor likely to affect the age at which infants identify a variable is how easy it is for them, after they are exposed to the relevant contrastive outcomes, to uncover the conditions that map onto the outcomes. To illustrate, consider the finding, discussed in the preceding section, that infants do not identify height as a containment variable until about 7½ months of age (Hespos, 1998; Hespos and Baillargeon, 2001a). In order to identify this variable, infants must be able to encode information about the heights of objects and containers. Prior research (e.g., see Baillargeon, 1991, 1994, 1995) suggests that, when infants begin to reason about a continuous variable in an event category, they can do so qualitatively, but not quantitatively: they cannot encode and remember information about absolute amounts.[7] To encode information about the heights of objects and containers qualitatively, infants must compare them as they stand side by side. Unfortunately, infants may witness relatively few instances in which objects are placed first next to and then inside containers; caretakers will more often insert objects directly into containers, allowing infants no opportunity to compare their heights. In the scenario outlined here, infants would thus notice that objects lowered inside containers are sometimes fully and sometimes only partly hidden. However, infants would have difficulty collecting data about the relative heights of the objects and containers, because they would have limited opportunities (perhaps until they themselves begin placing objects in containers) to see the objects standing next to the containers.

Additional Remarks The preceding speculations suggest possible explanations for the lags described earlier in infants' identification of similar variables across event categories. Consider, for example, the findings that infants identify height as an occlusion variable at about 3½ months of age (Baillargeon and DeVos, 1991), and as a containment variable at about 7½ months of age (Hespos, 1998; Hespos and Baillargeon,

2001a). It may be, of course, that in their daily lives infants observe many more occlusion than containment events, and hence can learn about occlusion events earlier. However, another possibility, related to the second factor discussed above, is that infants can more easily collect qualitative data about the relative heights of objects and occluders than of objects and containers. In the case of occlusion, infants will not only see objects being lowered from above behind occluders—they will also see objects being pushed from the side behind occluders (e.g., as when a parent slides a cup behind a box, or a sibling steps behind an armchair). In these side occlusions, it will usually be possible for infants to qualitatively compare the heights of the objects and their occluders; infants will then be in a position to begin mapping conditions onto outcomes.

The importance placed here on the availability of qualitative observations for the identification of continuous variables makes a number of interesting developmental predictions. For example, this approach suggests that, in containment events, infants should learn the variable width before height, because each time an object is lowered inside a container infants can compare their relative widths. And indeed, findings by Sitskoorn and Smitsman (1995) and Aguiar and Baillargeon (2000) indicate that infants do identify width before height as a containment variable, at some (still undefined) point between 4 and 6 months of age. Another prediction is that, in occlusion events, the variables height and width should be identified at about the same time, assuming that infants are exposed to occlusions from above and from the side about equally often. Preliminary results (Baillargeon and Brueckner, 2000) support this prediction.

What about the additional findings that infants do not consider height information in events involving tubes or covers until some point beyond 9 months of age (Wang and Paterson, 2000; see also Baillargeon, 1995, for similar results with events involving nonrigid covers)? One possibility is that young infants are not exposed to events involving tubes and covers often enough, and with sufficient opportunity for qualitative height comparisons, to be able to identify height as a relevant variable.

One way to test the general approach presented here would be to conduct observational studies to assess how often infants are presented with various occlusion, containment, and other events. The rationale of the studies would be to determine whether age of identification of variables can indeed be predicted from age of exposure to relevant condition-outcome data. A second way to test our general approach (and one we are actively pursuing) is to attempt to "teach" infants variables they have not yet acquired. Our view suggests that infants should acquire variables sooner than they would otherwise if exposed in the laboratory to appropriate condition-outcome observations. For example, infants should be able to identify the variable height in containment events prior to 7½ months of age if shown objects being placed next to and then inside containers of varying heights. Although we have not yet attempted to "teach" infants about height in containment, other experiments designed to teach 11-month-old infants the variable proportional distribution (described earlier) in support events have been highly successful (e.g., see Baillargeon, Fisher, and DeJong, 2000; for reviews, see Baillargeon, 1998, 1999). In addition, ongoing experiments in which Su-hua Wang and I are attempting to teach 9-month-old infants the variable height in covering events appear promising.[8]

Infants' Failures to Detect Continuity and Solidity Violations

If infants' interpretation of physical events is constrained from a very early age by continuity and solidity principles, as Spelke (e.g., see Spelke, 1994; Spelke et al., 1992, 1995) has suggested, we might expect infants to consistently detect all salient violations of these principles. However, this is not the case: infants often fail to detect even marked continuity and solidity violations. To illustrate, consider once again six of the results presented earlier: (1) 2½-month-olds are surprised when an object disappears behind one screen and reappears from behind another screen—but not when the two screens are connected at the top by a narrow strip (Aguiar and Baillargeon, 1999; Luo, 2000); (2) unlike 2½-month-olds, 3-month-olds are surprised when an object fails to appear between two screens that are connected at the top by a narrow strip; however, they are not surprised when the object fails to appear between two screens that are connected at the bottom by a narrow strip (Aguiar and Baillargeon, in press; Baillargeon and DeVos, 1991; Luo, 2000);

(3) 4-month-olds are not surprised when a wide object is lowered inside a container with a narrow opening (Sitskoorn and Smitsman, 1995); (4) 4½- to 6½-month-olds are not surprised when a tall object is fully hidden inside a short container (Hespos, 1998; Hespos and Baillargeon, 2001a); (5) 8½-month-olds are not surprised when an object that has been lowered inside a transparent container is not visible through the front of the container (Luo and Baillargeon, in preparation); and finally (6) 9-month-olds are not surprised when a tall object is fully hidden inside a short tube or under a short cover (Wang and Paterson, 2000).

How can we make sense of these results (see also Baillargeon, 1991, 1993, 1995)? If continuity and solidity principles constrain infants' interpretations of physical events, shouldn't they be able to readily detect all of these violations?

In this section, I first outline some of the assumptions my collaborators and I hold about infants' representations of physical events. Next, I discuss how limitations in infants' representations could lead to their failure to detect even salient continuity and solidity violations. Finally, I sketch out a possible experimental test of the approach proposed here.

How Do Infants Represent Physical Events?

My collaborators and I have developed a number of assumptions about infants' representations of physical events (e.g., see Aguiar and Baillargeon, in press; Hespos and Baillargeon, 2001b); three of these assumptions are described below.

A first assumption is that, when observing physical events, infants build *physical representations* that focus on the physical properties, displacements, and interactions of the objects within the events. (Infants no doubt build several representations simultaneously, for different purposes. For example, another representation might focus on the features of the objects in the events, and be used for recognition and categorization purposes—to ascertain whether these particular objects, or similar objects, have been encountered in the past; e.g., see Needham and Modi, 2000).

A second assumption is that infants' physical representations of events are by no means faithful copies of the events: they are abstract, functional descriptions that include some but not all of the physical information in the events.

Finally, a third assumption is that how much information infants include in their physical representations of events depends in part on their knowledge of the variables likely to affect the events. We suppose that, early in the representation process, infants categorize the event they are observing (e.g., as an occlusion or a containment event), and then access their knowledge of the event category selected. This knowledge specifies what variables should be attended to as the event unfolds—in other words, what information should be included in the physical representation of the event. To illustrate, this last assumption means that 3½-month-old infants who see an object being lowered behind a container (occlusion event) will include information about the relative heights and widths of the object and container in their physical representation of the event, because they have already identified height and width as occlusion variables. In contrast, 3½-month-old infants who see an object being lowered inside rather than behind a container (containment event) will not encode the relative heights and widths of the object and container, because they have not yet identified height and width as containment variables.[9]

A Case of Impoverished Physical Representations

If one accepts the assumptions discussed in the previous section, it becomes clear how infants might possess core continuity and solidity principles and still fail to detect salient violations of these principles. Infants' core principles, like all of their physical knowledge, can only operate at the level of their physical representations (i.e., infants do not apply their expectations directly to events, only to their representations of the events). It follows that, when infants bring to bear their continuity and solidity principles onto their physical representations of events, they will succeed in detecting violations of the principles *only* when the key information necessary to detect the violations is included in the representations. Infants' principles can only guide the interpretation of information that is included in their physical representations; information that has not been represented cannot be interpreted.

To illustrate how incomplete physical representations could lead infants to ignore violations of their continuity and solidity principles, consider one of the findings discussed earlier, that 3-month-old infants are not surprised when an object fails to appear between two screens connected at the bottom by a narrow strip (Aguiar and Baillargeon, in press; Baillargeon and DeVos, 1991; Luo, 2000). What is being suggested is that, when observing such an event, 3-month-old infants typically do not include information about the relative heights of the object and occluder in their physical representation of the event. Thus, when infants apply their continuity principle to their incomplete physical representation of the event, they have no basis for realizing that a portion of the object should be visible above the narrow strip between the screens.

To give another example, consider the finding that 4½- to 6½-month-old infants are not surprised when a tall object becomes fully hidden inside a short container (Hespos, 1998; Hespos and Baillargeon, 2001a). What is being suggested is that, when observing such an event, infants aged 6½ months and younger typically do not include information about the relative heights of the object and container in their physical representation of the event. Thus, when infants apply their continuity principle to their incomplete representation of the event, they cannot appreciate that a portion of the object should be visible above the container.

How Are Infants' Physical Representations Enriched?

I suggested in the previous section that young infants might possess continuity and solidity principles and still fail to detect violations of these principles because of incomplete physical representations. One important process by which infants' physical representations of events become more complete over time must be the identification of variables, as discussed in previous sections. After infants identify height as an occlusion variable, at about 3½ months of age (Baillargeon and DeVos, 1991), they begin to routinely include information about the heights of objects and occluders in their physical representations of occlusion events. Similarly, after infants identify height as a containment variable, at about 7½ months of age (Hespos, 1998; Hespos and Baillargeon, 2001a), they begin to routinely include information about the heights of objects and containers in their physical representations of containment events. (What makes it so certain that infants, once they have identified a variable, routinely include information about this variable in their physical representations, is that separate tests of sensitivity to a variable, conducted on different infants and often with different experimental events, consistently produce similar results; compare, for example, the positive results of Baillargeon and DeVos, 1991, and Hespos and Baillargeon, 2001a on height in occlusion events, and of Hespos and Baillargeon, 2001a and Wang and Paterson, 2000, on height in containment events).[10]

However, there might also be a process by which infants can be *temporarily* induced to include certain key information in their representations of physical events. What if, for example, $4^{1}/_{2}$- to $6^{1}/_{2}$-month-old infants could somehow be "primed" to include height information when representing containment events? This possibility is particularly intriguing because it suggests a direct test of the speculations advanced in the last section. According to these speculations, it should not really matter whether infants include information in a physical representation because (1) they have been primed to do so by the experimental context or (2) they have already identified the pertinent variable. In either case, the information, once represented, should be subject to infants' continuity and solidity principles, making it possible to detect violations of the principles. To return to our containment example, this means that $4^{1}/_{2}$- to $6^{1}/_{2}$-month-old infants who were induced to include height information in their physical representations of containment events *should* be surprised when shown a tall object being fully lowered inside a short container (recall that infants do not normally detect this violation until about $7^{1}/_{2}$ months of age; Hespos and Baillargeon, 2001a). The infants' continuity principle would guide the interpretation of their (artificially enriched) representation, resulting in an enhanced performance at a younger age.

Although no investigation has yet attempted to prime infants' physical representations in just the way described here, a recent series of experiments by Wilcox and her colleagues (Wilcox, 1999; Chapa and Wilcox, 1998) suggests that such attempts will be effective. In preliminary experiments (Wilcox, 1999), infants saw an object move behind one side of a screen; after a pause, a different object emerged from behind the opposite side of the screen. The screen was either too narrow or sufficiently wide to hide the two objects simultaneously. The results indicated that, by $9^{1}/_{2}$ months of age, infants showed surprise at the narrow-screen event when the objects on the two sides of the screen differed in size, shape, and pattern, but not color; only $11^{1}/_{2}$-month-old infants showed surprise at the narrow-screen event involving a red and a green ball (red-green event). In subsequent experiments, Chapa and Wilcox (1998) attempted to induce 9½-month-old infants to include color information in their physical representation of the red-green event. The infants received two pairs of priming trials. In the first, a red cup was used to pour salt, and a green cup was used to pound a wooden peg; the second pair of trials was similar except that different red and green containers were used. After receiving these priming trials, the infants showed surprise at the red-green event. One interpretation of these findings, in line with the speculations above, is that the infants were primed to include color information in their physical representation of the red-green event; this added information then became subject to the infants' continuity and solidity principles, allowing them to detect the violation in the event.

Of course, there may be several different ways of priming infants to include key information in their physical representations of events. Su-hua Wang and I have begun testing a very different approach, in which we capitalize on the fact that infants routinely include height or width information when representing some events (e.g., occlusion events), to induce them to include similar information when representing subsequent events (e.g., covering events) involving the same objects. For example, in one experiment, 8-month-old infants see a short or a tall cover standing next to a tall object. To start, the cover is pushed *in front of* the object; the tall cover occludes all of the object, the short cover only its bottom portion. Next, the cover is lifted and lowered *over* the object, until it is fully hidden. As mentioned earlier, Wang and Paterson (2000) found that 9-month-old infants are not surprised when a tall object becomes fully hidden under a short cover. This new experiment thus asks whether infants might detect this violation if first shown an occlusion event involving the same cover and object. Our reasoning is as follows: once infants have included the relative heights of the cover and object in their physical representation of the initial, occlusion event, they might be inclined to do the same in—or have this information available for—their physical representation of the subsequent, covering event. This information would then be subject to infants' core principles, making it possible to detect the violation in the short-cover event.

The preceding speculations hopefully make clear the potential interest of priming experiments. Assuming that priming effects can be produced, much research will be needed to find out, for example, what manipulations are helpful for priming variables and what manipulations are not; whether priming some variables improves infants' performance but priming others does not (i.e., priming variables not linked to core principles should have no immediate effect on infants' ability to detect violations); and finally, what are the long-term effects of successful priming experiences and how they compare to those of successful "teaching" experiences (as discussed earlier; Baillargeon, 1998, 1999). As a result of this research, we should learn a great deal more about the contents of infants' physical representations, the processes by which they can be enhanced, and the core principles that guide their interpretation.

Acknowledgments

The preparation of this manuscript was supported by a grant from the National Institute of Child Health and Human Development (HD-21104). I thank Jerry DeJong, Emmanuel Dupoux, Cynthia Fisher, Yu-yan Luo, Kristine Onishi, and Su-Hua Wang for many helpful comments and suggestions.

Notes

1. Preliminary data from experiments with Su-Hua Wang suggest that at 6½ months of age infants expect an object to be stable only if *over* half of its bottom surface is supported; by 8 months of age, infants have refined this rule and expect an object to be stable if *half or more* of its bottom surface is supported.

2. Recent data by Dan, Omori, and Tomiyasu (2000) suggest that, initially, infants expect an object whose middle section rests on a support to be stable, even when the section supported is very narrow (e.g., a pumpkin resting on a pencil-thin block). Over time, however, infants come to appreciate that a sufficient portion of the object's middle section must be supported for it to be stable.

3. It might be assumed that the lag reported here simply reflects the fact that young infants possess a concept of occlusion but not containment. However, this interpretation is unlikely. Recent findings (Hespos and Baillargeon, 2001b) indicate that, by 2½ months of age, infants already possess expectations about containment events. In particular, infants (1) believe that an object continues to exist

after it disappears inside a container and (2) expect the object to move with the container when displaced.

4. The phrase "when infants register these contrastive outcomes" is important because infants could of course be exposed to contrastive outcomes without actually registering the differences between them.

5. From the present perspective, a variable is thus akin to a dimension; conditions correspond to values on the dimension, with each value (or discernible range of values) being associated with a distinct outcome (hence the emphasis placed here on contrastive outcomes).

6. This discussion might lead readers to assume that the learning process as described here is primarily error-driven: infants notice that a rule (objects remain stable when released *on* supports) leads to incorrect predictions (objects do not always remain stable when released on supports), and set about correcting it. However, we mean our analysis to be more general. In some cases, infants will begin to notice contrastive outcomes from a different facet of an event, one they had largely ignored until then. For example, some time after infants realize that objects move when hit, they begin to notice that objects may move longer or shorter distances when hit; eventually, infants identify some of the variables responsible for these different outcomes (Kotovsky, 1994; Kotovsky and Baillargeon, 1998). A similar example has to do with the duration of occlusions—how long objects remain hidden when passing behind occluders (e.g., see Wilcox and Schweinle, submitted). The process of identifying variables is thus not always error-driven; in some cases, infants begin to notice new facets of events, and then identify the variables that contribute to them.

7. The distinction between qualitative and quantitative reasoning strategies is derived from computational models of everyday physical reasoning (e.g., Forbus, 1984).

8. Before leaving this section, I would like to address one common criticism of the notion that infants' learning mechanism is typically triggered by exposure to contrastive outcomes that cannot be explained or predicted by infants' current knowledge. This criticism is that infants are obviously capable of acquiring knowledge about objects in the absence of contrastive outcomes. For example, infants no doubt learn about the shapes and colors of bananas and carrots simply by repeated exposure to these objects. I fully agree that infants can learn facts about individual objects or categories of objects in the absence of contrastive outcomes (e.g., see Kotovsky and Baillargeon, 1998). What I would argue, however, is that (1) infants possess several different learning mechanisms, each with its own purpose and requirements for learning; and (2) the mechanism responsible for the acquisition of facts about specific objects and object categories (e.g., bananas are yellow) is different from the one responsible for the acquisition of facts about physical objects in general (e.g., objects typically fall when released in midair).

9. This discussion raises interesting questions about what basic information infants include in their physical representation of an event when they know no

variable about the event (or indeed possess no relevant event category). For example, what information do 2½-month-old infants, who know few if any variables, typically include in their physical representations of events? And what factors are responsible for these contents?

10. For a discussion of a situation in which infants who have identified a variable may nevertheless fail to reason correctly about it, see Aguiar and Baillargeon (2000) on perseveration and problem solving in infancy.

References

Aguiar, A., and Baillargeon, R. (1999). 2.5-month-old infants' reasoning about when objects should and should not be occluded. *Cognitive Psychology, 39,* 116–157.

Aguiar, A., and Baillargeon, R. (in press). Developments in young infants' reasoning about occluded objects. *Cognitive Psychology.*

Aguiar, A., and Baillargeon, R. (2000). Perseveration and problem solving in infancy. In H. W. Reese (Ed.), *Advances in child development and behavior.* Vol. 27 (pp. 135–180). San Diego: Academic Press.

Baillargeon, R. (1991). Reasoning about the height and location of a hidden object in 4.5- and 6.5-month-old infants. *Cognition, 38,* 13–42.

Baillargeon, R. (1993). The object concept revisited: New directions in the investigation of infants' physical knowledge. In C. E. Granrud (Ed.), *Visual perception and cognition in infancy* (pp. 265–315). Hillsdale, NJ: Erlbaum.

Baillargeon, R. (1994). How do infants learn about the physical world? *Current Directions in Psychological Science, 3,* 133–140.

Baillargeon, R. (1995). Physical reasoning in infancy. In C. Rovee-Collier and L. P. Lipsitt (Eds.), *Advances in infancy research.* Vol. 9 (pp. 305–371). Norwood, NJ: Ablex.

Baillargeon, R. (1998). Infants' understanding of the physical world. In M. Sabourin, F. Craik, and M. Robert (Eds.), *Advances in psychological science.* Vol. 2 (pp. 503–529). London: Psychology Press.

Baillargeon, R. (1999). Young infants' expectations about hidden objects: A reply to three challenges [article with peer commentaries and response]. *Developmental Science, 2,* 115–163.

Baillargeon, R., and Brueckner, L. (2000). 3.5-month-old infants' reasoning about the width of hidden objects. Presented at the Biennial International Conference on Infant Studies, Brighton, UK, July 2000.

Baillargeon, R., and DeVos, J. (1991). Object permanence in young infants: Further evidence. *Child Development, 62,* 1227–1246.

Baillargeon, R., Fisher, C., and DeJong, G. (2000). Teaching infants about support: What data must they see? Presented at the Biennial International Conference on Infant Studies, Brighton, UK, July 2000.

Baillargeon, R., Kotovsky, L., and Needham, A. (1995). The acquisition of physical knowledge in infancy. In D. Sperber, D. Premack, and A. J. Premack (Eds.), *Causal cognition: A multidisciplinary debate* (pp. 79–116). Oxford: Clarendon Press.

Baillargeon, R., Needham, A., and DeVos, J. (1992). The development of young infants' intuitions about support. *Early Development and Parenting, 1,* 69–78.

Chapa, C., and Wilcox, T. (1998). Object, color, and function in object individuation. Presented at the Biennial International Conference on Infant Studies, Atlanta, April 1998.

Dan, N., Omori, T., and Tomiyasu, Y. (2000). Development of infants' intuitions about support relations: Sensitivity to stability. *Developmental Science, 3,* 171–180.

Forbus, K. D. (1984). Qualitative process theory. *Artificial Intelligence, 24,* 85–168.

Hespos, S. J. (1998). Infants' physical reasoning about containment and occlusion: A surprising décalage. Presented at the Biennial International Conference on Infant Studies, Atlanta, April 1998.

Hespos, S. J., and Baillargeon, R. (2001a). Infants' knowledge about occlusion and containment events: A surprising discrepancy. *Psychological Science, 12,* 141–147.

Hespos, S. J., and Baillargeon, R. (2001b). Knowledge about containment events in very young infants. *Cognition, 78,* 207–245.

Karmiloff-Smith, A. (1992). *Beyond modularity: A developmental perspective on cognitive science.* Cambridge, MA: MIT Press.

Keil, F. C. (1995). The growth of causal understandings of natural kinds. In D. Sperber, D. Premack, and A. J. Premack (Eds.), *Causal cognition: A multidisciplinary debate* (pp. 234–262). Oxford: Clarendon Press.

Kotovsky, L. (1994). 2.5-month-old infants' reasoning about collisions. Presented at the Biennial International Conference on Infant Studies, Paris, June 1994.

Kotovsky, L., and Baillargeon, R. (1998). The development of calibration-based reasoning about collision events in young infants. *Cognition, 67,* 311–351.

Leslie, A. M. (1994). ToMM, ToBy, and Agency: Core architecture and domain specificity. In L. A. Hirschfeld and S. A. Gelman (Eds.), *Mapping the mind: Domain specificity in cognition and culture* (pp. 119–148). New York: Cambridge University Press.

Luo, Y. (2000). Young infants' knowledge about occlusion events. Presented at the Biennial International Conference on Infant Studies, Brighton, UK, July 2000.

Luo, Y., and Baillargeon, R. (in preparation). Infants' reasoning about transparency in occlusion and containment events.

Needham, A., and Baillargeon, R. (1993). Intuitions about support in 4.5-month-old infants. *Cognition, 47,* 121–148.

Needham, A., and Modi, A. (2000). Infants' use of prior experiences in object segregation. In H. W. Reese (Ed.), *Advances in child development and behavior.* Vol. 27 (pp. 99–133). New York: Academic Press.

Pauen, S. (1999). The development of ontological categories: Stable dimensions and changing concepts. In W. Schnotz, S. Vosniadou, and M. Carretero (Eds.), *New perspectives on conceptual change* (pp. 15–31). Amsterdam: Elsevier.

Rochat, P. (1992). Self-sitting and reaching in 5- to 8-month-old infants: The impact of posture and its development on early eye-hand coordination. *Journal of Motor Behavior, 24,* 210–220.

Sitskoorn, S. M., and Smitsman, A. W. (1995). Infants' perception of dynamic relations between objects: Passing through or support? *Developmental Psychology, 31,* 437–447.

Spelke, E. S. (1994). Initial knowledge: Six suggestions. *Cognition, 50,* 431–445.

Spelke, E. S., Breinlinger, K., Macomber, J., and Jacobson, K. (1992). Origins of knowledge. *Psychological Review, 99,* 605–632.

Spelke, E. S., Phillips, A., and Woodward, A. L. (1995). Infants' knowledge of object motion and human action. In D. Sperber, D. Premack, and A. J. Premack (Eds.), *Causal cognition: A multidisciplinary debate* (pp. 44–78). Oxford: Clarendon Press.

Wang, S., and Paterson, S. (2000). Infants' reasoning about containers and covers: Evidence for a surprising décalage. Presented at the Biennial International Conference on Infant Studies, Brighton, UK, July 2000.

Wilcox, T. (1999). Object individuation: Infants' use of shape, size, pattern, and color. *Cognition, 72,* 125–166.

Wilcox, T., and Schweinle, A. (submitted). Infants' use of speed information to individuate objects in occlusion events.

20

Learning Language: What Infants Know about It, and What We Don't Know about That

Peter W. Jusczyk

The Initial State

Speaking and understanding language is one of the defining features of human nature. It is no wonder, then, that philosophers and scientists have long been intrigued about how infants learn language. My own fascination with language development began during my undergraduate days at Brown University when I had the great good fortune to work with Peter Eimas on one of the very first studies of infants' perception of speech. During my graduate studies at the University of Pennsylvania, I continued to be interested in language acquisition and when I became an assistant professor at Dalhousie University, I set up my own laboratory for investigating infant speech perception capacities.

In the fall of 1976, I was invited to give a talk at the University of Texas. As it turned out, they already had another visiting speaker scheduled for that day who happened to be speaking on the same topic as I was: studies of infant speech perception. At that time, the field itself was but a few years old, and given the handful of individuals working in this area, it was certainly an interesting coincidence that Jacques Mehler and I were visiting at the same time. That day, as we listened to each other talk, we discovered that our views about infants and their capacities for perceiving speech had a lot in common. Jacques suggested that we keep in touch, and thus we began a dialogue that ultimately led to a fruitful research partnership, including three years working together in his laboratory.

The field of infant speech perception research was still struggling to learn about the range of infants' speech perception capacities. The groundbreaking research of Peter Eimas and his colleagues had established the fact that infants could perceive phonetic contrasts long before they ever produced them (Eimas, 1974, 1975; Eimas et al., 1971). Suddenly, language acquisition researchers began to take seriously the possibility that infants are biologically prepared with some specialized capacities for acquiring language. Jacques's laboratory was the first outside of North America to explore the abilities of infants exposed to a language other than English. At the time, Jacques had a very strong bias against developmental studies. He believed that cataloging developmental changes in language abilities would not be of much use in elucidating the mechanisms underlying language acquisition. Instead, he proposed a framework whereby one would study language competence at the two extreme endpoints. In the newborn, one would study the "initial state" (an idealization of what the cognitive apparatus is before it had any exposure to the environment), and in the adult, one would study the "stable state" (an idealization of a processing system that is functional and unchanging). Jacques believed that only when one understood these two extreme endpoints would studying the mapping between these two be possible. Consequently, much of Jacques's infant research has been devoted to studying infants of a few days to a few weeks old. Among other things, this work has established that even in the first few days of life, infants have the capacity to discriminate phonetic contrasts.

A critical objective of early research in this field was to learn more about the nature of the mechanisms underlying infant speech perception capacities. How widespread are infants' capacities at birth and to what extent are these capacities specialized for acquiring language? For example, although we both reported findings demonstrating that parallels exist in how young infants process speech and nonspeech sounds (Jusczyk et al., 1980, 1983; Mehler and Bertoncini, 1979), Jacques also showed that infants are better at discriminating a contrast in well-formed than in ill-formed syllables (Bertoncini and Mehler, 1981). Not only did these findings support the view that infants engage in some specialized processing of language but they also raised the prospect that syllables play a crucial role in early speech processing. In another series of studies, Jacques and his colleagues examined infants' abilities to process the types of cues available in longer utterances, namely the suprasegmental properties of speech. Whereas other investigators had demonstrated that infants can discriminate suprasegmental differences (Jusczyk and Thompson, 1978; Kaplan, 1969; Kuhl and Miller, 1982; Morse, 1972; Spring and Dale, 1977), Jacques and his colleagues showed that two-week-olds use such information in discriminating their own mother's voice from that of a stranger (Mehler et al., 1978).

Of course, delineating the initial state of speech and language capacities is critical for understanding just how these capacities develop. So, despite Jacques's antidevelopmentalist stance, his studies have contributed in an important way to our understanding of the development of speech perception capacities.

What We Have Learned

Much has been learned about infant speech perception capacities since those early studies. In fact, the current focus of research in the field has shifted toward understanding how these capacities develop during the first year and what their role is in acquiring a native language. One important impetus in this direction came from the discovery by Werker and Tees (1984) that toward the end of the first year, infants begin to show a decline in sensitivity to certain phonetic contrasts that do not appear in the native language that they are acquiring. These findings alerted other researchers to the fact that developmental changes were occurring in speech perception capacities in the context of acquiring a language.

Other developmental changes were discovered in the acquisition of phonetic categories, phonotactic constraints, and prosodic structure. For example, Kuhl and her colleagues (1992) have suggested that infants begin to develop native language vowel categories by six months of age (but see also Polka and Bohn, 1996). Other findings indicate that between six and nine months of age, infants have acquired information about phonotactic constraints, that is, what kinds of phonetic sequences can appear together in native language words (Friederici and Wessels, 1993; Jusczyk et al., 1993b), and further, that they are sensitive to the frequency with which these patterns appear in words (Jusczyk, Luce, and Charles-Luce, 1994). In addition to learning about the segmental features of their native language, infants acquire a great deal of information about the prosodic organization of language during their first year. For example, we have shown that, by nine months, infants are sensitive to the way that prosody marks clausal (Hirsh-Pasek et al., 1987) and phrasal units (Gerken, Jusczyk, and Mandel, 1994; Jusczyk et al., 1992) in native language utterances. At the same age, they also give evidence of recognizing the predominant stress patterns of native language words (Jusczyk, Cutler, and Redanz, 1993a).

In the meantime, Jacques and his colleagues continued their joint exploration of the initial state and the stable state. They wondered about the fact that children raised in multilingual environments do not seem to suffer any noticeable delays or trouble in language acquisition. Yet languages differ widely in their phonetic repertoires, syllabic structures, suprasegmental organization, and so on. Hence, one might have expected that multilingual infants would, at least initially, construct a linguistic mixture of all languages to which they were exposed. Jacques and his colleagues hypothesized that to avoid such confusions, infants must have a way of rapidly sorting out the different languages spoken in their environment. Mehler et al. (1988) started to investigate whether newborns could discriminate utterances in one language from those of another language. Not only did French newborns discriminate French utterances from Russian ones but they were able to do so even when they were forced to rely primarily on prosodic information (i.e., when the speech was low-pass–filtered). Thus, sensitivity to the prosodic features of utterances could provide young language learners with the means to keep utterances in one language separate from those in another language, a distinction that is critical in order to arrive at the correct generalizations about the linguistic organizations of the languages.

The notion that infants are attentive to the prosodic features of languages from a very early age meshed well with other findings from Jacques's studies of speech processing in adults. For many years, he had argued that the syllable constitutes the fundamental processing unit for French listeners (Mehler, 1981; Mehler et al., 1981). However, as he and his many collaborators began to explore the processing strategies used by listeners of other native languages, it became apparent that the basic processing unit was not the same for all languages. For instance, native English listeners appear to rely more heavily on stress cues in processing speech (Cutler et al., 1983, 1986), whereas native Japanese listeners seem to organize their processing strategies around moraic units (Cutler and Otake, 1995; Otake, Hatano, and Yoneyama, 1996; Otake et al., 1993). These findings led to the view that online processing strategies reflect the rhythmic properties of a particular language. However, it is not simply the case that listeners will shift their processing strategies when listening to a language from a rhythmic class that differs from the one that their native language belongs to (Cutler et al., 1986). Instead, when listening to languages from non-native rhythmic classes, listeners appear to rely on those strategies that they have developed for processing the native language.

One interesting possibility is that processing strategies tailored to the rhythmic properties of the native language originate in the early sensitivities that infants display in distinguishing utterances from different languages on the basis of prosodic features. Indeed, Jacques and his colleagues suggested that sensitivity to prosodic features such as rhythm were likely to play an important role in how infants learn to segment speech (Mehler, Dupoux, and Segui, 1990). The SARAH model, which they proposed, postulated a strong correspondence between the processes used by infants in word segmentation and those underlying lexical access by adults. SARAH posits that, even at the initial state, infants parse speech into elementary units that correspond roughly to the syllable. These elementary units provide the fodder for learning about the particular set of phonetic elements used in the native language and for retaining the type of word segmentation strategy that works best for that language. A crucial aspect of the latter is to use the syllabic representations and other acoustic information to compute elementary cues such as duration, stress, and so on, that serve as markers of word boundaries in the native language.

Subsequent findings from research with infants led Jacques and his colleagues to give greater emphasis to the role of vowels in enabling infants to discriminate utterances from different languages and in developing word segmentation strategies. In particular, although French newborns discriminated speech stimuli according to whether they contain two or three syllables, they did not discriminate stimuli that differed simply in how many phonetic segments that they contained (Bijeljac-Babic,

Bertoncini, and Mehler, 1993). Moreover, when the number of syllables is held constant, French newborns fail to discriminate stimuli that differ in the number of moras that they contain (Bertoncini et al., 1995). Other findings suggested that newborns are more sensitive to vowel changes than to consonantal changes (Bertoncini et al., 1988). On the basis of these findings, Mehler et al. (1996) proposed a new model, TIGRE, that is based on gridlike representation of vocalic nuclei in an utterance. In particular, the representation includes information about the amplitude and duration of the vowels and their relative spacing. Mehler et al. suggested that such a representation would allow infants to distinguish utterances in a syllable-timed language like French from those in either a stress-timed language like English or a mora-timed language like Japanese. Furthermore, they suggested that use of this type of representation might lead infants to fail to discriminate utterances from languages that belong to the same rhythmic class, such as English and Dutch.

In fact, recent findings from studies of the language discrimination abilities of infants support the view that, in the first few months of life, infants can discriminate utterances from languages belonging to different rhythmic classes, but not ones from languages belonging to the same rhythmic class. Thus, Nazzi, Bertoncini, and Mehler (1998) found that French newborns were able to discriminate stress-timed English utterances from ones in mora-timed Japanese. However, when utterances from two different stress-timed languages, English and Dutch, were presented, the infants failed to discriminate them. Moreover, Christophe and Morton (1998) showed that this failure to discriminate two languages from the same rhythmic class even extends to the one that includes the infants' native language. Thus, British two-month-olds also failed to distinguish utterances in English from ones in Dutch. In fact, infants do not show any capacity for distinguishing utterances from languages belonging to the same rhythmic class until four to five months of age. Even then, infants seem to be limited to discriminating utterances from their own native language from ones in other languages belonging to the same rhythmic class (Bosch, 1998; Bosch and Sebastián-Gallés, 1997; Nazzi, Jusczyk, and Johnson, 2000). For example, Nazzi et al. (2000) found that although American five-month-olds can distinguish American- and British-English dialects, they do not discriminate two unfamiliar languages

(Dutch and German) from the same rhythmic class. In general, then, there is good evidence that infants' sensitivity to prosodic features of languages, such as rhythm, is a crucial component of their abilities to discriminate utterances from different languages.

Is there evidence that sensitivity to prosodic characteristics also contributes to the development of word segmentation abilities in infants? When Jacques and his colleagues first raised this prospect (Mehler et al., 1990), virtually nothing was known about how or when infants begin to segment words from fluent speech. Investigations of infants' word segmentation abilities began with studies that focused on infants' capacities to discriminate utterances on the basis of potential word segmentation cues. For example, Jusczyk et al. (1993a) found that English-learning nine-month-olds display a preference for words with the predominant stress pattern over ones with a less frequent stress pattern. Friederici and Wessels (1993) showed that Dutch-learning nine-month-olds preferred utterances containing phonotactically legal sequences at syllable onsets and offsets to utterances with illegal sequences in these positions. Other investigations demonstrated that infants were capable of discriminating utterances on the basis of the kinds of allophonic and phonotactic cues that signal the presence or absence of word boundaries (Christophe et al., 1994; Hohne and Jusczyk, 1994). The results of these investigations showed that infants are sensitive to the kinds of acoustic markers that could signal word boundaries in fluent speech, but they did not show that infants actually used this information to segment speech.

The first indications of when infants show some ability to segment words from fluent speech came from an investigation by Jusczyk and Aslin (1995). We found that English-learning 7½-month-olds, familiarized with a pair of isolated words, listened significantly longer to passages containing these words than to ones without them. Furthermore, even when infants were first exposed to target words in fluent speech contexts, they gave evidence of segmenting the words from these contexts. By comparison, six-month-olds gave no evidence of segmenting words in these same situations. Thus, we concluded that English-learners begin to segment words between 6 and 7½ months of age.

Much of the subsequent research has focused on the kinds of information that infants use to segment words in fluent speech. Between 7½ and

10½ months, infants can use a number of different types of word boundary markers, including prosodic stress cues (Echols, Crowhurst, and Childers, 1997; Jusczyk, Houston, and Newsome, 1999b, Morgan, 1996), statistical cues (Saffran, Aslin, and Newport, 1996), phonotactic cues (Mattys and Jusczyk, in press, Mattys et al., 1999), and allophonic cues (Jusczyk, Hohne, and Bauman, 1999a). However, the developmental pattern of when infants are able to use each of these sources of information appears to give primacy to prosodic cues (Jusczyk, 1999a). Specifically, the available evidence suggests that infants can use prosodic cues by 7½ months (Jusczyk et al., 1999b) and statistical cues by 8 months (Saffran et al., 1996). However, they do not appear to use phonotactic cues until around 9 months (Mattys et al., 1999) nor allophonic cues until about 10½ months (Jusczyk et al., 1999a). Moreover, when stress cues conflict with phonotactic cues (Mattys et al., 1999) or statistical cues (Johnson and Jusczyk, in press), infants appear to favor the segmentation that is consistent with the stress cues. Finally, there are indications that infants who have learned a stress-based word segmentation strategy for one language, English, will apply this same strategy to another stress-based language, Dutch (Houston et al., 2000). Therefore, the findings to date are consistent with the view that infants' sensitivity to the prosodic features of language, especially rhythm, figures importantly in the way that they begin to segment speech.

What We Still Need to Know

Since the inception of the field, about thirty years ago, we have learned a great deal about infants' inborn speech perception capacities, as well as their development through exposure to a language. For example, we know that infants have some ability to discriminate many types of speech contrasts right from birth, and that they are able to deal with some of the variability that exists in the speech signal. We have also discovered that during the first year infants begin to adapt their capacities to maximize their efficiency for processing the sound structure of the native language that they are acquiring. Not only do they develop the ability to segment words from utterances but they also give evidence of responding to groupings of words into phrases and clauses. At this juncture, there is no longer an issue as to whether it is more important to study the initial state or the development of language capacities (see Mehler, Carey, and Bonati, in press). Instead, studies of the initial state have become integrated into those focusing on the development of speech and language processing. In particular, there is a great deal of interest in the possibility that infants may draw on information in the speech signal to help bootstrap the acquisition of the grammatical organization of the language (Gleitman and Wanner, 1982; Jusczyk, 1997; Morgan and Demuth, 1996; Peters, 1983). The basic premise is that one needs to understand how infants' speech-processing capacities may help them extract regularities from the linguistic input that provide clues about the grammatical organization of the language. As speech-processing abilities are modified to provide for more efficient processing of native language sound organization, this may enable learners to discover other potential clues to grammatical organization (Jusczyk, 1999b). In such a framework, studies of the initial state are necessarily intertwined with developmental studies. Still, much remains to be learned about how speech perception capacities develop and what role they play in the acquisition of language.

One critical issue that is yet to be resolved concerns linguistic diversity. Languages vary widely in their phonological, morphological, and syntactic organization. The sad fact remains that most studies to date have been carried out with English-learning infants. The consequence of this is that we have a clearer picture of how speech perception develops in English-learners, but no real conception of which aspects of development are language-general (or -universal) and which are language-specific. For instance, the fact that English-learning infants seem to identify word onsets with the occurrence of strong syllables in fluent speech may reflect their experience in listening to English. However, implausible as it may seem, one cannot yet rule out the possibility that all infants, regardless of what language they are learning, begin with a bias to identify word onsets with strong syllables. Were the latter true, then one would expect to see that infants acquiring languages with a different type of predominant word stress pattern might have to overcome such an initial bias. This is only one of many aspects of the acquisition of the sound structure of languages in which we need crosslinguistic data to understand how development occurs. Fortunately, there are many laboratories, including those in France, that are beginning to explore how language-processing strategies develop in infants learning languages other than English.

Another important domain for future research concerns brain organization and development. One obstacle that has hindered progress in this area has to do with the fact that current brain-imaging procedures cannot be used concurrently with the kinds of behavioral measures typically used in infant speech research. Although it may be a while before we can solve these difficulties, one step in the right direction would be to try and obtain both behavioral and brain measures on the same sets of infants, especially at different points in development. Two promising techniques are, in fact, being developed in Jacques's laboratory. One is the use of high-density evoked potentials, a technique that has been shown to be feasible in testing infants (Dehaene-Lambertz and Dehaene, 1994). The other is near-infrared spectroscopy (e.g., see Sato, Takeuchi, and Sakai, 1999). With these new procedures, many questions might be addressed. For instance, are there changes in the patterns of brain activation before and after infants begin to segment words, or before and after they distinguish native and non-native phonotactic patterns? In what ways might such changes in activation help explain why infants are able to carry out some particular process at one point but not at an earlier point?

A third domain that holds promise for understanding how language is acquired concerns the role that speech-processing abilities play in discovering the grammatical organization of one's native language. Does information in the speech signal serve only to delimit possible phrasal and clausal units in utterances or does it provide further information about the potential types of grammatical elements and their positioning within utterances? A related concern has to do with the consequences of achieving the early landmarks in the acquisition of the sound organization of one's native language. Are there individual differences in when such landmarks are achieved? If so, what if any consequences ensue when infants lag significantly behind in achieving these? Studies that track how the acquisition of sound structure relates to other types of achievements in language acquisition, such as vocabulary growth, complexity of word combinations, and so on, will yield a better understanding of the role that speech perception capacities play in language acquisition.

These are only a few of the bigger issues that future research in this area needs to address. Indeed, it is fair to say that one of the fundamental issues that early studies in the field set out to address—namely, the extent to which the capacities for learning language are highly specialized or very general mechanisms common to other perceptual and cognitive domains—still has not been adequately answered. Other issues loom just beyond the horizon. In fact, one characteristic of the progress that the field has made since Jacques and I first met is that new discoveries have often brought with them new sets of interesting questions to be addressed. Jacques's career in infant speech research provides a good example of why it is important to take the long view of how one's findings fit into the bigger picture. Even then, as I found out, sometimes it does help to be in the right place at the right time, as on that day in November of 1976 in Austin, Texas.

Acknowledgments

Preparation of this chapter was assisted by a Research Grant from NICHD (#15795) and a Senior Scientist Award from NIMH (#01490). I thank Ann Marie Jusczyk and Emmanuel Dupoux for helpful comments on previous versions of this manuscript.

References

Bertoncini, J., and Mehler, J. (1981). Syllables as units in infant speech perception. *Infant Behavior and Development, 4,* 247–260.

Bertoncini, J., Bijeljac-Babic, R., Jusczyk, P. W., Kennedy, L. J., and Mehler, J. (1988). An investigation of young infants' perceptual representations of speech sounds. *Journal of Experimental Psychology: General, 117,* 21–33.

Bertoncini, J., Floccia, C., Nazzi, T., and Mehler, J. (1995). Morae and syllables: Rhythmical basis of speech representations in neonates. *Language and Speech, 38,* 311–329.

Bijeljac-Babic, R., Bertoncini, J., and Mehler, J. (1993). How do four-day-old infants categorize multisyllabic utterances? *Developmental Psychology, 29,* 711–721.

Bosch, L. (1998). Bilingual exposure and some consequences on native language recognition processes at four months. Presented at the International Conference on Acoustics, Seattle, June 1998.

Bosch, L., and Sebastián-Gallés, N. (1997). Native-language recognition abilities in 4-month-old infants from monolingual and bilingual environments. *Cogniton, 65,* 33-69.

Christophe, A., and Morton, J. (1998). Is Dutch native English? Linguistic analysis by 2-month-olds. *Developmental Science, 1,* 215-219.

Christophe, A., Dupoux, E., Bertoncini, J., and Mehler, J. (1994). Do infants perceive word boundaries? An empirical approach to the bootstrapping problem for lexical acquisition. *Journal of the Acoustical Society of America, 95,* 1570-1580.

Cutler, A., and Otake, T. (1995). Mora or phonemes? Further evidence for language-specific listening. *Journal of Memory and Language, 33,* 824-844.

Cutler, A., Mehler, J., Norris, D. G., and Segui, J. (1983). A language-specific comprehension strategy. *Nature, 304,* 159-160.

Cutler, A., Mehler, J., Norris, D. G., and Segui, J. (1986). The syllable's differing role in the segmentation of French and English. *Journal of Memory and Language, 25,* 385-400.

Dehaene-Lambertz, G., and Dehaene, S. (1994). Speed and cerebral correlates of syllable discrimination in infants. *Nature, 370,* 292-295.

Echols, C. H., Crowhurst, M. J., and Childers, J. B. (1997). Perception of rhythmic units in speech by infants and adults. *Journal of Memory and Language, 36,* 202-225.

Eimas, P. D. (1974). Auditory and linguistic processing of cues for place of articulation by infants. *Perception and Psychophysics, 16,* 513-521.

Eimas, P. D. (1975). Auditory and phonetic coding of the cues for speech: Discrimination of the [r-l] distinction by young infants. *Perception and Psychophysics, 18,* 341-347.

Eimas, P. D., Siqueland, E. R., Jusczyk, P., and Vigorito, J. (1971). Speech perception in infants. *Science, 171,* 303-306.

Friederici, A. D., and Wessels, J. M. I. (1993). Phonotactic knowledge and its use in infant speech perception. *Perception and Psychophysics, 54,* 287-295.

Gerken, L. A., Jusczyk, P. W., and Mandel, D. R. (1994). When prosody fails to cue syntactic structure: Nine-month-olds' sensitivity to phonological vs syntactic phrases. *Cognition, 51,* 237-265.

Gleitman, L. R., and Wanner, E. (1982). The state of the state of the art. In E. Wanner and L. R. Gleitman (Eds.), *Language acquisition: The state of the art* (pp. 3-48). Cambridge, UK: Cambridge University Press.

Hirsh-Pasek, K., Kemler Nelson, D. G., Jusczyk, P. W., Wright Cassidy, K., Druss, B., and Kennedy, L. (1987). Clauses are perceptual units for young infants. *Cognition, 26,* 269-286.

Hohne, E. A., and Jusczyk, P. W. (1994). Two-month-old infants' sensitivity to allophonic differences. *Perception and Psychophysics, 56,* 613-623.

Houston, D. M., Jusczyk, P. W., Kuipers, C., Coolen, R., and Cutler, A. (2000). Cross-language word segmentation by 9-month-olds. *Psychonomic Bulletin and Review, 7,* 504–509.

Johnson, E. K., and Jusczyk, P. W. (in press). Finding words in fluent speech: How infants cope with different types of word segmentation cues. In I. Lasser (Ed.), *The process of language acquisition.* Potsdam: Peter Lang Verlag.

Jusczyk, P. W. (1997). *The discovery of spoken language.* Cambridge, MA: MIT Press.

Jusczyk, P. W. (1999a). How infants begin to extract words from fluent speech. *Trends in Cognitive Science, 3,* 323–328.

Jusczyk, P. W. (1999b). Narrowing the distance to language: One step at a time. *Journal of Communications Disorders, 32,* 207–222.

Jusczyk, P. W., and Aslin, R. N. (1995). Infants' detection of sound patterns of words in fluent speech. *Cognitive Psychology, 29,* 1–23.

Jusczyk, P. W., and Thompson, E. J. (1978). Perception of a phonetic contrast in multisyllabic utterances by two-month-old infants. *Perception and Psychophysics, 23,* 105–109.

Jusczyk, P. W., Cutler, A., and Redanz, N. (1993a). Preference for the predominant stress patterns of English words. *Child Development, 64,* 675–687.

Jusczyk, P. W., Friederici, A. D., Wessels, J., Svenkerud, V. Y., and Jusczyk, A. M. (1993b). Infants' sensitivity to the sound patterns of native language words. *Journal of Memory and Language, 32,* 402–420.

Jusczyk, P. W., Hirsh-Pasek, K., Kemler Nelson, D. G., Kennedy, L., Woodward, A., and Piwoz, J. (1992). Perception of acoustic correlates of major phrasal units by young infants. *Cognitive Psychology, 24,* 252–293.

Jusczyk, P. W., Hohne, E. A., and Bauman, A. (1999a). Infants' sensitivity to allophonic cues for word segmentation. *Perception and Psychophysics, 61,* 1465–1476.

Jusczyk, P. W., Houston, D. M., and Newsome, M. (1999b). The beginnings of word segmentation in English-learning infants. *Cognitive Psychology, 39,* 159–207.

Jusczyk, P. W., Luce, P. A., and Charles-Luce, J. (1994). Infants' sensitivity to phonotactic patterns in the native language. *Journal of Memory and Language, 33,* 630–645.

Jusczyk, P. W., Pisoni, D. B., Reed, M., Fernald, A., and Myers, M. (1983). Infants' discrimination of the duration of a rapid spectrum change in nonspeech signals. *Science, 222,* 175–177.

Jusczyk, P. W., Pisoni, D. B., Walley, A. C., and Murray, J. (1980). Discrimination of the relative onset of two-component tones by infants. *Journal of the Acoustical Society of America, 67,* 262–270.

Kaplan, E. L. (1969). *The role of intonation in the acquisition of language.* Unpublished PhD dissertation, Cornell University, Ithaca, NY.

Kuhl, P. K., and Miller, J. D. (1982). Discrimination of auditory target dimensions in the presence or absence of variation in a second dimension by infants. *Perception and Psychophysics, 31,* 279–292.

Kuhl, P. K., Williams, K. A., Lacerda, F., Stevens, K. N., and Lindblom, B. (1992). Linguistic experiences alter phonetic perception in infants by 6 months of age. *Science, 255,* 606–608.

Mattys, S. L., and Jusczyk, P. W. (in press). Do infants segment cues or recurring continguous patterns? *Journal of Experimental Psychology: Human Perception and Performance.*

Mattys, S. L., Jusczyk, P. W., Luce, P. A., and Morgan, J. L. (1999). Phonotactic and prosodic effects on word segmentation in infants. *Cognitive Psychology, 38,* 465–494.

Mehler, J. (1981). The role of syllables in speech processing: Infant and adult data. *Philosophical Transactions of the Royal Society of London. Series B: Biological Sciences, 295,* 333–352.

Mehler, J., and Bertoncini, J. (1979). Infants' perception of speech and other acoustic stimuli. In J. Morton and J. Marshall (Eds.), *Psycholinguistics 2: Structures and processes* (pp. 67–105). Cambridge, MA: MIT Press.

Mehler, J., Bertoncini, J., Barriere, M., and Jassik-Gerschenfeld, D. (1978). Infant recognition of mother's voice. *Perception, 7,* 491–497.

Mehler, Carey, and Bonati (in press).

Mehler, J., Dommergues, J. Y., Frauenfelder, U., and Segui, J. (1981). The syllable's role in speech segmentation. *Journal of Verbal Learning and Verbal Behavior, 20,* 298–305.

Mehler, J., Dupoux, E., Nazzi, T., and Dehaene-Lambertz, G. (1996). Coping with linguistic diversity: The infant's viewpoint. In J. L. Morgan and K. Demuth (Eds.), *Signal to syntax* (pp. 101–116). Mahwah, NJ: Erlbaum.

Mehler, J., Dupoux, E., and Segui, J. (1990). Constraining models of lexical access: The onset of word recognition. In G. T. M. Altmann (Ed.), *Cognitive models of speech processing* (pp. 236–262). Hillsdale, NJ: Erlbaum.

Mehler, J., Jusczyk, P. W., Lambertz, G., Halsted, N., Bertoncini, J., and Amiel-Tison, C. (1988). A precursor of language acquisition in young infants. *Cognition, 29,* 144–178.

Morgan, J. L. (1996). A rhythmic bias in preverbal speech segmentation. *Journal of Memory and Language, 35,* 666–688.

Morgan, J. L., and Demuth, K. (1996). *Signal to syntax.* Mahwah, NJ: Erlbaum.

Morse, P. A. (1972). The discrimination of speech and nonspeech stimuli in early infancy. *Journal of Experimental Child Psychology, 13,* 477–492.

Nazzi, T., Bertoncini, J., and Mehler, J. (1998). Language discrimination by newborns: Towards an understanding of the role of rhythm. *Journal of Experimental Psychology: Human Perception and Performance, 24,* 756–766.

Nazzi, T., Jusczyk, P. W., and Johnson, E. K. (2000). Language discrimination by English-learning 5-month-olds: Effects of rhythm and familiarity. *Journal of Memory and Language, 43,* 1–19.

Otake, T., Hatano, G., Cutler, A., and Mehler, J. (1993). Mora or syllable? Speech segmentation in Japanese. *Journal of Memory and Language, 32,* 258–278.

Otake, T., Hatano, G., and Yoneyama, K. (1996). Speech segmentation by Japanese listeners. In T. O. A. Cutler (Ed.), *Phonological structure and language processing: Cross-linguistic studies.* Berlin: Mouton de Gruyter.

Peters, A. (1983). *The units of language acquisition.* Cambridge, UK: Cambridge University Press.

Polka, L., and Bohn, O.-S. (1996). Cross-language comparison of vowel perception in English-learning and German-learning infants. *Journal of the Acoustical Society of America, 100,* 577–592.

Saffran, J. R., Aslin, R. N., and Newport, E. L. (1996). Statistical learning by 8-month-old infants. *Science, 274,* 1926–1928.

Sato, H., Takeuchi, T., and Sakai, K. L. (1999). Temporal cortex activation during speech recognition: An optical topography study. *Cognition, 73,* B55–66.

Spring, D. R., and Dale, P. S. (1977). Discrimination of linguistic stress in early infancy. *Journal of Speech and Hearing Research, 20,* 224–232.

Werker, J. F., and Tees, R. C. (1984). Cross-language speech perception: Evidence for perceptual reorganization during the first year of life. *Infant Behavior and Development, 7,* 49–63.

21

On Becoming and Being Bilingual

Núria Sebastián-Gallés and Laura Bosch

Thirty years ago, most of the research on language processing was conducted in English-speaking countries. As a consequence, models and hypotheses on its acquisition and use, though intended to be "universal," were in many cases language-specific: more precisely, English-specific. At that time the main goal of cognitive science was to uncover general mechanisms, and accordingly, differences between languages were considered secondary. Research into bilingualism was not central to the problems that cognitive science traditionally addressed. After all, what could research in bilingualism tell us about the very basic mechanisms of language acquisition and processing that monolingual research could not? Jacques Mehler realized both the importance of crosslinguistic research and of early learning experiences in determining how our language-processing system is attuned to processing our maternal language. By doing so, he helped put bilingual research firmly on the main agenda of cognitive science.

When we were first introduced to Professor Jacques Mehler, I also had the pleasure of meeting JM, a particularly interesting individual, with quite remarkable linguistic abilities. He speaks fluent Spanish, French, English, Italian, and Portuguese, and also writes these languages well; he speaks German, but for some reason he does not recognize this fact (a case of linguistic agnosia?). It is of interest that he speaks all these languages with a characteristic foreign accent, and it is difficult to determine which one is in fact his maternal language. Some researchers even claim that he no longer has a mother tongue. JM's experience with language exemplifies both the plasticity of the brain in dealing with several languages and the limits to this plasticity. Having continuous access to JM's language experiences allowed Mehler and his colleagues to develop a number of insightful hypotheses that have since shaped the field of language processing.

The Bilingual to Be: Opening Your Ears in a Multilingual World

JM often reports that when he overhears conversations in bars, he can identify the language being spoken, even when the words are unintelligible. He notes that his ability to identify languages is not perfect; once he thought that a conversation he was listening to was in Spanish, but later he discovered that it was in Greek, a language he does not understand. However, he has never reported confusing Spanish with English or Japanese. This, and similar observations, prompted Mehler and colleagues to investigate the hypothesis that languages fall into classes based on their rhythmic and prosodic properties (Cutler and Mehler, 1993). This hypothesis also helped to explain how infants that are raised in a multilingual environment might discover that there are several languages being spoken around them.

One of the most frequent questions asked about bilingual infants is how long it takes them to discover the existence of two languages. Mehler and coworkers (Mehler et al., 1996; Ramus and Mehler, 1999) have developed the theory that infants can notice at birth that two languages are not the same if they belong to different "rhythmic classes." Linguists (Abercrombie, 1967; Pike, 1946) proposed that languages could be classified into "syllable-timed" and "stress-timed" languages depending on whether syllables or stress-groups (from one stressed vowel to the next stressed vowel) were the basic timing units. In this typology Romance languages, for instance, are classified as syllabic languages, while Germanic and Slav languages are classified as stress languages. Subsequently, a further rhythmic group has been proposed to reflect the rhythmic properties of languages such as Japanese (Ladefoged, 1975). In such languages, the rhythm is based on the mora (a subsyllabic unit). Although this classification has been criticized as simplistic, recent work by Ramus and Mehler (1999) has shown that it reflects the vowel-consonant temporal ratio, that is, the proportion of vocalic intervals (as opposed to consonantal intervals) in the speech signal.

Significantly, early infant discrimination capacities can be explained by taking into consideration these rhythmic properties. Indeed, it has been shown that newborns can discriminate pairs of languages provided they belong to different rhythmic groups. Newborns can distinguish French from Russian or Italian from English (two between-group contrasts; Mehler and Christophe, 1995; Mehler et al., 1988; Nazzi, Bertoncini, and Mehler, 1998), though not Italian from Spanish or Dutch from English (two within-group contrasts; Mehler and Christophe, 1995; Mehler et al., 1988; Nazzi et al., 1998). Interestingly, this language discrimination capacity changes with age. On the one hand, it can be said to become less sensitive because by two months infants can no longer distinguish between two languages outside of their maternal language rhythmic group, even if they are quite distant from one another. For instance, English two-month-olds do not distinguish between French and Japanese (Christophe and Morton, 1998). But it also becomes more sensitive, because they notice differences within their maternal language rhythmic group. Spanish and Catalan monolingual infants aged 4½ months can distinguish Spanish from Catalan—two syllabic languages—(Bosch and Sebastián-Gallés, 1997; Bosch, Cortés, and Sebastián-Gallés, in press), and five-month-old American-English infants can distinguish English from Dutch—two stress languages (Nazzi, Juscyk, and Johnson, 2000). But these language discrimination patterns have been reported for monolingual infants. Do bilingual infants show a differential pattern of language discrimination behavior? How long does it take a bilingual infant to notice the existence of more than one language in his or her environment?

JM, as a newborn, probably had no difficulty in discovering that there was more than one language being spoken in his environment. He was born in Spain into a German-speaking family. German and Spanish show different rhythms, so although we do not have any direct report from him concerning his language confusions during the first days of his life (Mehler, personal communication), we can assume that the newborn JM already realized that more than one language was being spoken around him. However, not all infants are born into this type of environment. Many infants are born in a bilingual situation where two languages from the same rhythmic group are spoken. This constitutes quite a different starting point, because it may be several months before they can discover the existence of the two languages. If we consider that separating languages is a prerequisite of adequate language learning (Mehler and Christophe, 1995) and of the rapid language acquisition achievements taking place in the first year of life, any delay in separating languages may have important consequences for the way languages are represented in the brain of bilingual speakers.

In our laboratory we have studied the development of early language discrimination and language acquisition capacities of bilingual infants exposed to two rhythmically similar languages: Spanish and Catalan (incidentally, because JM was born in Barcelona, where Catalan and Spanish are commonly spoken, Catalan may have been one of the languages included in his early language exposure set). Spanish and Catalan are two Romance languages, and though they differ in some important phonological aspects, they represent a clear case of similar languages. Using a discrimination procedure, Bosch and Sebastián-Gallés (2001b) have obtained evidence that bilingual infants aged 4½ months do discriminate Catalan from Spanish. These data are evidence of early language discrimination capacities in bilingual infants, challenging some conclusions based on research in the production domain. For instance, Leopold (1954) and Volterra and Taeschner (1978) proposed that language differentiation in bilinguals did not generally take place before the third year of life, once functional categories have emerged. More recent studies (De Houwer, 1990; Genesee, Nicoladis, and Paradis, 1995) offer evidence of a somewhat earlier differentiation, but no previous study had addressed the discrimination issue in the prelinguistic child. What about newborns? Can they, too, discriminate Spanish from Catalan? A recent study suggests a negative answer to this question. French newborns were shown to be unable to differentiate Spanish from Catalan (F. Ramus, unpublished data). So, as suggested above, a developmental change arises between birth and four to five months that enables infants to make finer distinctions than those allowed by rhythmic classes alone. This is the time window inside which Spanish-Catalan bilingual infants can realize they are in a bilingual environment.

Does this mean, though, that there is no cost or delay in acquisition in bilinguals? In an initial series of experiments, we used a reaction time procedure to test language processing in young infants (Bosch and Sebastián-Gallés, 1997). In this procedure, infants fixate a screen, and are randomly presented with sentences from two languages from the left or the right side. This elicits an orientation movement toward the sound source, and visual orientation times (reaction times) are recorded. Previous results had shown that monolingual infants orient faster to a familiar (maternal) language than to a foreign language (Dehaene-Lambertz and Houston, 1998). We replicated this finding with monolingual Catalan and monolingual Spanish infants aged 4½ months. When presented with Catalan and Spanish sentences, they oriented faster to the maternal language than to the unfamiliar language. When the same materials were presented to bilingual Catalan-Spanish infants, they showed no differences between the two languages, suggesting that they treated both languages as equally familiar.

Therefore, it seems that monolingual and bilingual infants aged 4½ months have similar language discrimination capacities. However, other data in the study of Bosch and Sebastián-Gallés (1997) point in a different direction. In the same study, monolingual and bilingual infants were also presented with maternal (either Spanish or Catalan) vs. English sentences (a completely unknown language for all infants). While monolinguals showed the expected pattern of results of faster orientation times for the familiar language than for the unfamiliar language, bilingual infants showed exactly the opposite pattern: they oriented faster to the unfamiliar than to the familiar language. Orientation times to English were the same for monolingual and bilingual infants; this suggests that bilinguals are significantly slower than monolinguals to orient to their maternal language(s). This pattern of results was replicated with Italian instead of English as the foreign language, indicating that they were not due to some particular properties of the pairs of languages under study. Furthermore, bilinguals' slower orientation times for the maternal language have also been obtained with six-month-old infants (Bosch and Sebastián-Gallés, 2001a), suggesting that it is a rather robust finding.

Why are bilingual infants slower than monolinguals to orient to material in languages that are familiar to them? At this point, we can only offer speculations. Perhaps bilingual infants have to perform the extra task of deciding which of the two languages of their environment is the one they are listening to, and then switch to this language. Monolinguals would, of course, not have any extra processing to do, because they only know one language. Although we can only speculate on bilinguals' patterns of orientation times, there is initial evidence that bilingual infants do not treat language input in the same way as monolingual infants do. The long-term consequences of this differential pattern remain unknown.

Although being able to separate languages is a prerequisite of further linguistic acquisition, it does not tell us whether infants acquire both languages at the same time or not. Cutler et al. (1989) studied adult French-English early bilinguals, that is, bilinguals exposed to the two languages since birth, using a fragment detection task. This task had been shown to reflect the way in which monolinguals segment continuous speech in terms of prelexical units (Mehler et al., 1981; but see Cutler, McQueen, Norris, and Somejuan, chapter 10, for another interpretation). The results of these experiments were claimed to suggest that bilinguals segment the speech signal in a way that is consistent with only one of their two languages (as if they were monolinguals). Thus, although no infant data were collected in the study of Cutler et al., this result is compatible with the notion that bilingual infants prioritize learning one language over the other, and it is this which will become their dominant language. If this were the case, one would expect a bilingual infant's knowledge of the dominant language not to differ significantly from that of a monolingual infant, while the infant's knowledge of the nondominant language should be somehow more similar to that of an unknown language. We have tested this issue by comparing monolingual and bilingual Catalan-Spanish infants in their knowledge of some Catalan phonotactic properties that Spanish does not share.

Catalan, unlike Spanish, accepts final word complex consonant clusters (i.e., "tirk" is possible in Catalan, but it could not be a Spanish word). Yet, certain consonant clusters are impossible in Catalan (e.g., "tikf"). Previous work has shown that nine-month-old infants (Jusczyk, Luce, and Charles-Luce, 1994) prefer to listen to lists of stimuli that conform to the phonotactics of their maternal language than to lists of stimuli that do not. Thus, Catalan monolingual infants should prefer to listen to lists of legal stimuli like "tirk," rather than illegal ones like "tikf," whereas monolingual Spanish infants should not show any preference, since both types of stimuli are equally unfamiliar or illegal to them. This is the pattern we observed with monolinguals aged 10½ months (Sebastián-Gallés and Bosch, submitted, experiment 1). We then tested bilingual infants, that is, infants with one parent speaking mostly Spanish, and the other speaking mostly Catalan. Bilingual infants were divided into Catalan-dominant and Spanish-dominant according to the maternal language. The results demonstrated that only Catalan-dominant bilingual infants showed a pattern of listening time similar to that of monolingual Catalan infants. Spanish-dominant bilingual infants did not show any preference (like Spanish monolinguals).

These data are consistent with the hypothesis that bilingual infants give some priority to one of the languages spoken in their environment when acquiring phonotactic patterns. This result is most remarkable if we consider, as past research has shown (Jusczyk et al., 1994), that frequency of exposure plays a crucial role in infants' sensitivity to phonotactic patterns. When compared with Catalan monolinguals, Catalan-dominant bilingual infants had necessarily a reduced amount of exposure to properties that are specific to only one language (in our sample the average percentage of exposure to Catalan was 65 percent for Catalan-dominant bilinguals).[1] Thus, although there was a relatively large difference between monolingual Catalan and bilingual Catalan-dominant infants in their exposure to Catalan, they showed the same preference pattern. However, the relatively small difference between both bilingual populations in their exposure to Catalan (65 percent vs. 35 percent) was enough to show different patterns. The results challenge the hypothesis that bilingual infants build both language systems at similar rates, as far as phonotactic knowledge is concerned.

In short, bilingual infant research has shown that (a) bilingual infants, even those being raised in an environment where very similar languages are spoken, separate the two languages in the first months of their lives; (b) bilingual infants seem to suffer a cost compared to monolinguals in a task that requires them to orient toward one of their native languages (the exact origin of which is still to be fully understood); and (c) they seem to acquire the phonotactic properties of the dominant language, and do so more or less at the same rate as monolinguals.

Becoming Bilingual in Preschool: Too Late for Phonology

As with most second-language learners, the speech of JM, although very fluent in his several second languages, is full of slight phonological deviations that are characteristic of a foreign accent. This is so even though JM forcefully argues that his performance is fully native-like and has no accent at all in any of these languages. This prompted Mehler and colleagues to investigate whether a phenomenon similar to the foreign accent could also exist in perception: that is, non-native listening to speech sounds through the filter of native language phonology.

JM, unlike most human beings, learned not one, but several languages during the first two decades of his life. Spending several years of this period in South America, Europe, and the United States meant that he acquired different languages at different ages. Age of acquisition has always been a major issue in the second-language learning literature (see Flege, 1995; Johnson and Newport, 1989). The received knowledge has been that before a certain "critical" age, "perfect" second-language learning was possible—perfection meaning that "early" second-language speakers could not be distinguished from first-language speakers.

Recent research has shown that when learning a second language, not all aspects of the linguistic knowledge are equally hard to master (Flege, Yeni-Komshian, and Liu, 1999). Furthermore, age of acquisition seems to have different effects on different types of linguistic abilities, and phonology has been identified as a particularly difficult domain.

As we noted earlier, Catalan and Spanish differ in some important phonological dimensions. One of these is the phoneme repertoire: (1) while Spanish has only five vowels (a,e,i,o,u), Catalan has eight (a,e,ɛ,i,o,ɔ,u, and schwa); (2) while Catalan has voiced fricatives, Spanish has none (but it has some unvoiced fricatives that do not exist in Catalan, such as /x/ and /θ/). So segment perception seemed to be a good field in which to study first-language influences on second-language perception. This is a domain that had already received a great deal of attention and several models had been proposed to account for the fact that some foreign sounds are more easily learned than others (see, e.g., the proposals of Best, 1995, and Flege, 1995). What raised our interest was the puzzling observation that many native Spanish speakers who had been exposed to Catalan from the first years of their lives (normally between three and five years), who have lived in Catalonia for all of their lives (having received a bilingual education and living in a bilingual society), and who are extremely fluent in both languages seem to be unable to correctly perceive (and produce) some Catalan sounds. Similar observations have been made about immigrants arriving at a very early age in a foreign country. Although it is true that the earlier the exposure, the slighter the foreign accent, there seems to be a ceiling to our capacity to acquire foreign sounds. These observations are even more remarkable if we consider phoneme training studies claiming that even very difficult second-language segments can be learned through short (and intensive) training (see, e.g., Lively, 1994; McClelland et al., 1999).

In a series of studies we analyzed the segment perception capacities of highly skilled Catalan second-language learners (Spanish first-language), from the population described above. We tested these bilinguals in a wide range of perception tasks: identification, discrimination (ABX), and category goodness judgments (Pallier, Bosch, and Sebastián-Gallés, 1997); gating (Sebastián-Gallés and Soto-Faraco, 1999); and category goodness and discrimination (Bosch, Costa, and Sebastián-Gallés, 2000). In these studies a highly consistent pattern emerged. About 50 percent of the second-language learners failed to notice the differences between Catalan segment contrasts that do not exist in Spanish. In one of the studies (Sebastián-Gallés and Soto-Faraco, 1999) we were interested in analyzing the 50 percent of Spanish-dominant bilinguals who could perceive this contrast. We selected these bilinguals and asked them to perform a two-alternative gating task (participants had to guess at each gate which of two possible nonwords the fragment they had just heard corresponded to). The results showed that even if at the last gate (when the complete stimulus was presented), Spanish-dominant bilinguals did not differ from Catalan-dominant bilinguals; they performed worse at the preceding gates. Thus, even those Spanish-dominant bilinguals who have developed new phoneme categories for Catalan sounds do not treat Catalan stimuli as efficiently as native Catalan speakers do.

Recently, we have extended these results and, using an auditory repetition priming paradigm (with a lexical decision task), we have shown that the lack of differentiation between Catalan contrasts is not restricted to the phonemic representation, but projects into the lexicon (Pallier, Colomé, and Sebastián-Gallés, in press). Indeed, we observed that, unlike Catalan natives, Spanish-dominant bilinguals treated Catalan minimal pairs like /netə//nɛtə/ (meaning "granddaughter" and "clean" (feminine), respectively) as if they were homophones, that is, they showed a repetition priming effect, equivalent to a same-word repetition. Catalan natives did not show any repetition effect for these minimal pairs. In this experiment no significant overall differences in reaction times and error rates were observed between the two populations: their competence at the vocabulary level was equivalent.

The main conclusion of these experiments (see also Flege et al., 1999; Weber-Fox and Neville, 1996) is that even in the case of early, intensive exposure, it seems impossible to reach first-language proficiency, at least in some domains. Thus, when appropriate laboratory tools are used, the myth that early, intensive exposure is enough to achieve native speaker proficiency is untenable.

Then, what about studies claiming that with short, intense training procedures it is possible to acquire a second-language contrast? The answer to this puzzle may lie in considering the goal of the different studies. Research in training second-language contrasts considers it a major success (which, indeed, it is) if a significant improvement is achieved (usually around 10 to 20 percent increase in identification or discrimination tasks, when compared with pretraining tests). In this sense, learning new phonemic categories (and other aspects of language too) reflects the high degree of plasticity our brain has. However, the fact that even with very intense early exposure bilinguals' second-language performance does not reach the same competence levels as natives', and the fact that the first language continues to have a privileged status (in terms of processing efficiency), suggest that there are limits to this plasticity. Thus, for almost all second-language learners, even very early exposure is already too late for second-language sounds.

The Future of a Multilingual World

In modern Western societies, children raised in a multilingual environment are the norm. A monolingual environment is now becoming the exception; this trend seems quite irreversible as our world moves toward more and more interactions between cultures and economies. It is a fact that humans can learn several languages at once. Neither delay in acquisition nor serious problems in processing have been reported for the case of bilingual (or multilingual) infants. Such an important fact has to be taken into account by any serious model of language acquisition and processing. As we have seen, Jacques Mehler was one of the first to point this out to mainstream cognitive science, and the research that he set in motion points to a number of interesting directions to understand how a brain can sustain several languages at once.

How languages are represented in the brain of bilingual individuals is a question that has recently received some attention. Data coming from brain-imaging studies (Kim et al., 1997; Perani et al., 1996, 1998) and from intraoperative electrical cortical stimulation (Conesa et al., submitted) show parallel results. Indeed, the degree of overlapping between brain areas devoted to each language seems to be a function of language dominance and proficiency. Thus, while the first language tends to occupy similar areas in monolingual and bilingual individuals, second-language areas vary from one individual to another. The variation depends on the speaker's proficiency in the second language: the higher the proficiency, the greater the overlapping.

Yet there are still many open questions that will need to be explored. For instance, JM is a frequent traveler and because of the nature of his private life and his work commitments he has to switch languages very frequently. JM complains that, when moving from one country to another country in which another language is spoken, it takes him some time and effort to adapt his language perception system to the characteristics of the new language. This point raises the issue of the processes involved in language switching (see Price, Green, and von Studnitz, 1999, for a positron emission tomography study comparing brain areas activated by translation and language switching). It also addresses the issue of changes in brain and language-processing areas as a consequence of changes in the frequency of use of one language compared to the other(s). Is it the case that brain regions will reorganize whenever one language becomes predominant, as has been found in other domains, like the motor control domain? The research available does not allow us to disentangle these possibilities; in the studies mentioned above, the first language was always the most frequently used. Finally, are these reorganization processes similar in perception and production?

One can appreciate that these questions open up new and exciting research avenues regarding processing in bilinguals. These questions not only have fundamental implications concerning the functional and anatomical plasticity of higher functions in the brain but can also have very practical consequences regarding the early teaching of foreign languages in schools.

JM has been and still is the source of many enlightening remarks, which Jacques Mehler was able to shape into the form of interesting scientific questions. We thank both of them (the friend and the scientist) for all these years of inspiration.

Acknowledgments

The research described herein was supported by a research grant from the Spanish Ministerio de Educación y Cultura (PB97–0977) and the Catalan Government (1999SGR 083).

Note

1. One remote possibility could be that parents of bilingual infants speak more to their offspring. In this case, it could be that the total amount of exposure to Catalan would be fully comparable in both populations. However, we do not know of any data pointing in the direction that bilingual infants are spoken to more often than monolinguals.

References

Abercrombie, D. (1967). *Elements of general phonetics.* Edinburgh: Edinburgh University Press.

Best, C. T. (1995). A direct realist view of cross-language speech perception. In W. Strange (Ed.), *Speech perception and linguistic experience* (pp. 171–206). Baltimore: York Press.

Bosch, L., Cortés, C., & Sebastián-Gallés, N. (in press). El reconocimiento temprano de la lengua materna: Un estudio basado en la voz masculina. *Infancia y Aprendizaje.*

Bosch, L., Costa, A., and Sebastián-Gallés, N. (2000). First and second language vowel perception in early bilinguals. *European Journal of Cognitive Psychology, 12*, 189–222.

Bosch, L., and Sebastián-Gallés, N. (2001a). Early language differentiation in bilingual infants. In J. Cenoz and F. Genesee (Eds.), *Trends in bilingual acquisition*. Amsterdam: John Benjamins.

Bosch, L., and Sebastián-Gallés. (2001b). Evidence of early language discrimination abilities in infants from bilingual environments. *Infancy, 2*, 29–49.

Bosch, L., and Sebastián-Gallés, N. (1997). Native-language recognition abilities in four-month-old infants from monolingual and bilingual environments. *Cognition, 65*, 33–69.

Christophe, A., and Morton, J. (1998). Is Dutch native English? Linguistic analysis by two-month-olds. *Developmental Science, 1*, 215–219.

Conesa, G., Juncadella, M., Gabarrós, A., Sebastián Gallés, N., Marnov, A., Navarro, R., Busquets, N., Pujol, J., Deus, J., and Isamat, F. (submitted). Language cortical organization in bilingual Catalan and Spanish brain tumor patients.

Cutler, A., and Mehler, J. (1993). The periodicity bias. *Journal of Phonetics, 21*, 103–108.

Cutler, A., Mehler, J., Norris, D., and Seguí, J. (1989). Limits on bilingualism. *Nature, 320*, 229–230.

Dehaene-Lambertz, G., and Houston, D. (1998). Faster orientation latencies toward native language in two-month old infants. *Language and Speech, 41*, 21–43.

De Houwer, A. (1990). *The acquisition of two languages from birth: A case study*. Cambridge, UK: Cambridge University Press.

Flege, J. E. (1995). Second language speech learning: Theory, findings and problems. In W. Strange (Ed.), *Speech perception and linguistic experience* (pp. 233–272). Baltimore: York Press.

Flege, J. E., Yeni-Komshian, G. H., and Liu, S. (1999). Age constraints on second-language acquisition. *Journal of Memory and Language, 41*, 78–104.

Genesee, F., Nicoladis, E., and Paradis, J. (1995). Language differentiation in early bilingual development. *Journal of Child Language, 22*, 611–631.

Johnson, J., and Newport, E. (1989). Critical period effects in second language learning: The influence of maturational state on the acquisition of English as a second language. *Cognitive Psychology, 21*, 60–99.

Jusczyk, P. W., Luce, P. A., and Charles Luce, J. (1994). Infants' sensitivity to phonotactic patterns in the native language. *Journal of Memory and Language, 33*, 630–645.

Kim, K. H. S., Relkin, N. R., Lee, K.-M., and Hirsch, J. (1997). Distinct cortical areas associated with native and second languages. *Nature, 388*, 171–174.

Ladefoged, P. (1975). *A course in phonetics.* New York: Harcourt, Brace Jovanovich.

Leopold, W. F. (1954). A child's learning of two languages. In E. Hatch (Ed.) (1978), *Second language acquisition: a book of readings.* Rowley, Mass.: Newbury House.

Lively, S. E., Pisoni, D. B., Yamada, R. A., Tohkura, Y., and Yamada, T. (1994). Training Japanese listeners to identify English /r/ and /l/ III. Long-term retention of new phonetic categories. *Journal of the Acoustical Society of America, 96,* 2076–2087.

McClelland, J. L., Thomas, A., McCandliss, B. D., and Fiez, J. A. (1999). Understanding failures of learning: Hebbian learning, competition for representational space, and some preliminary experimental data. In J. Reggia, E. Ruppin, and D. Glanzman (Eds.), *Brain, behavioral and cognitive disorders: The neurocomputational perspective.* Oxford: Elsevier.

Mehler, J., and Christophe, A. (1995). Maturation and learning of language in the first year of life. In M. S. Gazzaniga (Ed.), *The cognitive neurosciences* (pp. 943–954). Cambridge, MA: MIT Press.

Mehler, J., Dommergues, J. Y., Frauenfelder, U., and Segui, J. (1981). The syllable's role in speech segmentation. *Journal of Verbal Learning and Verbal Behavior, 20,* 298–305.

Mehler, J., Dupoux, E., Nazzi, T., and Dehaene-Lambertz, G. (1996). Coping with linguistic diversity: The infant's viewpoint. In J. L. Morgan and K. Demuth (Eds.), *Signal to syntax* (pp. 101–116). Mahwah, NJ: Erlbaum.

Mehler, J., Jusczyk, P. W., Lambertz, G., Halsted, G., Bertoncini, J., and Amiel-Tison, C. (1988). A precursor of language acquisition in young infants. *Cognition, 29,* 143–178.

Nazzi, T., Bertoncini, J., and Mehler, J. (1998). Language discrimination by newborns: Towards an understanding of the role of rhythm. *Journal of Experimental Psychology: Human Perception and Performance, 24,* 756–766.

Nazzi, T., Juscyk, P. W., and Johnson, E. K. (2000). Language discrimination by English-learning 5-month-olds: Effects of rhythm and familiarity. *Journal of Memory and Language, 43,* 1–19.

Pallier, C., Bosch, L., and Sebastián-Gallés, N. (1997). A limit on behavioral plasticity in vowel acquisition. *Cognition, 64,* B9–B17.

Pallier, C., Colomé, A., and Sebastián-Gallés, N. (in press). The influence of native-language phonology on lexical access: Exemplar-based vs. abstract lexical entries. *Psychological Science.*

Perani, D., Dehaene, S., Grassi, F., Cohen, L., Cappa, S., Dupoux, E., Fazio, F., and Mehler, J. (1996). Brain processing of native and foreign languages. *Neuroreport, 7,* 2439–2444.

Perani, D., Paulesu, E., Sebastián-Gallés, N., Dupoux, E., Dehaene, S., Bettinardi, V., Cappa, S. F., Fazio, F., and Mehler, J. (1998). The bilingual brain: Proficiency and age of acquisition of the second language. *Brain, 121,* 1841–1852.

Pike, K. L. (1946). *The intonation of American English*. Ann Arbor: University of Michigan Press.

Price, C., Green, D., and von Studnitz, R. (1999). A functional imaging study of translation and language switching. *Brain, 122,* 2221–2235.

Ramus, F., and Mehler, J. (1999). Language identification with suprasegmental cues: A study based on speech resynthesis. *Journal of the Acoustical Society of America, 105,* 512–521.

Sebastián-Gallés, N., and Bosch, L. (submitted). The building of phonotactic knowledge in bilinguals: The role of early exposure.

Sebastián-Gallés, N., and Soto-Faraco, S. (1999). On-line processing of native and non-native phonemic contrasts in early bilinguals. *Cognition, 72,* 112–123.

Volterra, V., and Taeschner, T. (1978). The acquisition and development of language by bilingual children. *Journal of Child Language, 5,* 311–326.

Weber-Fox, C. M., and Neville, H. J. (1996). Maturational constraints on functional specialization for language processing: ERP and behavioral evidence in bilingual speakers. *Journal of Cognitive Neuroscience, 8,* 231–256.

V
Brain and Biology

On Language, Biology, and Reductionism

Stanislas Dehaene, Ghislaine Dehaene-Lambertz, and Laurent Cohen

"What relevance do neurophysiological findings have for psychological models?" This question was the lead sentence in a famously controversial article entitled "On Reducing Language to Biology" by Jacques Mehler, John Morton, and Peter Jusczyk (1984). The authors' answer was clearly "not much." In the carefully worded article, one learned that "constraints [between biological and psychological sciences] operate only under very restricted circumstances." It was stipulated that, at least in the domain of word recognition and speech perception, "cogent accounts of the psychological processes involved require the establishment of purely psychological constructs, which are to be judged in terms of their explanatory usefulness rather than their compatibility with . . . neurophysiological data." In private conversations and laboratory meetings, Jacques's view was even more provocative. "For all I know, language perception might be going on in the brain," he said in essence, "but my research would not be affected if it was found to be occurring in the left pinky!"

Mehler, Morton, and Jusczyk's (1984) pessimistic views on the contribution of biology to cognition was perhaps justified by the limited methods that were available at the time. In surveying the extant work, they ended up finding only two areas of potential interaction between biology and psychology: the study of brain-damaged patients and the activity of mapping psychological functions onto the brain. In both cases, it was easy for them to pinpoint the methodological difficulties and conceptual naivetés in the existing corpus of data, which was often based on clinical classifications of aphasic patients into wide and poorly defined categories, or on very general theories of hemispheric lateralization and the sometimes erratic results of the dichotic listening procedure.

Today, however, we know that the interactions between biology and psychology can be much richer than was foreseeable just a decade ago. The diversity of the contributions that follow testifies to the richness of the exchanges that take place across disciplines and justifies the christening of this previous no man's land with a new distinctive label, cognitive neuroscience. First and foremost, obviously, comes the revolution brought about by the new functional brain-imaging techniques. Michael Posner (chapter 22) tells how positron emission tomography (PET), and later functional magnetic resonance imaging (fMRI), in combination with the temporal sensitivity of event-related potentials (ERPs), brought about considerable progress in our understanding of cognitive functions such as attention and language. It is not simply a matter of being able to find precise localizations for the neural circuits underlying those functions. Rather, once they are localized, their modes of operation and coding principles become accessible to investigation with a directness that is lacking in behavioral measurements. Studies of the impact of attention orienting on early visual areas, for instance, have brought renewed support in favor of "early selection" views of attention (Moran and Desimone, 1985). Likewise, the finding that retinotopically organized areas can be activated during mental imagery provides the best evidence to date in favor of the theory that mental images are coded in an analogical visuospatial buffer (Kosslyn et al., 1995; but see Pylyshyn, chapter 4, for a different view).

While the basic concepts behind a common science of mind and matter arguably were laid down much earlier, perhaps as far back as ancient Greece (see Marshall, chapter 28), there is little doubt that modern technology provides unprecedented opportunities to look directly for bridges between the cognitive and neural spheres. Newport, Bavelier, and Neville (chapter 27) isolate a domain of research where this bridging process is particularly active: the understanding of critical periods in language acquisition. Here, psychological research informs neuroscientists about the limits of plasticity in language learning at various levels of processing, while neuroscience in turn informs psychologists about the neural mechanisms at the origins of those limits, thus providing new hypotheses for behavioral testing and intervention.

Cognitive neuroscience studies also contribute critical evidence to flesh out an important concept in cognitive psychology, the notion of modularity. Isabelle Peretz (chapter 24), for instance, describes how brain imaging and cognitive neuropsychological studies of music can reveal core systems of auditory and emotional processing with a fixed modular organization. José Morais and Régine Kolinsky (chapter 26) likewise describe how studies of illiterates, with both psychological and brain-imaging methods, can reveal the interplay of modular phonological representations and of learned alphabetical orthography in the human brain.

Evolution, the central concept of biology, is making a prominent comeback in the psychological arena. Whether or not one adheres to the varied, and sometimes extreme, claims of evolutionary psychology, there is no doubt that comparative approaches, aimed at separating the unique components of the human mind from those that are shared with other species, have much to contribute to cognitive neuroscience. In chapter 23, Marc Hauser explains how comparative studies of monkeys, preverbal infants, and human adults lead to a reevaluation of the "speech-is-special" hypothesis and may one day clarify the evolutionary origins of language. Galaburda (chapter 25) forcibly argues that there is no paradox in seeking biological explanations of dyslexia in such nonverbal species as mice and rats. He describes a rat model with brain anomalies strikingly similar to those found in the brains of human dyslexic patients, and in whom tests inspired by psychological research, such as the auditory mismatch negativity, reveal deficits in fast auditory processing.

Jacques Mehler himself has become one of the main players in the field of correlating—certainly not reducing!—language to biology. As soon as the techniques became available, he launched collaborative brain-imaging studies with the Milan and Orsay groups to reveal the neural basis of specialization for the mother tongue (Mazoyer et al., 1993) and of language representations in bilinguals (Dehaene et al., 1997; Perani et al., 1996, 1998). He also pioneered comparative work on the perception of continuous speech by nonhuman primates and young infants (Ramus et al., 2000). We, the authors of this introduction, owe a considerable debt to Jacques's openness and enthusiasm when it comes to discussing good experiments. After all, one of us (L.C.), although specializing in the disreputable discipline of clinical neurology, entered Jacques's laboratory in 1984, the publication date of Mehler, Morton, and Jusczyk's paper. The other two followed the next year, despite their questionable formation in neuropediatrics (G.D.) or association with a molecular neurobiologist (S.D.)!

How could it be the same Jacques who boldly stated that "as psychologists, we gain *nothing* by shifting from the purely information processing point of view to the neurologically centred point of view" (Mehler et al., 1984, p. 99; our emphasis)? One key can be found toward the end of the paper, where the authors state what they see as a precondition to any interaction between psychology and neurobiology: it is "the psychological analysis that constrains the nature of the mapping between psychology and neurophysiology" (p. 106), not the other way around. Over a decade of brain imaging and neuropsychology has confirmed this crucial insight. Cognitive neuroscience is not progressing independently of psychology. There might have been a time when brain-imaging techniques were so novel, or neuropsychological impairments so understudied, that any self-taught group could invent a task (say, playing chess), image it or study its impairment in a group of patients, and publish it in a top journal. That period is over. The best work in cognitive neuroscience nowadays is founded on a clear theoretical basis in information processing. Careful task design, chronometric analysis, and behavioral controls have become essential ingredients of any integrative cognitive neuroscience approach.

The prediction was made, at about the same time as the Mehler et al. paper, that a radical process of psychoneural reduction was under way and that psychological concepts were going to be eradicated from the future science of the mind-brain (Churchland, 1986). This sounds even more implausible today than it did at the time. Language will not be *reduced* to biology. However, both psycholinguistics and neuroscience will gain by working hand in hand.

References

Churchland, P. S. (1986). *Neurophilosophy: Toward a unified understanding of the mind/brain.* Cambridge, MA: MIT Press.

Dehaene, S. D., Dupoux, E., Mehler, J., Cohen, L., Paulesu, E., Perani, D., van de Moortele, P. F., Lehéricy, S., and Le Bihan, D. (1997). Anatomical variability in the cortical representation of first and second languages. *Neuroreport, 8,* 3809–3815.

Kosslyn, S. M., Thompson, W. L., Kim, I. J., and Alpert, N. M. (1995). Representations of mental images in primary visual cortex. *Nature, 378,* 496–498.

Mazoyer, B. M., Dehaene, S., Tzourio, N., Frak, V., Syrota, A., Murayama, N., Levrier, O., Salamon, G., Cohen, L., and Mehler, J. (1993). The cortical representation of speech. *Journal of Cognitive Neuroscience, 5,* 467–479.

Mehler, J., Morton, J., and Jusczyk, P. (1984). On reducing language to biology. *Cognitive Neuropsychology, 1,* 83–116.

Moran, J., and Desimone, R. (1985). Selective attention gates visual processing in the extrastriate cortex. *Science, 229,* 782–784.

Perani, D., Dehaene, S., Grassi, F., Cohen, L., Cappa, S. F., Dupoux, E., Fazio, F., and Mehler, J. (1996). Brain processing of native and foreign languages. *Neuroreport, 7,* 2439–2444.

Perani, D., Paulesu, E., Sebastián-Gallés, N., Dupoux, E., Dehaene, S., Bettinardi, V., Cappa, S. F., Fazio, F., and Mehler, J. (1998). The bilingual brain: Proficiency and age of acquisition of the second language. *Brain, 121,* 1841–1852.

Ramus, F., Hauser, M. D., Miller, C., Morris, D., and Mehler, J. (2000). Language discrimination by human newborns and by cotton-top tamarin monkeys [see comments]. *Science, 288,* 349–351.

22

Cognitive Neuroscience: The Synthesis of Mind and Brain

Michael I. Posner

In this chapter I try to examine what has happened in cognitive neuroscience starting with a meeting held in France in 1980 where, I believe, the first papers using the title were presented. The meeting itself was called by Jacques Mehler to discuss the topic of cognitive representation. It was then a dominant view in cognitive science that there was little that could be learned from attempting to determine where in the brain things happened. The predominant view was that cognitive science was about software and it just did not matter whether the hardware was silicon or protoplasm (see, e.g., Mehler, Morton, and Juszyck, 1984). The meeting Jacques Mehler arranged contained four papers that reflected efforts at the time to develop the neural basis of cognition. All four papers reflect issues still present in current studies, although they would all almost certainly be quite different if written today.

Edgar Zurif (1982) presented an account of Broca's aphasia in terms of processing of different types of vocabulary items. Today, owing to structural and functional imaging studies (Dronkers, 1996; Raichle et al., 1994) the close association of Broca's area with grammar and aphasia is less clear. José Morais (1982) argued for interpreting hemispheric function in terms of local and global processing, a view that has a current resonance (Ivry and Robertson, 1998) even though neuroimaging has reduced the tendency to view the hemispheres as functioning in isolation. Kean and Nadel (1982) argued for the advantages of a model system approach that might serve to correct the idea, then common to both behavioral and cognitive psychology, that the same principles will deal with all instances of a psychological phenomenon such as memory. The idea that all cognitive systems have memory, implicit to their function, has largely replaced the unified concept of memory which Kean and Nadel criticized.

I was amazed in reading our long-forgotten chapter how exactly it predicted what I and many of my colleagues have tried to do in the nearly twenty years that have elapsed since it was written (Posner, Pea, and Volpe, 1982). We argued that it was time for a physiology of human cognition. It was this phrase that raised the specter of reductionism which Jean Requin discussed in his commentary (Requin, 1982). However, we had in mind a true integration of the two sciences, not a reduction of one to the other. I believe that this is exactly what has happened in the last fifteen years. It is not that cognition has been reduced to neuroscience, but areas of overlap, such as the study of elementary mental operations, have been greatly aided by viewing them from the joint perspective of their function and structure. So much so, that in many neuroimaging laboratories each experiment is in itself a detailed synthesis of cognitive and neuroscience concepts.

Our 1982 paper began by defining spatial and temporal methods for relating cognition to underlying brain mechanisms. The main spatial method we emphasized was the use of computed tomography to locate brain lesions. We wrote that, "this allows the possibility of a quantitative analysis of the relation of brain tissue to detailed cognitive processes, such as might be involved in reading." The fine measurement of lesions, mostly by magnetic resonance imaging (MRI), has greatly expanded the utility of this kind of study (Rafal, 1998). There was only a brief discussion of the use of radioactive tracers to study areas of the brain active during reading passages based on a *Scientific American* article (Lassen, Ingvar, and Skinhoj, 1978). However, in the last decade, neuroimaging with positron emission tomography (PET) and functional MRI (fMRI) has certainly become central to cognitive neuroscience.

The section on temporal dynamics included reaction time (RT) studies, of course, but also cellular and scalp electrical recordings of brain electrical activity. The development of the current methods to record the time course and communication of specific anatomical areas active during cognitive tasks has taken up much of the last fifteen years. The story of PET and later fMRI and their combination with electroencephalography (EEG) and magnetoencephalography (MEG) has been examined in much detail in many publications (Hillyard and Anlo-Vento, 1998; Posner and Raichle, 1996, 1998; Posner and DiGirolamo, 2000).

In the 1982 paper we sought to apply the temporal and spatial methods then available to the study of attention and reading. In this chapter, I try to briefly summarize progress in the methods of measuring the time course of specific areas of brain anatomy under three general headings: localization, circuitry, and plasticity. After that, I seek to evaluate what has been learned in the areas of attention and reading to determine in what ways the science of synthesis has already developed.

Localization

Every psychologist who entered the field in the mid-twentieth century studied the history of efforts to localize higher-level mental activity in the brain. We were taught that Karl Lashley (1931) had discovered in his work with rats' learning mazes that the effect of a brain lesion depended on how much tissue was removed and not on what part of the cortex was taken out. Studies of human patients with frontal lesions were also interpreted as showing that the brain operated as a whole with respect to cognition. Although Broca and others, at the turn of the century, had argued strongly for localization, based upon lesions that caused aphasia, this effort had been so closely associated with the popularization of phrenology (see Finger, 1994, for a discussion) that Lashley's position, although based on animal work, was dominant. The somewhat vague principles of mass action and equipotentiality were used to explain failures to find any precise localization. We were also cautioned about efforts to separate cognitive processes into elements because the whole was more than the sum of the parts, or to use the subtractive method because tasks were completely altered when any part was changed.

Two events satisfied me that the facts were quite different. The first was evidence that unilateral lesions of the parietal lobe, thalamus, or colliculus interfered with attention to visual events on the side of space opposite the lesion. This general anatomy fit well with clinical findings. However, I found that the exact location of the lesion made a great difference in the particular mental operation that was damaged. If the lesion was in the parietal lobe I found a specific effect on RT that I associated with interference with the disengage operation. Subsequent research suggests that this effect depends upon the temporo-parietal junction (Corbetta et al., 2000). A lesion of the thalamus prevented attention from adequately filtering out surrounding distractors (engage operation). A lesion of the midbrain (superior colliculus) led to a difficulty in moving attention (move operation; Posner and Presti, 1987). Subsequent studies have suggested that the move operation involves the superior parietal lobe, probably in conjunction with the colliculus. It was as though each mental operation involved in the task was localized in a quite different brain region.

The second was a study employing PET to find brain areas active in processing words. Specific areas related to chunking letters into an orthographic unit, sounding things out, and determining meaning were revealed quite directly by subtracting less complex tasks (e.g., repeating a word) from more complex ones (generating a use for the word) to eliminate component operations (Petersen et al., 1987). Moreover, the brain areas were generally related to lesion data. These two events were enough to convince me that mental operations studied in cognition were localized in specific brain tissue (Posner et al., 1988).

It is hard to imagine as one reads journals devoted to cognitive neuroscience or mapping the human brain that there could ever have been doubts that there was a specific anatomy related to the performance of mental computations. The distributed nature of the activations in any realistic cognitive task helps explain why Lashley and others could have thought that the brain operated as a whole. Any task is represented by a network of brain areas. The more tissue that is removed, the more this network is likely to be affected and the poorer the overall performance. However, where we have been able to dissect tasks into plausible component operations, it is these operations not the tasks themselves, that are localized. As new generations of brain imagers are attracted to efforts to map human brain function, it will be important to keep in mind the difficulties of localization and the importance of careful task analysis.

Real-Time Analysis

In the 1960s cognitive psychologists showed that a wide variety of tasks could be described in terms of mental operations (subroutines) that took place over tens to hundreds of milliseconds (Posner, 1978). Since even quite early extrastriate visual areas can show top-down influences, it has become very important to document the time course of activity in neural substrates in the millisecond range to distinguish pure sensory from top-down influences.

While there has been improvement in the speed of measurement in imaging methods such as fMRI, the time course of changes in blood chemistry is delayed for some seconds after input. For several decades, event-related electrical potentials recorded from the EEG in humans have been important in tracing the real-time activity of cognitive processes. However, with the advent of imaging techniques, it is now possible to relate the scalp activity to generators obtained from blood flow measures in humans and neurophysiological measures in animals. While the earliest parts of the event-related potential are primarily driven from the sensory event, within 100 ms, top-down attentional influences are clearly evident in human cortex. During the last ten years the link between electrical recording and hemodynamic imaging has been securely made. The most impressive studies came in the area of visual attention (Hillyard and Anlo-Vento, 1998). The importance of this work is illustrated in one recent study that traced the increased activity in primary visual cortex (Martinez et al., 1999). When complex visual displays are used, a number of cellular recording and fMRI studies have shown increased neural activity within primary visual cortex when subjects (monkeys or animals) attend to visual input (Posner and Gilbert, 1999). However, it appears from time course data that these activations of primary visual cortex (V1) are fed back from attentional influences in extrastriate regions that occur earlier in time (Martinez et al., 1999).

In high-level skills such as reading, timing is constrained by the natural task of moving the eyes. A skilled reader takes only about 275 ms for an individual fixation. There is clear evidence that the length of the saccade that follows the fixation is influenced by the meaning of the fixated word (Sereno, Rayner, and Posner, 1998). Thus, 275 ms sets an upper bound to achieve the word meaning sufficiently to convey appropriate information to the saccadic eye movement system. Studies of visual word processing using scalp electrical recording have shown specific areas of the occipital lobe and of the frontal lobe that are activated within the first 200 ms of input (Posner et al., 1999).

Plasticity

There are many time scales on which changes in neural circuits can be studied. In this section we consider rapid changes in neurocircuitry that take place following a few minutes of practice on a list of associations. Priming also involves fast changes in brain areas involved in processing input. However, some forms of plasticity may involve many weeks or years of practice. One of these is the development of the visual word system during the acquisition of literacy discussed in the next section.

Automaticity

In one study using PET (Raichle et al., 1994), subjects were required to generate a use for a read or heard noun (e.g., pound as a use for a hammer). When a new list of words was presented there was activity in left frontal and posterior cortex, in the anterior cingulate, and in the right cerebellum. Activity in the anterior insula was reduced over what was found in simply reading the words aloud. A few minutes of practice at generating associations for the same list of words shifted activation so that the left frontal and posterior areas, important in generating a new use, dropped away, and the anterior insula, strongly activated during reading aloud, increased. When a new list was introduced, the original activation returned. When generating the associations for a word became automated with practice, the same circuit was used as when skilled readers speak words aloud.

There appeared to be one neural circuit associated with the thought needed to generate a familiar, but unpracticed use of a noun, and a different neural circuitry when the task was automated, as in reading aloud or generating a just-practiced association. The circuit used for thought includes brain areas known to be involved in attention and effortful control, while the automated circuit does not involve these areas (Posner and McCandliss, 1999).

Priming

A major area of brain imaging has been to understand changes with repetition. Neurophysiological studies in animals that present the same item on successive occasions (repetition priming) show that a single repetition of an object can lead to dropping out of many of the neurons active during its first presentation (Desimone and Duncan, 1995). Reductions in blood flow (Demb et al., 1995) indicate tuning or selection of the neural activity needed to process the items. The discovery of how priming takes places in ensembles of neurons fits well with the finding that as repetitions allow faster RTs to identify the primed item, the degree of attention given the item is reduced.

Where Are We Now?

The idea of synthesis between brain and mind represents a whole spectrum of possibilities, as was pointed out by the late J. Requin (1982) in commenting on our 1982 paper. At one end is the idea that psychological and physiological studies should proceed separately and may, one distant day, find an appropriate relationship. At the other end is a naive reductionism in which function disappears to be replaced by a true understanding of cell and molecular biology. Requin cautioned against these extremes and in his later career he carried out studies requiring a close integration of ideas from psychology and neurobiology.

This integrative empirical direction dominates the current scene in which neuroimaging is a tool providing day-to-day interaction between brain structure and cognitive function. In most of the world's neuroimaging centers cognitive psychologists work closely with neuroscientists and with physicists and chemists. Societies and journals with the name "cognitive neuroscience" encourage the daily interaction that is the reality of life in leading centers in Europe and America. The McDonnell summer institute on cognitive neuroscience attract young researchers from many fields. In addition, it is now the goal of many young physicists and engineers to enhance methods to measure the activity of the human brain noninvasively and with increased precision. In this sense the science of synthesis we envisaged has taken place. However, one can also ask if it has yielded the new principles required to support a true science.

Attention

The synthesis has been most impressive in the study of visuospatial attention (Posner and DiGirolamo, 2000). The cognitive facts of neglect are that patients with right parietal lesions show a permanent deficit in their ability to detect targets in the left visual field if they are already attending to a target in the right visual field or in any location rightward of the target. How is it that these patients can learn to voluntarily shift attention to the left and sometime appear to do so normally when a target occurs, but at other times miss the identical target? These results seem to depend upon the fact that quite separate circuits are involved in moving attention when it is not otherwise engaged from those that are required to disengage the target from a current focus of attention.

Imaging data have shown that an area of the right superior parietal lobe is important in shifting attention leftward, but this brain area does not serve to disengage a person from an already existing focus of attention (Corbetta et al., 2000). For that purpose, it appears to be necessary to access the temporoparietal junction. Thus a novel event first breaks the current focus of attention, if one exists, and this depends upon just the part of the parietal lobe that is most frequently lesioned in neglect (Rafal, 1998). In addition, lesions of the corpus callosum break the coordination between the two parietal lobes, leaving a patient with increased ability to search the two visual fields in parallel (Luck et al., 1989).

These physical findings help us explain some of the most puzzling aspects of normal orienting and its disorders. Why not leave the cognitive theory to stand alone? After all, it would be possible to separate the function of disengaging and moving attention based only on an analysis of the task. However, the same cognitive predictions can result from a number of underlying models some of which require no localization of operation, in fact no specific operations at all (Cohen et al., 1994). It is possible to account for the results of RT experiments with normal people with either model. However, since one of the models has no anatomical suggestion, it tells us nothing about the influence of specific damage to some part of the system. Moreover, the close connection of the superior parietal lobe with the eye movement system suggests that moving attention is closely related to voluntary eye movements, but that the disengage operation is more closely related to what happens when novel events occur regardless of location. Thus, knowing the anatomy suggests direction for cognitive research, as well as providing a basis for understanding brain damage.

Reading

In their paper, Mehler et al. (1984) suggest that too tight a linkage between biology and psychological explanations would retard the independent development of each of them. While I share the concern for the continued importance of functional models of the brain arising from purely cognitive studies, I do not see how in the future such models can be constructed accurately without consideration of the biological results.

Consider the visual input logogen, a device that chunks the letters of a word into a unit (Morton, 1982). What information does such a device use? The finding of an area of the left ventral occipital lobe that responds to words and orthographically regular nonwords but not to nonsense material (Petersen et al., 1990) is of obvious potential importance. There have been disputes about this area, possibly because the importance of a visual input logogen may depend in part on whether the orthography of the language is shallow or deep and exactly by what methods one learns to read. It seems possible that in some persons and perhaps all persons on some occasions visual letter strings are translated very swiftly into a phonological form with little evidence of a separate visual word form. In accord with this possibility, recent studies have distinguished between a more ventral occipitotemporal system involved in chunking letters into a unit and a more dorsal area more closely related to phonology. The ventral visual word-form system appears to be relatively late-developing (Posner and McCandliss, 1999). The visual word-form area must be influenced by the specific learning experiences that children have during their early education. Indeed, this area appears to start in childhood with responding to specific words the child has learned and only after training does it become more related to the orthography of the language than to the familiarity of learned exemplars (Posner and McCandliss, 1999).

Future Goals

Given these developments, have we reached the long-sought-after synthesis of mind and brain, and if not, what remains to be done? One direction that is currently being pursued is a firmer connection between the level of analysis given by neuroimaging, which is at its best in the millimeter range, and the study of cellular, synaptic, and genetic mechanisms of neuronal activation.

For example, studies of alert monkeys have shown that alerting depends upon the neurotransmitter norepinepherine, whereas spatial switches of attention involve the cholinergic system (Marrocco and Davidson, 1998; Davidson and Marrocco, 2000). Not only do these findings support some cognitive views but they also provide suggestions as to therapies following lesions.

Another important step will be to determine the plasticity of brain circuits that serve cognition functions. When in infant or child development do particular circuits come online? New adaptations of magnetic imaging may be able to trace noninvasively myelination of specific pathways between brain areas (Le Bihan, 1995). We could then be able to try to predict when particular behaviors should emerge. I suspect the visual system, and particularly visually guided eye movements, may be studied first, but higher-level cognitive activity is sure to follow. We would then have a disciplined approach to understanding when a human brain becomes ready to learn a cognitive skill.

Development is a particularly important and obvious way to study plasticity. Jacques Mehler is a world leader in showing the importance of studies of infants and children in constructing an adequate view of human cognition. We now have available methods to observe changes in brain circuits with educational experiences (Posner and McCandliss, 1999). Another important application will be in studying what various brain injuries or pathological changes do to the circuitry that supports cognitive processes and how they recover spontaneously or with drugs, practice, or insight therapies. We are just at the very start of guiding our therapeutic interventions with imaging methods.

A quite different direction for the use of imaging is to begin to understand the reasons why certain computations occur in particular brain areas. There has been spectacular progress in efforts to relate imaged human brain areas to the visual maps obtained from cellular recording in primates (Tootell et al., 1998).

A more distant goal for the study of imaging, but one that needs careful thought, is the development of general principles of how closely related computations are expressed both in brain tissue and in performance.

Kinsbourne and Hicks (1978) proposed that the more densely connected two brain areas were, the more interference they would show when simultaneously active and the more priming when successively active. At the time this was proposed we were not able to test these ideas except by experiments using different motor outputs. Now, if one assumes that being closer in brain space relates to connectivity, there are many opportunities to test links between performance and neural distance.

The progress made in mapping cognitive functions in the human brain has been swift and startling. However, there is still a long way to go to exploit these maps into a more general understanding that would support a genuine science of synthesis. I believe the integration of brain and mind will come closer as issues of the circuitry and plasticity of the networks of anatomical areas underlying cognition become ever-stronger features of neuroimaging studies. Outstanding psychologists, at first and perhaps still very skeptical about integration, are participating very actively in these developments. From my perspective, the leadership that Mehler provided in supporting the 1982 conference and since then in much of his work and that of his students (Dehaene, 1996; Mazoyer et al., 1993) has been of central importance in suggesting what might be possible.

References

Cohen, J. D., Romero, R. D., Servan-Schreiber, D., and Farah, M. J. (1994). Mechanisms of spatial attention: The relationship of macrostructure to microstructure in parietal neglect. *Journal of Cognitive Neuroscience, 6,* 377–387.

Corbetta, M., Kincade, J. M., Ollinger, J. M. McAvoy, M. P., and Shulman, G. L. (2000). Dynamics of visual spatial expectancies from target detection in human posterior parietal cortex. *Nature and Neuroscience, 3,* 292–297.

Davidson, M. C., and Marrocco, R. T. (2000). Local infusion of scopolamine into intraparietal cortex slows covert orienting in rhesus monkeys. *Journal of Neurophysiology, 83,* 1536–1549.

Dehaene, S. (1996). The organization of brain activations in number comparison: Event-related potentials and the additive-factors method. *Journal of Cognitive Neuroscience, 8,* 47–68.

Demb, J. B., Desmond, J. E. Wagner, A. D., Vaidya, C. T., Glover, G. H., and Gabrieli, J. D. (1995). Semantic encoding and retrieval in the left inferior prefrontal cortex: A functional MRI study of task difficulty and process specificity. *Journal of Neuroscience, 15,* 5870–5878.

Desimone, R., and Duncan, J. (1995). Neural mechanisms of selective attention. Annual Review of Neuroscience, 18, 193–222.

Dronkers, N. (1996). A new brain region for coordinating speech articulation. *Nature, 384,* 159–161.

Finger, S. (1994). *Origins of neuroscience.* New York: Oxford Univesity Press.

Hillyard, S. A., and Anlo-Vento, L. (1998). Event related brain potentials in the study of selective attention. *Proceedings of the National Academy of Sciences of the United States of America, 95,* 781–787.

Ivry, R. and Robertson, L. (1998). *The two sides of perception.* Cambridge, MA: MIT Press.

Kean, M-L., and Nadel, L. (1982). On the emergence of cognitive neuroscience. In J. Mehler, E. Walker, and M. Garrett (Eds.), *Perspectives on mental representations* (pp. 317–325). Hillsdale, NJ: Erlbaum.

Kinsbourne, M., and Hicks, H. (1978). Functional cerebral space: A model for overflow, transfer and interference effects in human performance. In J. Requin (Ed.), *Attention and performance VII* (pp. 345–363). New York: Wiley.

Lashley, K. (1931). Mass action in cerebral function. *Science 73,* 245–254.

Lassen, N. A., Ingvar, D. H., and Skinhø, E. (1978). Brain function and blood flow. *Scientific American, 239*(Oct.), 62–71.

LeBihan, D. (1995). *Diffusion and perfusion magnetic resonance imaging.* New York: Raven Press.

Luck, S. J., Hillyard, S. Q., Manguan, G. R., and Gazzaniga, M. S. (1989). Independent hemispheric attentional systems mediate visual search in split brain patients. *Nature, 342,* 543–545.

Marrocco, R. T., and Davidson, M. C. (1998). The neurochemistry of attention. In R. Parasuraman (Ed.), *The attentive brain.* Cambridge, MA: MIT Press.

Martinez, A., Anlo-Vento, L., Sereno, M. I., Frank, L. R., Buxton, R. B., Dubowitz, D. J., Wong, E. C., Hinrichs, H., Heinze, H. J., and Hillyard, S. A. (1999). Involvement of striate and extrastriate visual cortical areas in spatial attention. *Nature and Neuroscience, 2,* 364–369.

Mazoyer, B. M., Tzouria, N., Frak, V., Syrota, A., Murayama, N., Levrier, N., Salamon, G., Dehaene, S., Cohen, L., and Mehler, J. (1993). The cortical representation of speech. *Journal of Cognitive Neuroscience, 5,* 467–479.

Mehler, J., Morton, J., and Juszyck, P. W. (1984). On reducing language to biology. *Journal of Cognitive Neuropsychology, 1,* 83–11.

Morais, J. (1982). The two sides of cognition. In J. Mehler, E. Walker, and M. Garrett (Eds.), *Perspectives on mental representations* (pp. 277–309). Hillsdale, NJ: Erlbaum.

Morton, J. (1982). Disintegrating the lexicon: An information processing approach. In J. Mehler, E. Walker, and M. Garrett (Eds.), *Perspectives on mental representations.* Hillsdale, NJ: Erlbaum.

Petersen, S. E., Fox, P. T., Posner, M. I., Mintun, M., and Raichle, M. E. (1987). Positron emission tomographic studies of the cortical anatomy of single word processing. *Nature, 331,* 585–589.

Petersen, S. E., Fox, P. T., Snyder, A., and Raichle, M. E. (1990). Activation of extrastriate and frontal cortical areas by words and word-like stimuli. *Science, 249,* 1041–1044.

Posner, M. I. (1978). *Chronometric explorations of mind.* Hillsdale, NJ: Erlbaum.

Posner, M. I. (1988). Structures and functions of selective attention. In T. Boll and B. Bryant (Eds.), *Master lectures in clinical neuropsychology and brain function: Research, measurement, and practice* (pp. 171–202). Washington, DC: American Psychological Association.

Posner, M. I., and DiGirolamo, G. J. (2000). Attention in cognitive neuroscience: An overview. In M. S. Gazzaniga (Ed.), *The new cognitive neurosciences,* 2d ed. (pp. 621–632). Cambridge, MA: MIT Press.

Posner, M. I., and Gilbert, C. D. (1999). Attention and primary visual cortex. *Proceedings of the National Academy of Sciences of the U.S.A., 96,* 2585–2587.

Posner, M. I., and McCandliss, B. D. (1999). Brain circuitry during reading. In R. Klein and P. McMullen (Eds.), *Converging methods for understanding reading and dyslexia* (pp. 305–337). Cambridge, MA: MIT Press.

Posner, M. I., and Raichle, M. E. (Eds.) (1998). Overview: The neuroimaging of human brain function. *Proceedings of the National Academy of Sciences of the U.S.A., 95,* 763–764.

Posner, M. I., and Presti, D. (1987). Selective attention and cognitive control. *Trends in Neuroscience, 10,* 12–17.

Posner, M. I., and Raichle, M. E. (1996). *Image of mind.* New York: Scientific American Library.

Posner, M. I., Abdullaev, Y., McCandliss, B. D., and Sereno, S. E. (1999). Neuroanatomy, circuitry and plasticity of word reading. *Neuroreport, 10,* R12–23.

Posner, M. I., Pea, R., and Volpe, B. (1982). Cognitive neuroscience: Developments toward a science of synthesis. In J. Mehler, E. Walker, and M. Garrett (Eds.), *Perspectives on mental representations* (pp. 251–275). Hillsdale, NJ: Erlbaum.

Posner, M. I., Petersen, S. E., Fox. P. T., and Raichle, M. E. (1988). Localization of cognitive functions in the human brain. *Science, 240,* 1627–1631.

Rafal, R. D. (1998). Neglect. In R. Parasuraman (Ed.), *The attentive brain* (pp. 489–525). Cambridge, MA: MIT Press.

Raichle, M. E., Fiez, J. A., Videen, T. O., Pardo, J. V., Fox, P. T., and Petersen, S. E. (1994). Practice-related changes in human brain functional anatomy during nonmotor learning. *Cerebral Cortex, 4,* 8–29.

Requin, J. (1982). The meaning of an experiment combining cognitive and neurobiological approaches. In J. Mehler, E. Walker, and M. Garrett (Eds.), *Perspectives on mental representations* (pp. 457–464). Hillsdale, NJ: Erlbaum.

Sereno, S. C., Rayner, K., and Posner, M. I. (1998). Establishing a time line of word recognition: Evidence from eye movements and event-related potentials. *Neuroreport, 9,* 2195–2200.

Tootell, R. B. H., Mendola, J. D., Hadjikhani, N. K., Liu, A. K., and Dale, A. M. (1998). The representation of the ipsilateral visual field in human cerebral cortex. *Proceedings of the National Academy of Sciences of the United States of America, 95,* 818–824.

Zurif, E. B. (1982). Language and the brain: Some points of connection. In J. Mehler, E. Walker, and M. Garrett (Eds.), *Perspectives on mental representations* (pp. 311–316). Hillsdale, NJ: Erlbaum.

23

What's So Special about Speech?

Marc Hauser

Humans have long been obsessed with their uniqueness. Early on, it was "man the hunter." But studies of cooperative hunting in chimpanzees put an end to this view (Boesch, 1994; Goodall, 1986). Then it was "man the toolmaker." Once again, studies of chimpanzees and, more recently, of several other animals have shown that humans are by no means unique in their capacity to make and use tools (Alp, 1997; Hauser, 1997; Hunt, 1996; Matsuzawa, 1996; Visalberghi, 1990; reviewed in Hauser, 2000). As recent studies of chimpanzees have clearly shown, the diversity of tools used for solving both social and ecological problems is extraordinary (Whiten, et al., 1999). Then it was "humans, the sexy species." But observations of bonobos, also known as pygmy chimpanzees, have demonstrated that we are not alone in our sexual extravaganzas (de Waal, 1988, 1989, 1996; Dixson, 1999). Bonobo females are continuously receptive to the sexual advances of males, as well as females. There are homo- and heterosexual encounters with individuals of all ages, including oral sex and tongue kissing. What, then, of our verbal abilities, the capacity to speak about quantum mechanics or syntactic structure, the beauty of nature, the passion of romantic love, and the batting record of a Babe Ruth or Sammy Sosa? What, more precisely, is so special about our capacity for speech? My aim in this essay is to explore the "speech-is-special" problem using what I consider to be the only viable empirical approach. It is an approach that Charles Darwin championed, and that I and several colleagues have tapped over the past few years: the comparative method (Kluender, Lotto, and Holt, in press). Here, I compare human infants and adults on the one hand, with nonhuman animals (hereafter animals) on the other, exploring the extent to which the mechanisms underlying speech perception in humans are evolutionarily ancient, inherited from a vertebrate ancestor.

The Speech-Is-Special Debate

Historical Background

First, a brief history. In the 1960s, following Chomsky's (1957a) famous attack on Skinner's *Verbal Behavior,* Al Liberman and his colleagues (Liberman et al., 1957, 1967) at the Haskins laboratory began to explore in detail the mechanisms underlying speech perception in humans. What is particularly interesting about the claims emerging from Haskins at this time is that they were intellectually allied with Chomsky's (1957b, 1966) position concerning the special nature of human language. In particular, it is clear that Chomsky thought of the "language organ" as a uniquely human organ, and that its capacity for generating syntactic structure evolved for reasons that had nothing to do with communication. Although one can certainly challenge this claim, what is important for biologists about Chomsky's position is that it sets up a testable hypothesis about the nature of the comparative database (see Hauser, 1996). Specifically, if humans are truly unique with respect to the language organ, then we should see little to no evidence of a precursor mechanism in other animals. For the past 40 years, biologists, psychologists, and anthropologists have been chasing Chomsky's particular version of the human uniqueness claim by looking at the capacities of animals to acquire some form of a human natural language under intensive training environments, or for animals to use their natural, species-typical vocalizations in ways that are similar to spoken language.

Thus, for example, studies have focused on the capacity of human-reared apes to string symbols together to form sentences or comprehend them (Gardner, Gardner, and Van Cantfort, 1989; Savage-Rumbaugh and Lewin, 1996; Savage-Rumbaugh et al., 1993; Terrace, 1979), and of wild monkey populations to use vocalizations to refer to objects and events in the external environment (Fischer, 1998; Gouzoules, Gouzoules, and Marler, 1984; Hauser, 1998; Marler, Duffy, and Pickert, 1986; Seyfarth, Cheney, and Marler, 1980; Zuberbuhler, Noe, and Seyfarth, 1997). Though these studies have met with mixed success, especially when viewed from the perspective of linguists looking at such comparative data for insights into the evolution of language (Bickerton, 1990; Lieberman 1991; Pinker, 1994), there has been another approach, one that brings us back to Liberman and the Haskins laboratory. In particular, rather than a focus on the semantics and syntax of language, much of the early work on speech perception was aimed at identifying particular signatures of an underlying, specialized mechanism. Perhaps one of the most important, and early, entries into this problem was Liberman's discovery of the phenomenon of categorical perception.

Categorical Perception: Uniquely Human, Special to Speech?

When we perceive speech, we clearly create categories. Using an artificially created acoustic continuum running from /ba/ to /pa/, human adults show a categorical discrimination and labeling function. More precisely, discrimination of exemplars is excellent for between-category exemplars, but not for within-category exemplars. To determine whether the mechanism underlying categorical perception is specialized for speech, uniquely human, and fine-tuned by the linguistic environment, new methods were required, as were subjects other than human adults. In response to this demand, the phenomenon of categorical perception was soon explored in (1) adult humans, using nonspeech acoustic signals as well as visual signals; (2) human infants, using a non-nutritive sucking technique together with the presentation of speech stimuli; and (3) animals, using operant techniques and the precise speech stimuli used to first demonstrate the phenomenon in adult humans. Results showed that categorical perception could be demonstrated for nonspeech stimuli in adults (Bornstein, 1987; Remez, 1979), and for speech stimuli in both human infants (Eimas et al., 1971) and nonhuman animals (Kuhl and Miller, 1975; Kuhl and Padden, 1982). Although the earliest work on animals was restricted to mammals (i.e., chinchilla, macaques), more recent studies have provided comparable evidence in birds (Dent et al., 1997; Kluender, Diehl, and Killeen, 1987). This suggests that the mechanism underlying categorical perception in humans is shared with other animals, and may have evolved at least as far back as the divergence point with birds. Although this finding does not rule out the importance of categorical perception in speech processing, it does indicate that the underlying mechanism is unlikely to have evolved for speech.

Categorical perception has also been demonstrated in animals using tasks involving their own, species-typical vocalizations (reviewed in Kuhl, 1989; Hauser, 1996; Wyttenbach and Hoy, 1999). And here, the breadth of species tested is truly extraordinary, including field crickets (Wyttenbach, May, and Hoy, 1996), swamp sparrows (Nelson and Marler, 1989), mice (Ehret and Haack, 1981), pygmy marmosets (Snowdon, 1987), and Japanese macaques (May, Moody, and Stebbins, 1989). Perhaps one of the best examples, based on methodological elegance as well as functional and ecological considerations, comes from Wyttenbach and Hoy's work on the field cricket. In this species, individuals emit a contact call of 4 to 5 kHz. When conspecifics hear this call, they often approach. In contrast, predatory bats produce ultrasonic signals in the range of 25 to 80 kHz, and when crickets hear such sounds, they move away. The perceptual task, therefore, involves a discrimination between two ecologically meaningful acoustic signals, one that elicits approach and a second that elicits avoidance. Laboratory experiments had already indicated a transition between approach and avoidance in the range of 10 to 20 kHz. In the labeling task, crickets were presented with signals that varied from 2.5 to 40 kHz. Results showed an abrupt transition from approach to avoid between 13 and 16 kHz, providing strong evidence of a categorical boundary. In the discrimination task, crickets were habituated to 20-kHz pulses (i.e., a signal that elicits escape), and a photocell used to measure the movement of the subject's hind leg. Once subjects habituated (i.e., showed little to no escape response), they then received one test stimulus from a different frequency and one 20-kHz stimulus. Of the frequencies tested, only stimuli falling below 16 kHz caused dishabituation; no stimuli falling in the ultrasound range caused dishabituation, providing strong evidence of between-category discrimination.

The Next Generation

The history of work on categorical perception provides an elegant example of the comparative method. If you want to know whether a mechanism has evolved specifically for a particular function, in a particular species, then the only way to address this question is by running experiments on a broad array of species. With respect to categorical perception, at least, we can confidently claim that the underlying mechanism did not evolve for processing speech. A question, however, arises from such work: What, if anything, is special about speech, especially with respect to processing mechanisms? Until the early 1990s, animal scientists pursued this problem, focusing on different phonemic contrasts as well as formant perception (Lotto, Kluender, and Holt, 1997; Sinnott, 1989; Sinnott and Brown, 1997; Sinnott, Petersen, and Hopp, 1985; Sommers et al., 1992); most of this work suggested common mechanisms, shared by humans and nonhuman primates. In 1991, however, Patricia Kuhl (1991) published an important paper showing that human adults and infants, but not rhesus monkeys, perceive a distinction between so-to-speak good and bad exemplars of a phonemic class. The good exemplars, which Kuhl described as prototypes, functioned like perceptual magnets, anchoring the category, and making it more difficult to distinguish the prototype from sounds that are acoustically similar; nonprototypes function in a different way, and are readily distinguished from more prototypical exemplars. In the same way that robins and sparrows, but not penguins or storks, are prototypical birds because they possess the most common or salient visual features (e.g., feathers, beak, wings) within the category bird, prototypical phonemes consist of the most common or salient acoustical features.

At present, there is some controversy in the literature concerning the validity of Kuhl's work for thinking about the perceptual organization and development of speech (Kluender et al., 1998; Lotto, Kluender, and Holt, 1998; Sussman and Lauckner-Morano, 1995). My concern here, however, is with the comparative claim. Because Kuhl failed to find evidence that rhesus monkeys distinguish prototypical from nonprototypical instances of a phonetic category, she argued that the perceptual magnet effect represents a uniquely human mechanism, specialized for processing speech. Moreover, because prototypes are formed on the basis of experience with the language environment, Kuhl further argued that each linguistic community will have prototypical exemplars tuned to the particular morphology of their natural language.

To address the comparative claim, Kluender and colleagues (1998) attempted a replication of Kuhl's original findings, using European starlings and the English vowels /i/ and /I/, as well as the Swedish vowels /y/ and /u/; these represent the stimuli used in Kuhl's original work on the prototype effect. Based on a mel scale of the first and second formants, these vowels have distinctive prototypes that are, acoustically, nonoverlapping. Once starlings were trained to respond to exemplars from these vowel categories, they readily generalized to novel exemplars. More important, the extent to which they classified a novel exemplar as a member of one vowel category or another was almost completely predicted by the F1 and F2 values, as well as by the exemplar's distance from the prototype or centroid of the vowel sound. Because the starlings' responses were graded, and matched human adult listeners' ratings of goodness for a particular vowel class, Kluender and colleagues concluded, contra Kuhl, that the perceptual magnet effect is not uniquely human, and can be better explained by general, perceptual learning mechanisms.

In contrast to the extensive comparative work on categorical perception, we have only two studies of the perceptual magnet effect in animals. One study of macaques claims that animals lack such capacities, whereas a second study of starlings claims that animals have such capacities. If starlings perceive vowel prototypes, but macaques do not, then this provides evidence of a homoplasy—a character that is similar between species because of convergent evolution. Future work on this problem must focus on whether the failure with macaques is due to methodological issues (e.g., would a different testing procedure provide different results?) or to an absence of a capacity.

New Approaches
To date, every time a claim has been made that a particular mechanism X is special to speech, animal studies have generally shown that the claim is false. Speech scientists might argue, however, that these studies are based on extensive training regimens, and thus fail to show what animals spontaneously perceive. Although experiments involving training show what an animal's brain is capable of computing, they do not allow us to understand how animals and humans compare on tasks involving spontaneous methods. Over the past few years, my students and I have been pushing the development of methodological tools that involve no training, and thus may provide a more direct approach to comparing the be-

havioral, perceptual, and cognitive mechanisms that are either shared or distinctive across species (Hauser, 1996, 1997, 1998; Santos and Hauser, 1999; Santos, Frieson, and Hauser, 1999); we are certainly not alone in this endeavor (Diamond, 1990; Diamond, Zola-Morgan, and Squire, 1989; Terrace, 1993; Tomasello and Call, 1997; Tomasello, Savage-Rumbaugh, and Krugerf, 1993). In the remainder of this chapter, I describe several recent results using spontaneous techniques, focusing specifically on how nonhuman primates perceive speech and whether they tap mechanisms that are shared with human primates.

Over the past 15 or so years, my students, colleagues, and I have conducted observations and experiments on the vocal behavior of three nonhuman primate species, each from a different branch of the primate phylogeny (figure 23.1). Specifically, we have conducted studies of cotton-top tamarins (*Saguinus oedipus*) in the laboratory, rhesus monkeys (*Macaca mulatta*) on an island off the coast of Puerto Rico, and chimpanzees (*Pan troglodytes*) in a rainforest in Uganda. A powerful technique for exploring spontaneous perceptual distinctions is the habituation-dishabituation technique (Cheney and Seyfarth, 1988; Hauser, 1998), briefly mentioned above. Given the variety of conditions in which our animals live, each situation demands a slightly different use of this technique. The logic underlying our use of the procedure is, however, the same. In general, we habituate a subject to different exemplars from within an acoustic class and then present it with a test stimulus. A response is scored if the subject turns and orients in the direction of the speaker. We consider the subject to be habituated if it fails to orient toward the speaker on at least two consecutive trials; as such, all subjects enter the test trial having failed to respond on the previous two trials. The advantage of this approach is that we can not only score whether or not the subject responds to the test stimulus but in some cases the magnitude of the response; that is, we can score the amount of time spent looking in the direction of the speaker. In the case of speech stimuli, duration is not a reliable measure, whereas in the case of conspecific vocalizations it is.

The first playback experiment on speech perception was run in collaboration with Franck Ramus, Cory Miller, Dylan Morris, and Jacques Mehler (2000). The goal of these experiments was twofold. Theoretically,

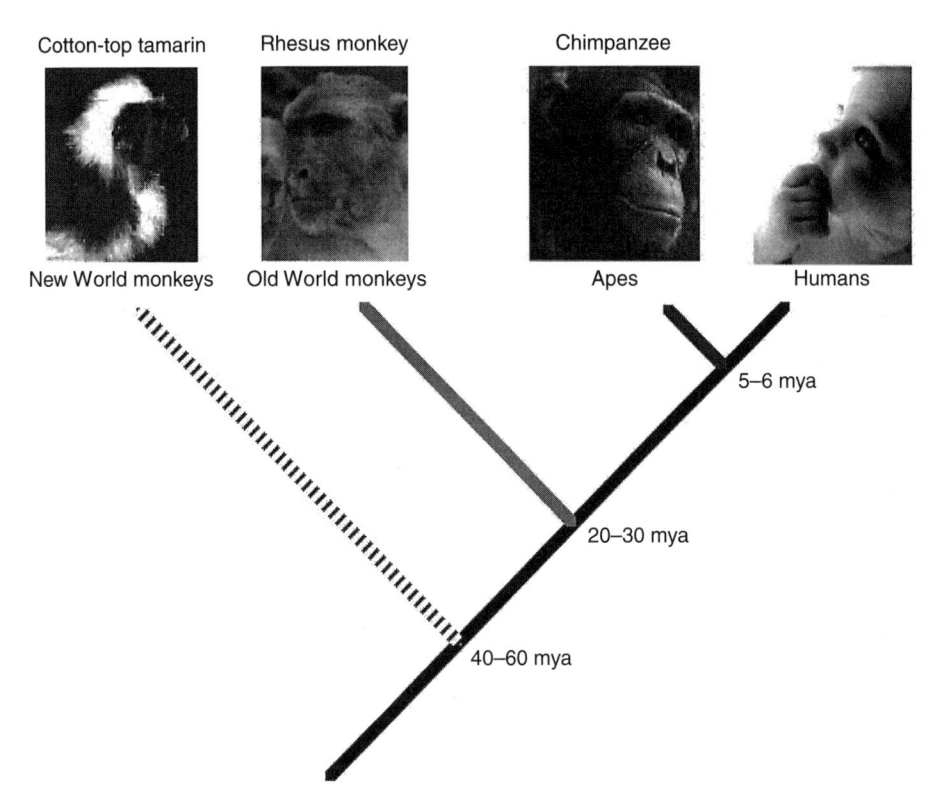

Figure 23.1
Phylogeny of four primate groups, with approximate divergence times listed at each node.

we wanted to understand whether the capacity of human infants to both discriminate and subsequently acquire two natural languages is based on a mechanism that is uniquely human or shared with other species. Though animals clearly lack the capacity to produce most of the sounds of our natural languages, their hearing system is such (at least for most primates; Stebbins, 1983) that they may be able to hear some of the critical acoustic features that distinguish one language from another. To explore this problem, we asked whether human neonates and cotton-top tamarins can discriminate sentences of Dutch from sentences of Japanese, and whether the capacity to discriminate these two languages depends on whether they are played in a forward (i.e., normal) or backward direction; given the fact that adult humans process backward speech quite differently from forward speech, we expected to find some differences, though not necessarily in both species. Methodologically, we wanted to determine whether tests of speech processing could be run on neonates and captive cotton-top tamarins using the same stimuli and procedure. Specifically, would tamarins attend to sentences from a natural language, and could we implement the habituation-dishabituation technique to ask questions about discrimination?

Neonates and adult tamarins were tested in four different conditions involving naturally produced sentences of Dutch and Japanese. In the first language change condition, subjects were habituated to one language played in the normal/forward direction, and then tested with sentences from the second language played in the normal/forward direction. In the second language change condition, all sentences were played backward, but with the same shift from one language to the other. In the first speaker change condition—run as a control for the language change condition—subjects were habituated to normal/forward sentences of one language spoken by two speakers, and then tested with normal/forward sentences of the same language, but spoken by two new speakers. The second speaker change condition was the same, but with the sentences played backward.

There were a few differences in the testing procedures used for neonates and tamarins. The behavioral assay for neonates was a high-amplitude sucking response, whereas for tamarins, we used a head-orienting response in the direction of the concealed speaker. For neonates, habituation stimuli were played back until the sucking response attenuated to 25 percent less than the previous minute, and then maintained this level for two consecutive minutes. Once habituated, test stimuli were repeatedly played back. For tamarins, in contrast, we played back exemplars from the habituation category until the subject failed to orient on two consecutive trials. Following habituation, we played back sentences of the test category. If subjects failed to respond in the test trial, we played a post-test stimulus, specifically, a tamarin scream. The logic behind the post-test was to ensure that the tamarins had not habituated to the entire playback setup. Thus, if they failed to respond in the post-test, we assumed that they had habituated to the setup, and reran the entire session a few weeks later.

Neonates failed to discriminate the two languages played forward, and also failed to discriminate the two speakers. Rather than run the backward condition with natural speech, we decided to synthesize the sentences and run the experiment again, with new subjects. One explanation for the failure with natural speech was that discrimination was impaired by the significant acoustic variability imposed by the different speakers. Consequently, synthetic speech provides a tool for looking at language discrimination, while eliminating speaker variability. When synthetic speech was used, neonates dishabituated in the language change condition, but only if the sentences were played forward; in the backward speech condition, subjects failed to dishabituate.

In contrast to the data on neonates tested with natural speech, tamarins showed evidence of discrimination in the forward language change condition, but failed to show evidence of discrimination in any of the other conditions (figure 23.2). When the synthetic stimuli were used, the results were generally the same (see figure 23.2). Only the forward language change condition elicited a statistically significant level of discrimination, though the backward speaker change was nearly significant; thus, there was a nonsignificant difference between the language and speaker change condition. When the data from the natural and synthetic stimuli are combined, tamarins showed a highly significant discrimination of the forward language change condition, but no other condition.

These results allow us to make five points with respect to studying the speech-is-special problem. First, the same method can be used with human infants and nonhuman animals. Specifically, the habituation-dishabituation paradigm provides a powerful tool to explore similarities and differences in perceptual mechanisms, and avoids the potential problems associated with training. Second, animals such as cotton-top tamarins not only attend to isolated syllables as previously demonstrated in studies of categorical perception but also attend to strings of continuous speech. Consequently, it is now possible to ask comparative questions about some of the higher-order properties of spoken languages, including some of the relevant prosodic or paralinguistic information. Third, given the fact that tamarins discriminate sentences of Dutch from sentences of Jap-

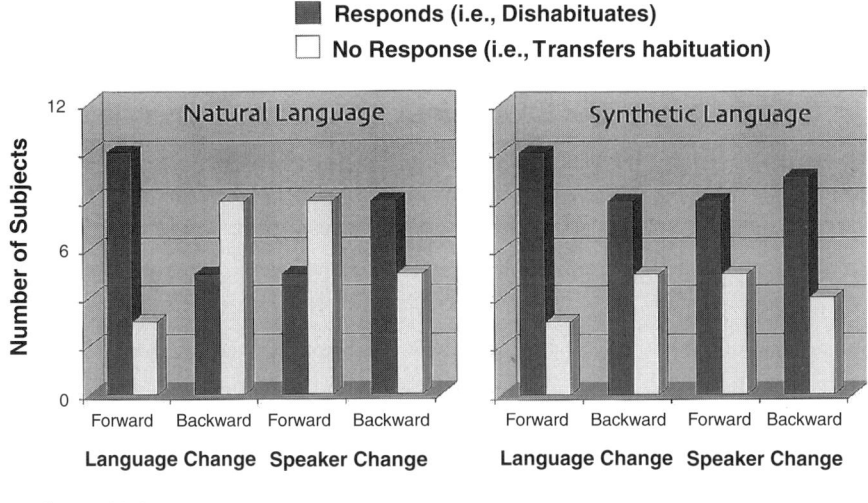

Figure 23.2
Results from habituation-dishabituation experiments with cotton-top tamarins using natural and synthetic sentences of Dutch and Japanese. The y-axis plots the number of subjects responding (dark bars) by turning toward the speaker, or not responding (white bars), following the habituation series. Each subject was tested in four conditions: forward sentences with a language change, backward sentences with a language change, forward sentences with a speaker change, and backward sentences with a speaker change

anese in the face of speaker variability, they are clearly able to extract acoustic equivalence classes. This capacity is not present in the human neonate, coming online a few months after birth (Jusczyk, 1997; Oller, 2000). Fourth, because tamarins fail to discriminate sentences of Dutch from sentences of Japanese when played backward, their capacity to discriminate such sentences when played forward shows that they must be using specific properties of speech as opposed to low-level cues. Fifth, given that the tamarins' capacity to discriminate Dutch from Japanese was weaker in the second test involving synthetic speech, it is possible that newborns and tamarins are responding to somewhat different acoustic cues. In particular, newborns may be more sensitive to prosodic differences (e.g., rhythm), whereas tamarins may be more sensitive to phonetic contrasts. Future research will explore this possibility.

In addition to our collaborative work with Mehler, we have also begun tests involving other aspects of speech processing. In particular, as several recent papers suggest, a real-world problem facing the human infant is how to segment the continuous acoustic stream of speech into functional units, such as phonemes, words, and phrases. Work by Saffran, Aslin, and Newport (1996) suggests that infants may be equipped with mechanisms that enable them to extract the statistical regularities of a particular language. Similarly, Marcus and colleagues (1999) have suggested that infants are equipped with the capacity to extract abstract rules that, subsequently, may form the foundation upon which grammars are constructed. In collaboration with these two groups, we have used our tamarin colony to determine whether other animals are capable of computing transitional probabilities as well as other statistical inferences. Using the original material of Saffran et al., we have recently replicated the findings with tamarins (Hauser, Newport, and Aslin, in press). Specifically, having been exposed to a continuous acoustic stream of syllables, where the transitional probabilities provide the only relevant information for discriminating words (i.e., three syllable sequences with high transitional probabilities) from nonwords (i.e., three-syllable sequences with low transitional probabilities), tamarins were able to compute the relevant statistics. Thus, like human infants, tamarins oriented to playbacks of nonwords (novel) more often than to words (familiar). This result is powerful, not only because tamarins show the same kind of capacity as do human infants but because the methods and stimuli are the same, and involve no training.

What can be said about our verbal abilities? Unique or not? If I had to place a wager, I would bet that humans share with other animals the core mechanisms for speech perception. More precisely, we inherited from animals a suite of perceptual mechanisms for listening to speech, ones that are quite general, and did not evolve for processing speech. Whether the similarities across species represent cases of homology or homoplasy cannot be answered at present and will require additional neuroanatomical work, tracing circuitry, and establishing functional connectivity. What is perhaps uniquely human, however, is our capacity to take the units that constitute spoken and signed language, and recombine them into an infinite variety of meaningful expressions. Although much work remains, my guess is that animals will lack the capacity for recursion, and their capacity for statistical inference will be restricted to items that are in close, temporal proximity. With the ability to run animals and human infants on the same tasks, with the same material, we will soon be in a strong position to pinpoint when, during evolution and ontogeny, we acquired our specially designed system for language.

References

Alp, R. (1997). "Stepping-sticks" and "seat-sticks": New types of tools used by wild chimpanzees (*Pan troglodytes*) in Sierra Leone. *American Journal of Primatology, 41,* 45–52.

Bickerton, D. (1990). *Language and species.* Chicago: Chicago University Press.

Boesch, C. (1994). Cooperative hunting in wild chimpanzees. *Animal Behaviour, 48,* 653–667.

Bornstein, M. H. (1987). Perceptual categories in vision and audition. In S. Harnad (Ed.), *Categorical perception* (pp. 287–300). Cambridge, UK: Cambridge University Press.

Cheney, D. L., and Seyfarth, R. M. (1988). Assessment of meaning and the detection of unreliable signals by vervet monkeys. *Animal Behaviour, 36,* 477–486.

Chomsky, N. (1957a). A review of B.F. Skinner's *Verbal Behavior. Language, 35,* 26–58.

Chomsky, N. (1957b). *Syntactic structures.* The Hague: Mouton.

Chomsky, N. (1966). *Cartesian linguistics.* New York: Harper & Row.

de Waal, F. B. M. (1988). The communicative repertoire of captive bonobos (*Pan paniscus*), compared to that of chimpanzees. *Behaviour, 106,* 183–251.

de Waal, F. B. M. (1989). *Peacemaking among primates.* Cambridge, UK: Cambridge University Press.

de Waal, F. B. M. (1996). *Good natured.* Cambridge, MA: Harvard University Press.

Dent, M. L., Brittan-Powell, F., Dooling, R. J., and Pierce, A. (1997). Perception of synthetic /ba/-/wa/ speech continuum by budgerigars (*Melopsittacus undulatus*). *Journal of the Acoustical Society of America, 102,* 1891–1897.

Diamond, A. (1990). Developmental time course in human infants and infant monkeys, and the neural bases of higher cognitive functions. *Annals of the New York Academy of Sciences, 608,* 637–676.

Diamond, A., Zola-Morgan, S., and Squire, L. R. (1989). Successful performance by monkeys with lesions of the hippocampal formation on AB and object retrieval, two tasks that mark developmental changes in human infants. *Behavioral Neuroscience, 103,* 526–537.

Dixson, A. F. (1999). *Primate sexuality*. New York: Oxford University Press.

Ehret, G., and Haack, B. (1981). Categorical perception of mouse pup ultrasounds by lactating females. *Naturwissenshaften, 68*, 208.

Eimas, P. D., Siqueland, P., Jusczyk, P., and Vigorito, J. (1971). Speech perception in infants. *Science, 171*, 303–306.

Fischer, J. (1998). Barbary macaques categorize shrill barks into two call types. *Animal Behaviour, 55*, 799–807.

Gardner, R. A., Gardner, B. T., and Van Cantfort, E. (1989). *Teaching sign language to chimpanzees*. Albany: State University of New York Press.

Goodall, J. (1986). *The chimpanzees of Gombe: Patterns of behavior*. Cambridge, MA: Harvard University Press.

Gouzoules, S., Gouzoules, H., and Marler, P. (1984). Rhesus monkey (*Macaca mulatta*) screams: Representational signalling in the recruitment of agonistic aid. *Animal Behaviour, 32*, 182–193.

Hauser, M. D. (1996). *The evolution of communication*. Cambridge, MA: MIT Press.

Hauser, M. D. (1997). Artifactual kinds and functional design features: What a primate understands without language. *Cognition, 64*, 285–308.

Hauser, M.D. (1998). Functional referents and acoustic similarity: Field playback experiments with rhesus monkeys. *Animal Behaviour, 55*, 1647–1658.

Hauser, M.D. (2000). *Wild minds: What animals really think*. New York: Henry Holt.

Hauser, M. D., Newport, E. L., and Aslin, R. N. (in press). Sequestation of the speech stream in a nonhuman primate: Statistical learning in cotton-top tamarins. *Cognition*.

Hunt, G. R. (1996). Manufacture of hook-tools by New Caledonian crows. *Nature, 379*, 249–251.

Jusczyk, P. W. (1997). *The discovery of spoken language*. Cambridge, MA: MIT Press.

Kluender, K. R., Diehl, R. L., and Killeen, P. R. (1987). Japanese quail can learn phonetic categories. *Science, 237*, 1195–1197.

Kluender, K. R., Lotto, A. J., & Holt, L. L. (in press). Contributions of nonhuman animal models to understanding human speech. In S. Greenberg and W. Ainsworth (Eds.), *Listening to speech: An auditory perspective*. New York: Oxford University Press.

Kluender, K. R., Lotto, A. J., Holt, L. L., and Bloedel, S. L. (1998). Role of experience for language-specific functional mappings of vowel sounds. *Journal of the Acoustical Society of America, 104*, 3568–3582.

Kuhl, P. (1989). On babies, birds, modules, and mechanisms: A comparative approach to the acquisition of vocal communication. In R. J. Dooling and S. H. Hulse (Eds.), *The comparative psychology of audition* (pp. 379–422). Hillsdale, NJ: Erlbaum.

Kuhl, P. (1991). Human adults and human infants show a "perceptual magnet effect" for the prototypes of speech categories, monkeys do not. *Perception and Psychophysics, 50,* 93–107.

Kuhl, P. K., and Miller, J. D. (1975). Speech perception by the chinchilla: Voiced-voiceless distinction in alveolar plosive consonants. *Science, 190,* 69–72.

Kuhl, P. K., and Padden, D. M. (1982). Enhanced discriminability at the phonetic boundaries for the voicing feature in macaques. *Perception and Psychophysics, 32,* 542–550.

Liberman, A. M., Cooper, F. S., Shankweiler, D. P., and Studdert-Kennedy, M. (1967). Perception of the speech code. *Psychological Review, 74,* 431–461.

Liberman, A. M., Harris, K. S., Hoffman, H. S., and Griffith, B. C. (1957). The discrimination of speech sounds within and across phoneme boundaries. *Journal of Experimental Psychology, 54,* 358–368.

Lieberman, P. (1991). *Uniquely human.* Cambridge, MA: Harvard University Press.

Lotto, A. J., Kluender, K. R., and Holt, L. L. (1997). Perceptual compensation for coarticulation by Japanese quail (*Coturnix coturnix japonica*). *Journal of the Acoustical Society of America, 102,* 1134–1140.

Lotto, A. J., Kluender, K. R., and Holt, L. L. (1998). Depolarizing the perceptual magnet effect. *Journal of the Acoustical Society of America, 103,* 3648–3655.

Marcus, G., Vijayan, S., Bandi Rao, S., and Vishton, P. M. (1999). Rule learning by seven-month-old infants. *Science, 283,* 77–80.

Marler, P., Dufty, A., and Pickert, R. (1986). Vocal communication in the domestic chicken. II. Is a sender sensitive to the presence and nature of a receiver? *Animal Behaviour, 34,* 194–198.

Matsuzawa, T. (1996). Chimpanzee intelligence in nature and in captivity: Isomorphism of symbol use and tool use. In W. C. McGrew, L. F. Marchant, and T. Nishida (Eds.), *Great ape societies* (pp. 196–209). Cambridge, UK: Cambridge University Press.

May, B., Moody, D. B., and Stebbins, W. C. (1989). Categorical perception of conspecific communication sounds by Japanese macaques, *Macaca fuscata. Journal of the Acoustical Society of America, 85,* 837–847.

Nelson, D. A., and Marler, P. (1989). Categorical perception of a natural stimulus continuum: Birdsong. *Science, 244,* 976–978.

Oller, D. K. (2000). *The emergence of the speech capacity.* Mahwah, NJ: Erlbaum.

Pinker, S. (1994). *The language instinct.* New York: Morrow.

Ramus, F., Hauser, M. D., Miller, C., Morris, D., and Mehler, J. (2000). Language discrimination by human newborns and by cotton-top tamarin monkeys. *Science, 288,* 349–351.

Remez, R. E. (1979). Adaptation of the category boundary between speech and nonspeech: A case against feature detectors. *Cognitive Psychology, 11,* 38–57.

Saffran, J. R., Aslin, R. N., and Newport, E. L. (1996). Statistical learning by 8-month-old infants. *Science, 274,* 1926–1928.

Santos, L. R., and Hauser, M. D. (1999). How monkeys see the eyes: Cotton-top tamarins' reaction to changes in visual attention and action. *Animal Cognition, 2,* 131–139.

Santos, L. R., Ericson, B., and Hauser, M. D. (1999). Constraints on problem solving and inhibition: Object retrieval in cotton-top tamarins. *Journal of Comparative Psychology, 113,* 1–8.

Savage-Rumbaugh, E. S., and Lewin, R. (1996). *Kanzi.* New York: Basic Books.

Savage-Rumbaugh, E. S., Murphy, J., Sevcik, R. A., Brakke, K. E., Williams, S. L., and Rumbaugh, D. M. (1993). Language comprehension in ape and child. *Monographs of the Society for Research in Child Development, 58,* 1–221.

Seyfarth, R. M., Cheney, D. L., and Marler, P. (1980). Monkey responses to three different alarm calls: Evidence of predator classification and semantic communication. *Science, 210,* 801–803.

Sinnott, J. M. (1989). Detection and discrimination of synthetic English vowels by Old World monkeys (*Cercopithecus, Macaca*) and humans. *Journal of the Acoustical Society of America, 86,* 557–565.

Sinnott, J. M., and Brown, C. H. (1997). Perception of the English liquid /ra-la/ contrast by humans and monkeys. *Journal of the Acoustical Society of America, 102,* 588–602.

Sinnott, J. M., Petersen, M. R., and Hopp, S. L. (1985). Frequency and intensity discrimination in humans and monkeys. *Journal of the Acoustical Society of America, 78,* 1977–1985.

Snowdon, C. T. (1987). A naturalistic view of categorical perception. In S. Harnad (Ed.), *Categorical perception* (pp. 332–354). Cambridge, UK: Cambridge University Press.

Sommers, M. S., Moody, D. B., Prosen, C. A., and Stebbins, W. C. (1992). Formant frequency discrimination by Japanese macaques (*Macaca fuscata*). *Journal of the Acoustical Society of America, 91,* 3499–3510.

Stebbins, W. C. (1983). *The acoustic sense of animals.* Cambridge, MA: Harvard University Press.

Sussman, J. E., and Lauckner-Morano, V. J. (1995). Further tests of the perceptual magnet effect in the perception of [i]: Identification and change/no change discrimination. *Journal of the Acoustical Society of America, 97,* 539–552.

Terrace, H. S. (1979). *Nim.* New York: Knopf.

Terrace, H. S. (1993). The phylogeny and ontogeny of serial memory: List learning by pigeons and monkeys. *Psychological Science, 4,* 162–169.

Tomasello, M., and Call, J. (1997). *Primate cognition.* Oxford: Oxford University Press.

Tomasello, M., Savage-Rumbaugh, E. S., and Kruger, A. (1993). Imitative learning of actions on objects by children, chimpanzees, and enculturated chimpanzees. *Child Development, 64,* 1688–1706.

Visalberghi, E. (1990). Tool use in Cebus. *Folia Primatologica, 54,* 146–154.

Whiten, A., Goodall, J., McGrew, W. C., Nishida, T., Reynolds, V., Sugiyama, Y., Tutin, C. E. G., Wrangham, R. W., and Boesch, C. (1999). Cultures in chimpanzees. *Nature, 399,* 682–685.

Wyttenbach, R. A., and Hoy, R. R. (1999). Categorical perception of behaviorally relevant stimuli by crickets. In M. D. Hauser and M. Konishi (Eds.), *The design of animal communication* (pp. 559–576). Cambridge, MA: MIT Press.

Wyttenbach, R. A., May, M. L., and Hoy, R. R. (1996). Categorical perception of sound frequencies by crickets. *Science, 273,* 1542–1544.

Zuberbuhler, K., Noe, R., and Seyfarth, R. M. (1997). Diana monkey long-distance calls: Messages for conspecifics and predators. *Animal Behaviour, 53,* 589–604.

24
The Biological Foundations of Music

Isabelle Peretz

All human societies have music. To our knowledge, this has always been so. Unlike other widespread human inventions, such as the writing systems, music was not created by a few individuals and then transmitted to others. Instead, music seems to have emerged spontaneously in all forms of human societies. Moreover, this emergence is not recent in human evolution. Music apparently emerged as early as 40,000 to 80,000 years ago, as suggested by the recent discovery of a bone flute attributed to the Neanderthals (Kunej and Turk, 2000). Thus, not only is music ubiquitous to human societies, it is also old in evolutionary terms. Accordingly, music may pertain to the areas covered by human biology.

Surprisingly, the notion that music might have biological foundations has only recently gained legitimacy.[1] Over the past thirty years, music has mostly been studied as a cultural artifact, while a growing body of research, well represented by the work of Jacques Mehler, contributed to the building of the neurobiology of language (following the pioneering work of Eric Lenneberg, 1967, *Biological Foundations of Language*). For most musicologists, each musical system could only be understood in the context of its specific culture. Moreover, according to most neuroscientists and psychologists, music served as a convenient window to the general functioning of the human brain. However, recent evidence suggests that music might well be distinct from other cognitive functions, in being subserved by specialized neural networks under the guidance of innate mechanisms. The goal of this chapter is to briefly review the current evidence for the premise that music is a part of the human biological endowment.

I begin with the best available evidence, or at least the evidence that convinces me, that music cannot be reduced to an ephemeral cultural product. This material relies, in particular, on the detailed examination of music-specific disorders that can occur either after brain damage at the adult age or as an acquisition failure. Next, I briefly discuss the issue of brain localization with special attention to expertise effects, since these are related and enduring questions. Afterward, I mention some recent and fundamental discoveries made on infants' musical abilities and suggest a few candidates for musical universal principles. Lastly, I conclude with a few speculations on the biological functions of music.

Specialized Neural Networks for Music Processing

If music is biologically determined, then it is expected to have functional and neuroanatomical specialization, as do all major cognitive functions, such as language. Support for the existence of such specialized neural networks is presently compelling. The major source of evidence is coming from the functional examination of individuals who happen to suffer from some brain anomaly and who, as a consequence, exhibit disorders that selectively either disturb or spare musical abilities. Presently, there is no evidence for music-specificity derived from the study of normal brains. However, it will not be long in coming, since the patient-based approach converges on the notion that music is subserved by neural networks that are dedicated to its processing. By "dedicated neural structures," I am referring to neural devices that process musical information selectively and exclusively. Support for their existence comes from essentially three types of neuropsychological findings.

The first type of evidence lies in the classic neuropsychological dissociations that can be observed after an accidental focal damage in a mature brain. For example, we have been able to document several cases whose characteristic symptom was the loss of the ability to recognize and memorize music. These patients retained the ability to recognize and understand speech, as well as to identify common environmental sounds normally (Peretz, 1996; Peretz, Belleville, and Fontaine, 1997; Peretz et al., 1994; replicated by Griffith et al., 1997). The deficit can be remarkably selective. For example, CN was unable to recognize hummed melodies coming from familiar songs above chance. Yet, she could perfectly recognize the lyrics accompanying the melodies that she failed to recognize (e.g., see Peretz, 1996). Moreover, CN was also able to recognize the voice of speakers (Peretz et al., 1994) and the intonation of speech (Patel et al., 1998). The existence of such a specific problem with music alongside normal functioning of other auditory abilities, including speech comprehension, suggests damage to processing components that are both essential to the normal process of music recognition and specific to the musical domain. The reverse condition, which corresponds to a selective sparing of music recognition, has also been reported (e.g., see Godefroy et al., 1995). In this latter condition, the lesion compromises both speech comprehension and recognition of familiar environmental sounds. Such cases, suggesting isolated sparing of music recognition abilities, complement the music-specific deficits, and provide the key evidence of double dissociation. This neuropsychological pattern is the prototypical signature of the presence of specialized brain circuits.

Further neuropsychological evidence that is indicative of the domain-specificity of music is apparent in studies of autistic people. The autistic individual is generally more apt in the area of music in contrast to other cognitive activities, such as verbal communication (e.g., see Heaton, Hermelin, and Pring, 1998). Of note is that certain autistic individuals become *musical savants,* a term which refers to the observation of high achievements in musical competence in individuals who are otherwise socially and mentally handicapped (e.g., see Miller, 1989). The mirror image of this condition would consist of individuals who are musically totally inept, despite normal exposure to music and normal intelligence. Such individuals exist and are commonly called *tone-deaf* (see Grant-Allen, 1878, for the first report). Affected individuals are born without the essential wiring elements for developing a normally functioning system for music, hence experiencing music as noise at the adult age while achieving a high degree of proficiency in their professional and social life. A well-known figure who was known to be tone-deaf was Che Guevara (Taibo, 1996). We have been able to confirm the existence of this music-specific learning disability in at least twelve other adults (Ayotte et al., in preparation). The selectivity of their musical failure is striking. All these individuals match the case of CN (the brain-damaged patient previously described) with no evidence of brain lesions. Accordingly, these individuals should be referred to as *congenital amusics.* These amusic individuals have above-average language skills, being able to speak several languages without accent. Moreover, developmental amusics have never been able to sing, dance, or to recognize music as simple as their national anthem, despite sterile efforts to learn music during childhood. Their condition is the reverse condition of that of the musical savant syndrome, hence illustrating exceptional isolation of the musical modules in the developing brain.

The third source of evidence that speaks for the existence of neural networks that are dedicated to music comes from the examination of epileptic patients. It is well known that in a few individuals, the pathological firing of neurons, conductive to the epileptic crisis, will be elicited by music exclusively. This form of epilepsy is called "musicogenic epilepsy" (Wieser et al., 1997) and suggests that the epileptogenic tissue lie in a music-specific neural region. This suggestion is consistent with the content of the descriptions elicited in vivo by electric stimulation of the brain of epileptic patients before brain surgery (Penfield and Perot, 1963). Direct electric stimulation of particular areas of the auditory associative cortex of awake patients often produces highly vivid musical hallucinations. Again, these music-specific experiences support the existence of neural networks specialized for music.

Localization of the Music-Specific Networks

Taken together, neuropsychological explorations provide compelling evidence for the existence of brain specialization for music. One important implication of this observation is that music does not look like a parasite or a byproduct of a more important brain function, such as language. Although such a conclusion supports the notion that music is a biological function, it is not sufficient. Brain specialization does not entail prewiring. Music may simply recruit any free neural space in the infant's brain and modify that space to adjust it to its processing needs. These needs may be computationally complex to satisfy and hence require free and plastic neural tissue. This opportunistic scenario of brain organization for music may respond to cultural pressure and not biological requirements. However, if this were true, a highly variable distribution of the musical networks should be observed across individuals. Depending on the moment, quality, and quantity of exposure, various brain spaces might be mobilized. Thus, if music is a brain "squatter," localization should vary capriciously across members of the same culture.

A prewired organization must exhibit consistency in localization. The primary auditory areas are systematically located in the Heschl's gyri, buried in the sylvian fissure; this holds for all humans including individuals who are born deaf. Similarly, brain regions that are dedicated to music processing should correspond to a *fixed* arrangement. That is, brain implementation of music networks should be similar in the vast majority of humans, nonmusicians and musicians alike. Moreover, this organization is not expected to vary as a function of the musical culture considered. Musical functions are expected to be similarly implemented in the brain of an isolated Pacific islander, a Chinese opera singer, and a Western fan of rap music. This is a very strong prediction, perfectly suited to the exploitation of the new brain imagery techniques.

Although clear and straightforward, the demonstration of a similar brain organization for music in all humans remains elusive. The only consensus that has been reached today concerns only one component of the music-processing system: the pitch contour extraction mechanism that involves the superior temporal gyrus and frontal regions on the right side of the brain (see Peretz, 2000, for a recent review). However, it remains to be determined if this processing component is music-specific, since the intonation patterns of speech seem to recruit similar, if not identical, brain circuitries (Zatorre et al, 1992; Patel et al., 1998). Clearly, what is needed at the present stage is a grid that allows specification of the processing mechanisms that are essential to music appreciation, an ability shared by all humans. Once these essential ingredients have been identified, their respective localization could be tracked down in the brain of musicians and nonmusicians of different musical cultures. The research agenda involved is dense and will only be briefly sketched in the next section.

What Is the Content of the Music-Specific Neural Networks?

The common core of musical abilities, which is acquired by all individuals of the same culture and which forms the essence of the musical competence acquired by members of other cultures, must be organized around a few essential processing components that are the core of the brain specialization for music. It is these highly specialized mechanisms that are probably embedded in the neural networks specifically associated with music, and which may be detected in neurological practice. Thus, there is no need for all musical abilities to have initial specialization. Brain specialization for a few mechanisms that are essential to the normal development of musical skills should suffice.

I am proposing that the two anchorage points of brain specialization for music are the encoding of pitch along musical scales and the ascribing of a regular pulse to incoming events. The notion that a special device exists for pitch processing in music has been developed in previous papers (Peretz and Morais, 1989, 1993) and will thus not be elaborated on further here. Similarly, the notion that regularity might be fundamental to music appreciation is slowly emerging (e.g., see Drake, 1998), although its specificity to music is rarely addressed. It is worth mentioning that the universality of musical scales and of regular pulse faces difficulties in finding a niche in ethnomusicologist circles (which are more concerned about understanding each musical system in its context). Yet the plausibility of considering these two principles as music universals has increased in recent years (e.g., see Arom, 2000).

While musicologists generally remain reluctant to envision biological determinism in music and, consequently, do not engage in the active search of relevant evidence in the various types of music around the world, developmental psychologists have been more courageous. Infants have been shown to possess precocious sensitivity to musical scales and for temporal synchronicity. For example, six- to nine-month-old infants process consonant intervals better than dissonant intervals (e.g., see Schellenberg and Trehub, 1996) and exhibit learning preferences for musical scales (Trehub et al., 1999). In most musical cultures, musical scales make use of unequal pitch steps. Infants already show a sensitivity bias toward musical scales, since they have been shown to be better at detecting a small pitch change in an unequal-step scale than in an equal-step scale. On the time dimension, infants prefer music that is subject to an isochronous temporal pulse. For instance, like adults, four-month-old infants are biased toward perceiving regularity; they exhibit sensitivity to slight disruptions of temporal isochrony (see Drake, 1998, for a review). All of these aspects of auditory pattern processing suggest the presence of innate learning preferences.

Surprisingly, preference biases are rarely explored in adults. Adults are studied as information-processing machines, not having emotional biases. Part of this situation may be attributed to the widely held belief that preferences, or emotional interpretations of music, are highly personal and variable, hence preclude scientific examination. This belief is false. A recent study of ours (Peretz, Gagnon, and Bouchard, 1998) showed that emotional appreciation of music appears highly consistent across individuals, to have immediacy, and to be available to the layperson without conscious reflection and with little effort. Therefore, emotional appreciation of music fits well both with the product of a specialized cortical arrangement and the purpose of music. Emotional purposes are the first reasons offered by people to explain why they listen to music (Panksepp, 1995; Sloboda, 1999).

It is probably the study of music as an emotional language that is the most likely to tap the universal principles that are responsible for its ubiquity. Until the 1960s it was believed that languages could vary arbitrarily and without limit. Today, there is a consensus among linguists that there is a universal grammar underlying diversity. Similarly, it was once thought that facial expressions of emotion could vary arbitrarily across cultures, until Ekman (Ekman et al., 1987) showed that a wide variety of emotions are expressed cross-culturally by the same facial movements. Likewise, certain aspects of music are culture-specific, although the general rules and processes subserving musical encoding of pitch and timing can be universal. These universal principles may in turn be subserved by neural networks that are shaped by natural selection.

What Is Music For?

If indeed music corresponds to a musical propensity that emerged early in human evolution, that is universal and functional early in human development, and that resides in a dedicated neural system, then the key question becomes "why?" What adaptive function was served by music in ancestral activities so as to provide its practitioners with a survival advantage in the course of natural selection? There are two main explanations.[2] The initial account was provided by Darwin himself (1871) who proposed that music serves to attract sexual partners. This view has been recently revived by Miller (2000) who reminds us that musicmaking is still a young male trait. However, the dominant view is that the adaptive value of music lies at the group level rather than at the individual level, with music helping to promote group cohesion. Music is present at all kinds of gatherings—dances, religious rituals, ceremonies—thereby strengthening interpersonal bonds and identification with one's group. The initial step for this bonding effect of music could be the mother-infant interactive pattern created through singing and motherese (or baby talk, which refers to the musical way adults talk to infants), thereby favoring emotional communion. These two different adaptive roles attributed to music do not need to be mutually exclusive. As pointed out by Kogan (1994), individuals taking the lead in ceremonies by virtue of their musical and dance prowess can achieve leadership status in the group, a factor that contributes to reproductive success.

In support of the contention that music has adaptive value, particularly for the group, is the fact that music possesses two design features which reflect an intrinsic role in communion (as opposed to communication, which is the key function of speech). Pitch intervals or frequency ratios allow harmonious voice blending, and regularity favors motor synchronicity or grace. These two musical features are highly effective at promoting simultaneous singing and dancing while admitting some degree of autonomy between voices and bodies (Brown, 2000). This design appears specific to music; it is certainly not shared with speech, which requires individuality for its intelligibility. These special features fit with the important criterion, discussed by Buss and collaborators (1998), that for a system to qualify as adaptive it must have a "special design" in order to offer effective solutions to a problem. The bonding problem in the case of music is to override selfish genes for the benefit of the group.

The Vogue Caveat

My major reservation toward the biological account of music is that it subscribes to the biological vogue. Nowadays, all human phenomena, from humor to rape, are explained in biological terms. Thus, the risk that future research on music will lose track of the essential role of cultural diversity, aesthetic freedom, and education is high. Fortunately, there is one notorious discrepant voice that reminds us to advance cautiously on the biological path. This discrepant voice is that of Steven Pinker (1997) who argues that music "is useless. . . . Music could vanish from our species and the rest of our lifestyle would be virtually unchanged. Music appears to be a pure pleasure technology. . . . All this suggests that music is quite different from language and that it is a technology, not an adaptation" (pp. 528–529). The merit in this position is that it provides us with the necessary incentive to prove that music is more than just a game for our mind or for our senses.[3]

. . . that music is a language which is understood by the immense majority of mankind, although only a tiny minority of people are capable of expressing it, and that music is the only language with the contradictory attributes of being at once intelligible and untranslatable, make the musical creator comparable to the gods, and music itself the supreme mystery of the science of man, a mystery that all the various disciplines come up against and which holds the key to their progress.
—Claude Lévi-Strauss (1969)

Acknowledgments

This chapter is based on research supported by grants from the Medical Research Council of Canada and the Natural Science and Engineering Research Council of Canada. I am grateful to Serge Larochelle, Emmanuel Dupoux, and Carol Krumhansl for their valuable comments on a previous draft.

Notes

1. As attested, for example, by the recent meeting entitled "Biological Foundations of Music" that was sponsored by the New York Academy of Sciences and held in New York in May 2000.
2. One notable exception that is worth the detour is David Huron's recent lecture entitled "An Instinct for Music: Is Music an Evolutionary Adaptation?" The lecture is available on the Web @ http://dactyl.som.ohio-state.edu/music220/Bloch.lectures/2.Origi ns.html
3. Certainly, the vacuity of music is helpless to explain how military music succeeds in leading sexually productive adults to death. The adaptive group-level explanation does, albeit in its darkest effects.

References

Arom, S. (2000). Prolegomena to a biomusicology. In N. Wallin, B. Merker, and S. Brown (Eds.), *The origins of music* (pp. 27–29). Cambridge, MA: MIT Press.

Ayotte, J., Peretz, I., and Hyde, K. Congenital amusia: A music-specific disorder in adults. In preparation.

Brown, S. (2000) The "musilanguage" model of music evolution. In N. Wallin, B. Merker, and S. Brown (Eds.), *The origins of music* (pp. 271–300). Cambridge, MA: MIT press.

Buss, D., Haselton, M., Shackelford, T., Bleske, A., and Wakefield, J. (1998). Adaptations, exaptations, and sprandels. *American Psychologist, 53,* 533–548.

Darwin, C. (1871). *The descent of man, and selection in relation to sex.* 2 vols. London: Murray.

Drake, C. (1998). Psychological processes involved in the temporal organization of complex auditory sequences: Universal and acquired processes. *Music Perception, 16,* 11–26.

Ekman, P., Friesen, W., O'Sullivan, M., Chan, A., Diacoyanni-Tarltzis, I., Heider, K., Krause, R., LeCompte, W., Pitcairn, T., Ricci-Bitti, P., Sherer, K., Tomita, M., and Tzavaras, A. (1987). Universals and cultural differences in the judgments of facial expressions of emotion. *Journal of Personality and Social Psychology, 53,* 712–717.

Godefroy, O., Leys, D., Furby, A., De Reuck, J., Daems, C., Rondepierre, P., Dabachy, B., Deleume, J-F., and Desaulty, A. (1995). Psychoacoustical deficits related to bilateral subcortical hemorrhages: A case with apperceptive auditory agnosia. *Cortex, 31,* 149–159.

Grant-Allen (1878). Note-deafness. *Mind, 10,* 157–167.

Griffith, T., Rees, A., Witton, C., Cross, P., Shakir, R., and Green, G. (1997). Spatial and temporal auditory processing deficits following right hemisphere infarction: A psychophysical study. *Brain, 120,* 785–794.

Heaton, P., Hermelin, B., and Pring, L. (1998). Autism and pitch processing: A precursor for savant musical ability? *Music Perception, 15,* 291–305.

Kogan, N. (1994). On aesthetics and its origins: Some psychobiological and evolutionary considerations. *Social Research, 61,* 139–165.

Kunej, D., and Turk, I. (2000). New perspectives on the beginnings of music: Archeological and musicological analysis of a middle paleolithic bone "flute." In N. Wallin, B. Merker, S. Brown (Eds), *The origins of music* (pp. 235–269). Cambridge, MA: MIT Press.

Lenneberg, E. (1967). *Biological foundations of language.* New York: Wiley.

Lévi-Strauss, C. (1969) *The raw and the cooked.* Translated by John and Doreen Weightman. New York: Harper & Row.

Miller, G. (2000). Evolution of human music through sexual selection. In N. Wallin, B. Merker, and S. Brown (Eds.), *The origins of music* (pp. 329–360). Cambridge: MIT Press.

Miller, L. (1989). *Musical savants. Exceptional skill in the mentally retarded.* Hillsdale, NJ: Erlbaum.

Panksepp, J. (1995). The emotional sources of "chills" induced by music. *Music Perception, 13,* 171–207.

Patel, A. D., Peretz, I., Tramo, M. and Labrecque, R. (1998). Processing prosodic and musical patterns: A neuropsychological investigation. *Brain and Language, 61,* 123–144.

Penfield, W., and Perot, P. (1963). The brain's record of auditory and visual experience. *Brain, 86,* 595–696.

Peretz, I. (1996). Can we lose memories for music? The case of music agnosia in a nonmusician. *Journal of Cognitive Neurosciences, 8,* 481–496.

Peretz, I. (2000). Music perception and recognition. In B. Rapp (Ed.), *The handbook of cognitive neuropsychology.* Hove: Psychology Press.

Peretz, I., and Morais, J. (1989). Music and modularity. *Contemporary Music Review, 4,* 279–293.

Peretz, I., and Morais, J. (1993). Specificity for music. In F. Boller and J. Grafman (Eds.), *Handbook of Neuropsychology.* Vol. 8 (pp. 373–390). Amsterdam: Elsevier.

Peretz, I., Belleville, S., and Fontaine, F. S. (1997). Dissociations entre musique et langage après atteinte cérébrale: Un nouveau cas d'amusie sans aphasie. *Revue canadienne de psychologie expérimentale, 51,* 354–367.

Peretz, I., Gagnon, L., and Bouchard, B. (1998). Music and emotion: Perceptual determinants, immediacy and isolation after brain damage. *Cognition, 68,* 111–141.

Peretz, I., Kolinsky, R., Tramo, M., Labrecque, R., Hublet, C., Demeurisse, G., and Belleville, S. (1994). Functional dissociations following bilateral lesions of auditory cortex. *Brain, 117,* 1283–1301.

Pinker, S. (1997). *How the mind works.* New York: Norton.

Schellenberg, E. G., and Trehub, S. (1996). Natural musical intervals: Evidence from infant listeners. *Psychological Science, 7,* 272–277.

Sloboda, J. (1999). Music—where cognition and emotion meet. *The Psychologist, 12,* 450–455.

Taibo, P. I. II (1996). *Ernesto Guevara, también conocido como el Che.* Buenos Aires: Planeta.

Trehub, S., Schellenberg, G., and Kamenetsky, S. (1999). Infants' and adults' perception of scale structure. *Journal of Experimental Psychology: Human Perception and Performance, 25,* 965–975.

Wieser, H. G., Hungerbühler, H., Siegel, A., and Buck, A. (1997). Musicogenic epilepsy: Review of the literature and case report with ictal single photon emission computed tomography. *Epilepsia, 38,* 200–207.

Zatorre, R. J., Evans, A. C., Meyer, E., and Gjedde, A. (1992). Lateralization of phonetic and pitch processing in speech perception. *Science, 256,* 846–849.

25

Brain and Sounds: Lessons from "Dyslexic" Rodents

Albert M. Galaburda

Dyslexia is a disorder of language that affects primarily reading and writing (Vellutino, 1987). It is perhaps the best known of the specific learning disabilities. Affected children have trouble learning written language and do not achieve expected levels of performance. Dyslexia becomes evident when a child first encounters reading and writing in kindergarten or first grade, although in some cases there are problems with speech development that are seen earlier, between two and three years of age. The latter may take the form of speech-onset delay or frequent mispronunciations of words. In most, however, there are no other clues than the knowledge that other members of the family have been diagnosed with dyslexia, which by itself increases the odds that the child in question is also at risk.

Dyslexia implies normal intelligence and emotional makeup, neurological health, and cultural and educational opportunities. However, mild problems in one or more of these areas are often present. It is not the case that a mildly retarded child cannot also be dyslexic, or that all dyslexics are bright, but when reading problems occur together with other cognitive problems, such as attention deficit hyperactivity disorder or poor memory, it is more difficult to attribute the reading retardation to a specific failure.

Reading involves seeing the correspondence between the visual symbols and sounds of language. Dyslexic children have problems with conscious awareness of the sounds of the native language, which they show on tests of phonological awareness, such as rhyming, pig Latin, word segmentation tasks, phoneme deletion tasks, and so on (Bradley and Bryant, 1981; Liberman and Shankweiler, 1985; others). This form of dyslexia is known as *phonological dyslexia* and it is the most common form of developmental dyslexia. Autopsy studies indicate that brain areas involved in phonological and other language processes contain minor malformations that originate during fetal life (Drake, 1968; Galaburda and Humphreys 1989; Galaburda and Kemper 1979; Galaburda et al., 1985). Similarly, functional imaging studies, such as functional magnetic resonance imaging (fMRI), by which it is possible to learn about the functioning, living brain in subjects performing language or other cognitive tasks, also show dysfunction in the parts of the cerebral cortex that are involved in phonological processing and auditory-visual association (Paulesu et al., 1996; Rumsey et al., 1987; Shaywitz et al., 1998). All of this points to an innate, constitutional problem arising during early brain development, well before a child is exposed to reading.

Probably a small proportion of dyslexics (the exact numbers are not known) do not have problems handling speech sounds and can read pseudowords without difficulty. Instead, these dyslexics have trouble only with irregular words, such as "enough," "yacht," and "naive." This form of dyslexia is known as *surface dyslexia* (Castles and Coltheart, 1993). All dyslexics have problems with irregular words, but this subset do not have problems with pseudowords. In stroke and acquired brain injury, the site of the lesion that produces problems with reading irregular words is different from that accounting for problems with pseudowords. It should then be the case that the two forms of developmental dyslexia, phonological and surface, have separate brain mechanisms, too. This, however, has not as yet been determined.

Dyslexics may also exhibit problems with processing nonlanguage rapidly changing sounds (Tallal, 1977) and rapidly changing visual images (Lovegrove, Heddle, and Slaghuis, 1980). A part of the visual system, called the magnocellular pathway, is directly implicated in the visual system (Livingstone et al., 1991), because this pathway is used for stabilizing images as the eyes move across the page. Eye discomfort in reading may be result when the magnocellular pathway is dysfunctional, and the use of tinted lenses has attempted to deal with this problem, with variable and unclear results. The significance of these visual and auditory deficits for acquiring competence in reading is not widely accepted, but they could contribute to the reading problem. Because sound processing is an element in developmental dyslexia of the phonological type, in this chapter we will focus on sound-processing deficits, as modeled in laboratory animals.

Animals usually employed in the laboratory process sound. It is less likely that they process language, at least in the way humans process language. Primates perhaps share some characteristics with humans, even with respect to learning the kinds of sounds that form part of language. Rodents, on the other hand, are simpler creatures. It would be foolhardy to propose that laboratory rodents be considered as useful models for the linguistic anomalies seen in dyslexias. However, it may be far less farfetched to explore laboratory rodents with a class of cortical malformations in order to obtain a clearer understanding of the type of sound-processing deficits dyslexic children exhibit. This is what we have done.

Some of our work has consisted in modeling developmental cortical anomalies in laboratory rodents for the purpose of shedding light on the problem of sound processing in developmental dyslexia. For obvious reasons, we have focused in these animal models on the temporal processing deficit of nonlinguistic sounds, as seen in phonological dyslexics. It is possible to induce cortical malformations in rats and mice identical to those found in developmental dyslexia (Humphreys et al., 1991). Furthermore, there exist several lines of mouse mutants that develop equivalent malformations spontaneously (Denenberg et al., 1988; Sherman, Galaburda, and Geschwind, 1985). Both the rat and mouse models have shown deficits in discrimination of rapidly changing sounds, so we have looked in these models for the relationship between the anatomical changes in the cortex and the behavioral deficits. Whether the problem with processing of rapidly changing sounds has anything to do with the phonological problem seen in dyslexia is still an open question, which cannot be answered in the rodent model. On the assumption that the problems are related, it is reasonable to learn more about the sound-processing deficit—hence the animal model.

In the laboratory rat and mouse we have modeled a neuronal migration anomaly we have called "ectopia" (Galaburda and Kemper, 1979; Galaburda et al., 1985) (figure 25.1). In the mouse we study the genetics and neurophysiology of this disorder. Genetics is not this chapter's concern, and will not be considered further. In the rat we look at the anatomical

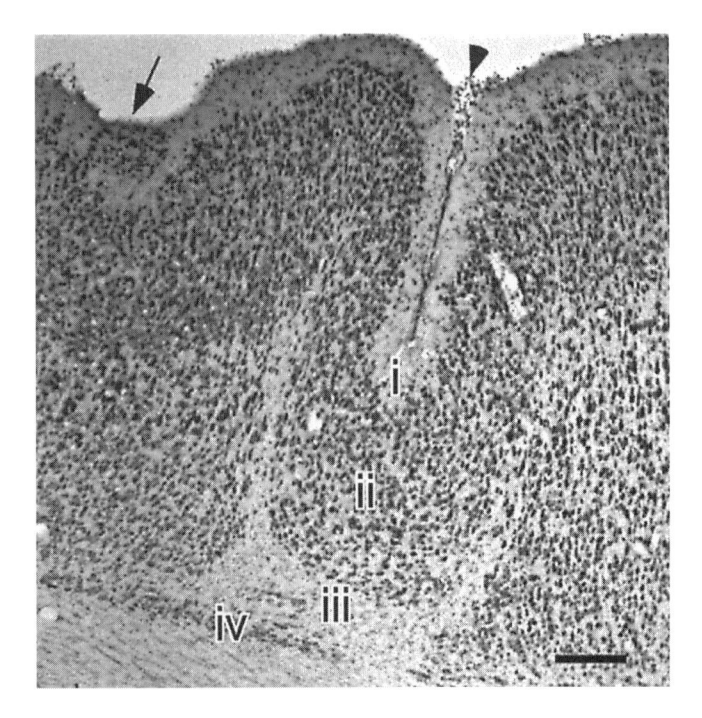

Figure 25.1
Example of induced cortical malformation in the rat, which mimics the two malformations seen in the cortex of the dyslexic brain. A freezing probe is placed on the skull of the newborn rat pup damaging the underlying cortex. Directly below the probe, a classic four-layer microgyrus is formed, resulting from the destruction of neurons present in the cortex at the time and fusion of subsequent formed layers. The arrowhead points to a microsulcus which is formed under the probe. Layer i corresponds to normal layer I; layer ii corresponds to fused normal layers II to IV; cell free layer iii corresponds to normal layers IV and V; and layer iv corresponds to layer VI in the normal. At the edge of the microgyrus, where the damage is more superficial, an ectopic nest of neurons and glia is formed in the molecular layer (layer I; arrow) Bar = 800 µm.

organization of the ectopia and some neurophysiological and behavioral consequences.

Our interest in developing these models sprang from the finding that ectopias are found in large numbers in the brains of persons with developmental dyslexia (Galaburda et al., 1985). In eight male dyslexic brains we found these ectopias in perisylvian cortex, affecting frontal opercular areas, superior and middle temporal gyri, and parietal operculum. In most of the specimens, there were more ectopias in the left than the right hemisphere. In two female brains we did not find ectopias. Instead we found small prenatal gliotic scars of roughly the same size as the ectopias and in roughly the same locations. This led us to postulate that they were indeed the product of damage in both cases, but affecting a more mature brain in the female case. This hypothesis was later confirmed experimentally (Humphreys et al., 1991).

We assumed that ectopias had something to do with the language behavior displayed by dyslexics, because, other than the excessive tendency for the planum temporale to be symmetrical in the dyslexic brains, we found no other anomalies. However, except to say that the ectopias probably scrambled cortical areas involved in language, we could not get any closer to the neural mechanism associated with the ectopias. The rat, and in a few cases the mouse, offered the opportunity to learn more about the anatomical, physiological, and behavioral effects ectopias might have on neural networks.

Anatomical and Functional Features of Malformations

Ectopias consist of 50 to 100 neurons and glia nested quite compactly in the molecular layer of the neocortex. Ectopias appear in the cortex soon after young neurons born in the germinal zones begin to migrate to the cortex to find their laminar positions. Analysis of the neuronal types in the ectopias using Golgi stains and radioactive dating methods demonstrates neurons of different birth dates, beginning with the earliest-born neurons destined to populate layer VI and ending with neurons destined for layer II (Rosen et al., 1992a; Sherman et al., 1992). These neurons escape into the molecular layer through a breach in the external glial limiting membrane, which ordinarily serves as a barrier to keep neurons from entering this layer. We assume that the breach is there early because of the presence of multiaged neurons. Furthermore, we can cause an ectopia by poking a small hole in the membrane before the end of neuronal migration in the rat (Rosen et al., 1992b).

Whereas ectopias are produced by mild injury to the developing cortex, focal microgyria occurs when the injury is more severe (Rosen and Galaburda, in press; see figure 25.1). Microgyria disturbs the organization of all the layers including and underlying the molecular layer. Ectopias and microgyria can occur together when adjacent areas of the cortex are exposed to different degrees of injury. Focal microgyria has also been seen in the brains of dyslexics (Galaburda and Kemper, 1979; Galaburda et al., 1985). Microgyria is easier to induce in the rats, so we have studied them in more detail, although many of the findings in microgyria apply to ectopias too.

Neurons in the mycrogyria are mainly excitatory (Frenkel et al., 2000). Initially there is a complete paucity of GABAergic inhibitory neurons (G. D. Rosen, unpublished observations), and excessive excitability is demonstrated physiologically in and near the microgyria (Frenkel et al., 2000). Placement of axonal tracers in microgyria or ectopias shows that the abnormal region is connected both locally and widely (Jenner, Galaburda, and Sherman, in press; Rosen, Burstein, and Galaburda, 2000). Barrel field microgyria, for instance, sends projections to the ventrobasal complex of the thalamus and diffusely to the contralateral hemisphere and receives reciprocal connections. There may be inappropriate connections, too, because some developmentally transient connections are not eliminated or new ones are formed by abnormal sprouting. The diffuse nature of the contralateral projections, for instance, is an example of lack of withdrawal of transient connections. This pattern of organization suggests that ectopias are in a position to affect the development of areas to which they connect, possibly through the mechanism of excitotoxicity and with functional consequences.

We found that induction of microgyria in the frontal lobe, some synapses away from the thalamus, causes histological changes in the medial geniculate nucleus (MGN) (Herman et al., 1997). Animals with microgyria, compared with animals with sham injury, showed an excess of small neurons in the MGN. Dyslexic brains, too, have an excess of small neurons in the MGN, in this case only the left MGN (Galaburda, Menard, and Rosen, 1994). Because, in general, smaller neurons are slower, we sought to demonstrate whether animals with microgyria exhibited slowing of auditory processing.

Microgyria and molecular layer ectopias slow auditory processing speed in the rat, and this can be demonstrated in two different ways. Rats were taught to discriminate between two tones presented sequentially, and the interstimulus separation was varied. Animals with microgyria could not discriminate the tones at shorter interstimulus separations (Galaburda, et al., 1994). In another experiment, mice with ectopias were implanted with electrodes and tested on an oddball paradigm using tones. Short interstimulus distances failed to produce mismatch negativity in animals with ectopias (Frenkel et al., 2000). These findings indicate that ectopias interfere with rapid auditory processing for tones. They do not, however, implicate a specific stage of processing, whether in the thalamus or the cortex.

We observed sexual dimorphism in the phenomena described above (Rosen, Herman, and Galaburda, 1999). Female rats with microgyria did not show any deficits in rapid auditory processing once examined separately from the males. We investigated the microgyria in the females and found it to be identical to that of the males, both in size and location. On the other hand, examination of the thalamus did not show changes in cell size in the females as it did in the males. This showed that the microgyria itself was not directly responsible for the behavioral deficit, but rather the changes in the MGN, which it stimulated in the males but not in the females. Furthermore, this finding helped to causally link the slowing of sound processing to the earliest of the forebrain processors for sound—the MGN of the thalamus. Additional experiments indicated that the sex difference was caused by sex hormones (Rosen et al., 1999). Thus, exposure of fetal female rats to male sex steroids changed their pattern of response to microgyria to the male pattern.

Discussion

The animal model might explain why dyslexics have problems with processing rapidly changing sounds. Induction of cortical malformations led to changes in the thalamus, which are in turn associated with the sound discrimination deficits. When the changes in the thalamus do not occur, the cortical changes are not themselves associated with discrimination deficits. What the research does not and cannot address is whether cortical malformations like those in the dyslexic brain and in the laboratory animals are capable of also changing the processing of speech sounds. Current research is looking at the animals' ability to make phonetic distinctions based on voice-onset time or place of articulation. Although these results will be important to gather, they may not address the fundamental question, which is whether humans have special systems for speech sounds as compared to other sounds. If rats with malformations fail at these phonological tasks, it may simply mean that they use one system for all. However, if they succeed where they fail with nonspeech sounds, assuming equivalent psychophysical demands, the possibility exists that even in rodents speech sounds are processed separately, an unlikely case.

There are other reasons for which dyslexics may fail at speech sounds separately from nonspeech sounds, and this has to do with other anatomical changes associated with the ectopias and microgyria. As mentioned, dyslexic brains show symmetry of the planum temporale (Galaburda et al., 1985; Steinmetz, 1996). Such symmetry is found also in two thirds of unselected autopsy brains (Geschwind and Levitsky, 1968), but all dyslexic brains analyzed thus far were symmetrical in this region. We looked at the relationship between symmetry and the presence of ectopias. We did not find that symmetry is more common in the presence of ectopias, but we did find that the nature of symmetry changes from the normal pattern (Rosen et al., 1989b). Symmetrical cortical areas are larger in total than asymmetrical cortical areas. In the asymmetrical cases one of the sides is smaller than a symmetrical side rather than larger, as if the asymmetrical case is a unilaterally curtailed form of a symmetrical case (Galaburda et al., 1987). There is an inverse linear relationship between total amount of cortex (left plus right) and the coefficient of asymmetry of a cortical area. This is seen in human and animal brains where the degree of asymmetry varies from individual to individual (Galaburda et al., 1986, 1987). When ectopias are present, however, this relationship breaks down and no relationship between asymmetry and symmetry vis-à-vis total area emerges (Rosen et al., 1989b). Symmetry and asymmetry differ in the number of neurons (Rosen, Sherman, and Galaburda, 1991) and in the pattern of callosal connections (Rosen, Sherman, and Galaburda, 1989a). It is therefore possible that areas in the planum temporale having to do with phonological representation and processing are rendered anomalous by alterations associated with ectopias, although the exact mechanism for this is unknown.

Another possibility for involvement of phonological systems directly in dyslexics may have to do with the local effects of ectopias and microgyria, separately from their effects on the thalamus and asymmetry. We find alterations in the numbers of GABAergic neurons (Rosen, Jacobs, and Prince, 1998) and neuronal excitability in cortex adjacent to microgyria (J. J. LoTurco, unpublished observations). Frontal opercular involvement by malformations is perhaps the most common finding in dyslexic brains, leading to the supposition that inferolateral frontal functional capacities are changed. The inferolateral frontal region has been implicated in phonological processing (Fiez et al., 1995; Fujimaki et al., 1999; Poldrack et al., 1999) and one of the areas that shows abnormal activation in dyslexics is the inferior prefrontal cortex (Shaywitz et al., 1998). One aspect of phonological processing is critically abnormal in dyslexia—phonological awareness. Phonological awareness is required for learning to read and for reading unknown words and pseudowords. It is reasonable to propose that it is the frontal lobe that participates in phonological awareness, more so than the auditory cortices themselves. It is further reasonable to suppose that malformation in this region in dyslexics may be behind the difficulty with phonological awareness. This is over and above the issue of plain sound processing, which probably implicates lower-level systems, including the thalamus.

A problem arises when examining this type of brain evidence. Injury to the cortex early in development is accompanied by reaction close by and far afield. Microgyria in the frontal cortex produces changes in neuronal sizes in the thalamus, and probably in all intervening processing stations along the way. Changes may not stop at the thalamus and may continue instead to stations in the brainstem. We suspect that changes occur downstream, too, affecting cortical areas further along in the processing stream. In fact, a whole pathway may be affected by the ectopia (figure 25.2).

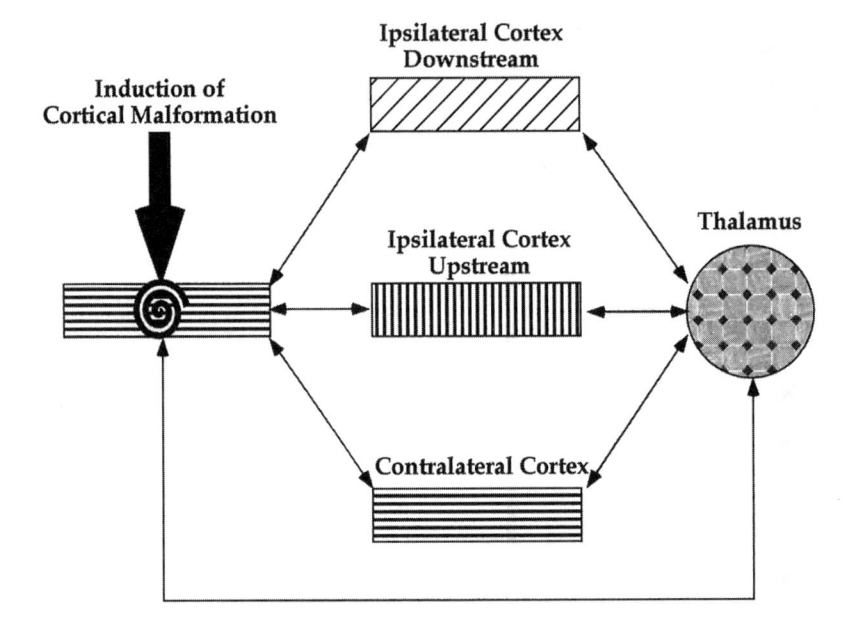

Figure 25.2
Schematic diagram of the developmental cascade thought to follow production of a cortical malformations. Spontaneous or induced malformations in the cortex before completion of neuronal migration leads to changes downstream to cortical areas farther away from input channels (ipsilateral and contralateral cortex) and upstream (ipsilateral cortex and thalamus, the latter directly and indirectly) toward input channels. Cortical changes, ipsi- and contralateral, are thought to produce cognitive deficits, whereas thalamic changes are thought to underlie deficits in sensory processing, both of which are noted in dyslexia. In this model, the disorders of sensory and cognitive processing need not be causally related to each other by both result from the initial cortical malformation.

We have suggested that some of the changes—namely, those in the thalamus—could account for sound-processing abnormalities, whereas others—those in the frontal lobe—could explain problems with phonological awareness. However, it is not clear why all aspects of speech processing would not be affected. So, for instance, this corruption in the total network does not explain normal speech production and speech comprehension, which is by and large the case in dyslexia. The female case may be enlightening in this regard. It shows that malformation in the cortex is not always associated with plasticity changes in connectionally related areas. At least, if there is plasticity, it is not of the type that can be measured in cell sizes and does not affect temporal processing. It is therefore possible that phonological representations and the areas that accommodate them may remain intact, or may react adaptively to the initial event and remain functional.

One of the characteristics of processing modules is that they degrade gracefully. Developmental injury may lead to an exception to this rule. Connectivity is demonstrably anomalous after cortical malformation. Units that normally do not connect in adult brains are poised to connect in the event of abnormal influences coming to bear upon them during development. Thus, somatosensory cortices are functionally connected to the visual system in congenitally blind subjects who read braille (Sadato et al., 1998). Cortical ectopias and microgyria lead to alterations in the patterns of connections within and between the hemispheres. Could this result in the elimination of normal functional boundaries between modular systems? Alternatively, systems having separate genetic backgrounds, which may apply to modules, may never interconnect, irrespective of cause or timing of brain damage. This question cannot be answered at present. If intersystem connectivity is indeed possible, then it is also possible that in some abnormal developmental states, systems that remain isolated in the normal state become intertwined, with resultant degradation of multiple functions.

Neuroscience may help resolve some of these questions in ways not possible only a decade ago. It is possible, therefore, to map function onto selected areas of the cortex using instruments such as magnetoelectroencephalography (MEG) and fMRI. The functionality of these areas of cortex identified by these tools can be checked with transcranial magnetic stimulation (Flitman et al., 1998). Some, albeit not enough, of these approaches are now available for the study of children and infants. We need to know for certain whether phonemes and other sounds are mapped separately in the auditory cortex and whether the maps can be changed one independently of the other. We could perhaps learn about whether dyslexics follow the rules vis-à-vis mapping sounds and speech early in the processing pathway. It may even be possible to find out whether behavioral improvement in dyslexics through the use of a given therapeutic approach exerts any change, even normalization, of the maps. The question in dyslexia is still primarily about sounds and speech, and its resolution will help further define theories regarding the uniqueness of language in the brain.

Conclusion

The laboratory rodent model has disclosed a probable but as yet unspecified mechanism for the problems dyslexics have with processing nonspeech sounds—this is the thalamic change associated with cortical malformations. Although malformation of the cortical language areas itself probably lies behind the phonological deficits dyslexics also show, additional research needs to be done to explain how the cortical scrambling produced by the malformation leads to the language deficits.

I have presented research that helps to bridge neurobiology to psychology. Cognitive psychologists, including Jacques Mehler, whom we celebrate in this book, have traditionally doubted the usefulness of this type of bridge building. There is a fundamental chasm separating neurobiological description from psychological description. How can an infinitely detailed account of anatomy explain a cognitive process? We do not know a way, yet, certainly not for anything interesting and complex, like language, or mental imagery, or consciousness. However, bridges are useful because they help guide research across levels. What Jacques learned about babies learning language helped me to focus on what was interesting to study about the brain. Hopefully, research on brain development can help choose interesting questions and constrain hypotheses, even if entirely within the field of cognition.

References

Bradley, L., and Bryant, P. (1981). Visual memory and phonological skills in reading and spelling backwardness. *Psychological Research, 43,* 193–199.

Castles, A., and Coltheart, M. (1993). Varieties of developmental dyslexia. *Cognition, 47,* 149–180.

Denenberg, V. H., Sherman, G. F., Rosen, G. D., and Galaburda, A. M. (1988). Learning and laterality differences in BXSB mice as a function of neocortical anomaly. *Society for Neuroscience Abstracts, 14,* 1260.

Drake, W. E. (1968). Clinical and pathological findings in a child with a developmental learning disability. *Journal of Learning Disabilities, 1,* 9–25.

Fiez, J. A., Raichle, M. E., Miezin, F. M., Petersen, S. E., Tallal, P., and Katz, W. F. (1995). PET studies of auditory and phonological processing: Effects of stimulus characteristics and task demands. *Journal of Cognitive Neuroscience, 7,* 357–375.

Flitman, S. S., Grafman, J., Wassermann, E. M., Cooper, V., O'Grady, J., Pascual-Leone, A., and Hallett, M. (1998). Linguistic processing during repetitive transcranial magnetic stimulation. *Neurology, 50,* 175–181.

Frenkel, M., Sherman, G. F., Bashan, K. A., Galaburda, A. M., and LoTurco, J. J. (2000). Neocortical ectopias are associated with attenuated neurophysiological responses to rapidly changing auditory stimuli. *Neuroreport, 11,* 575–579.

Fujimaki, N., Miyauchi, S., Putz, B., Sasaki, Y., Takino, R., and Sakai, K. (1999). Functional magnetic resonance imaging of neural activity related to orthographic, phonological, and lexico-semantic judgments of visually presented characters and words. *Human Brain Mapping, 8,* 44–59.

Galaburda, A. M., and Humphreys, P. (1989). Developmental dyslexia in women: Neuropathological findings in two cases. *Neurology, 39* (Suppl. 1), 317.

Galaburda, A. M., and Kemper, T. L. (1979). Cytoarchitectonic abnormalities in developmental dyslexia: A case study. *Annals of Neurology, 6,* 94–100.

Galaburda, A. M., Aboitiz, F., Rosen, G. D., and Sherman, G. F. (1986). Histological asymmetry in the primary visual cortex of the rat: Implications for mechanisms of cerebral asymmetry. *Cortex, 22,* 151–160.

Galaburda, A. M., Corsiglia, J., Rosen, G. D., and Sherman, G. F. (1987). Planum temporale asymmetry: Reappraisal since Geschwind and Levitsky. *Neuropsychologia, 25,* 853–868.

Galaburda, A. M., Menard, M., and Rosen, G. D. (1994). Evidence for aberrant auditory anatomy in developmental dyslexia. *Proceedings of the National Academy of Sciences of the United States of America, 91,* 8010–8013.

Galaburda, A. M., Sherman, G. F., Rosen, G. D., Aboitiz, F., and Geschwind, N. (1985). Developmental dyslexia: Four consecutive cases with cortical anomalies. *Annals of Neurology, 18,* 222–233.

Geschwind, N., and Levitsky, W. (1968). Human brain: Left-right asymmetries in temporal speech region. *Science, 161,* 186–187.

Herman, A. E., Galaburda, A. M., Fitch, H. R., Carter, A. R., and Rosen, G. D. (1997). Cerebral microgyria, thalamic cell size and auditory temporal processing in male and female rats. *Cerebral Cortex, 7,* 453–464.

Humphreys, P., Rosen, G. D., Press, D. M., Sherman, G. F., and Galaburda, A. M. (1991). Freezing lesions of the newborn rat brain: A model for cerebrocortical microgyria. *Journal of Neuropathology and Experimental Neurology, 50,* 145–160.

Jenner, A. R., Galaburda, A. M., and Sherman, G. F. (2000). Connectivity of ectopic neurons in the molecular layer of the somatosensory cortex in autoimmune mice. *Cerebral Cortex, 10,* 1005–1013.

Liberman, I. Y., and Shankweiler, D. (1985). Phonology and the problems of learning to read and write. *Remedial and Special Education, 6,* 8–17.

Livingstone, M., Rosen, G., Drislane, F., and Galaburda, A. (1991). Physiological and anatomical evidence for a magnocellular defect in developmental dyslexia. *Proceedings of the National Academy of Science of the United States of America, 88,* 7943–7947.

Lovegrove, W., Heddle, M., and Slaghuis, W. (1980). Reading disability: Spatial frequency specific deficits in visual information storage. *Neuropsychologia, 18,* 111–115.

Paulesu, E., Frith, U., Snowling, M., Gallagher, A., Morton, J., Frackowiak, R. S. J., and Frith, C. D. (1996). Is developmental dyslexia a disconnection syndrome? Evidence from PET scanning. *Brain, 119,* 143–157.

Poldrack, R. A., Wagner, A. D., Prull, M. W., Desmond, J. E., Glover, G. H., and Gabrieli, J. D. (1999). Functional specialization for semantic and phonological processing in the left inferior prefrontal cortex. *Neuroimage, 10,* 15–35.

Rosen, G. D., and Galaburda, A. M. (in press). Single cause, polymorphic neuronal migration disorder: An animal model. *Developmental Medicine and Child Neurology.*

Rosen, G. D., Burstein, D., and Galaburda, A. M. (2000). Changes in efferent and afferent connectivity in rats with cerebrocortical microgyria. *Journal of Comparative Neurology, 418,* 423–440.

Rosen, G. D., Herman, B. A., and Galaburda, A. M. (1999). Sex differences in the effects of early neocortical injury on neuronal size distribution of the medial geniculate nucleus in the rat are mediated by perinatal gonadal steroids. *Cerebral Cortex, 9,* 27–34.

Rosen, G. D., Jacobs, K. M., and Prince, D. A. (1998). Effects of neonatal freeze lesions on expression of parvalbumin in rat neocortex. *Cerebral Cortex, 8,* 753–761.

Rosen, G. D., Richman, J. M., Sherman, G. F., and Galaburda, A. M. (1992a). Birthdates of neocortical neurons in induced microgyria in the rat. *Society for Neuroscience Abstracts, 18,* 1446.

Rosen, G. D., Sherman, G. F., and Galaburda, A. M. (1989a). Interhemispheric connections differ between symmetrical and asymmetrical brain regions. *Neuroscience, 33,* 525–533.

Rosen, G. D., Sherman, G. F., and Galaburda, A. M. (1991). Ontogenesis of neocortical asymmetry: A [^{3}H]thymidine study. *Neuroscience, 41,* 779–790.

Rosen, G. D., Sherman, G. F., Mehler, C., Emsbo, K., and Galaburda, A. M. (1989b). The effect of developmental neuropathology on neocortical asymmetry in New Zealand Black mice. *International Journal of Neuroscience, 45,* 247–254.

Rosen, G. D., Sherman, G. F., Richman, J. M., Stone, L. V., and Galaburda, A. M. (1992b). Induction of molecular layer ectopias by puncture wounds in newborn rats and mice. *Developmental Brain Research, 67,* 285–291.

Rumsey, J. M., Berman, K. F., Denckla, M. B., Hamburger, S. D., Kruesi, M. J., and Weinberger, D. R. (1987). Regional cerebral blood flow in severe developmental dyslexia. *Archives of Neurology, 44,* 1144–1150.

Sadato, N., Pascual-Leone, A., Grafman, J., Deiber, M. P., Ibañez, V., and Hallett, M. (1998). Neural networks for Braille reading by the blind. *Brain, 121,* 1213–1229.

Shaywitz, S. E., Shaywitz, B. A., Pugh, K. R., Fulbright, R. K., Constable, R. T., Mencl, W. E., Shankweiler, D. P., Liberman, A. M., Skudlarski, P., Fletcher, J. M., Katz, L., Marchione, K. E., Lacadie, C., Gatenby, C., and Gore, J. C. (1998). Functional disruption in the organization of the brain for reading in dyslexia. *Proceedings of the National Academy of Sciences of the United States of America, 95,* 2636–2641.

Sherman, G. F., Galaburda, A. M., and Geschwind, N. (1985). Cortical anomalies in brains of New Zealand mice: A neuropathologic model of dyslexia? *Proceedings of the National Academy of Sciences of the United States of America, 82,* 8072–8074.

Sherman, G. F., Stone, L. V., Walthour, N. R., Boehm, G. W., Denenberg, V. H., Rosen, G. D., and Galaburda, A. M. (1992). Birthdates of neurons in neocortical ectopias of New Zealand Black mice. *Society for Neuroscience Abstracts, 18,* 1446A.

Steinmetz, H. (1996). Structure, function and cerebral asymmetry: In vivo morphometry of the planum temporale. *Neuroscience and Biobehavioral Reviews, 20,* 587–591.

Tallal, P. (1977). Auditory perception, phonics and reading disabilities in children. *Journal of the Acoustical Society of America, 62,* S100.

Vellutino, F. R. (1987). Dyslexia. *Scientific American, 256* (March), 34–41.

26

The Literate Mind and the Universal Human Mind

José Morais and Régine Kolinsky

In a chapter written in honor of Paul Bertelson, Dupoux and Mehler (1992) alerted psycholinguists that they "will have to include considerations about literacy in the interpretation of on-line studies. Indeed, literacy adds representations that make it possible for subjects to perform in tasks like phoneme-monitoring experiment which they would otherwise be unable to do. Given this observation, it is surprising that psycholinguists have neglected the potential effects of cultural representations" (p. 60). Eight years after these remarks were made, the impact of literacy on the human cognitive mind remains relatively ignored. In this chapter we question the implicit assumption of most cognitive psychology that the literate mind is an adequate model of the universal human mind.

We are the only animals that have acquired speech and literacy. Speech is ubiquitous in the human species after a necessary developmental period. It is only absent in a few pathological conditions, including extreme deprivation of linguistic input from the environment. Literacy, on the contrary, is not universal. If the universal mind exists, then the acquisition of literacy leads to two possible cases: either literacy represents an additional system of information processing, changing in no significant way the universal mind; or it modifies what are considered to be the universal properties. The latter hypothesis admits of many possibilities, depending on the number and nature of the properties of the mind that cease to be universal as a result of literacy acquisition. Concretely, one needs to ascertain whether the data, generalizations, and models derived from the study of literate people are or are not valid for illiterate people.

What Is Literacy?

Literacy is highly correlated with education. However, the two notions are distinct and can be separated empirically. We define *literacy* as the ability to read and write, and education as the whole corpus of knowledge acquired, for a large part, through the exercise of literacy.

Literacy is also associated with schooling, both contributing to the formation of the educated mind. However, one can isolate the specific effects of literacy either by comparing illiterate adults to "ex-illiterate" adults, that is, people who never attended school in childhood but became literate later on in special classes, or by comparing people who learned to read and write in different literacy systems.

We used the first method to demonstrate that awareness of phonemes does not develop spontaneously. Illiterates and ex-illiterates were compared on the ability to manipulate phonemes intentionally (Morais et al., 1979). Contrary to ex-illiterates, illiterates were unable to either delete the initial consonant of a verbal item or add one at the onset.

Read et al. (1986) used the second method to demonstrate that what entails phoneme awareness is not literacy in general but alphabetic literacy. Using tests adapted from ours, they compared alphabetized to nonalphabetized literate Chinese adults. The scores of the nonalphabetized Chinese readers were similar to those of illiterates, whereas the scores of the alphabetized ones were similar to those of ex-illiterates.

In this chapter, we focus mainly on the first method. A specific effect of literacy is demonstrated whenever, all other things being equivalent (especially social and cultural level), illiterates display lower performance than ex-illiterates do. These populations are rarely examined. Due to practical reasons but also to the belief that the investigated capacities are literacy-independent, the portrait of the literate mind is in most cases the portrait of the educated mind.

Literacy as a Biological or a Cultural Function

As said above, literacy is not universal. This indisputable fact leaves open, however, the question of whether literacy is mainly a biological (in the strong sense of genetically determined) or a cultural function. Indeed, exactly as the universality of a function does not mean that is biological, its lack of universality does not mean that it is cultural. The few examples below may illustrate these two claims.

Cross-cultural studies have shown that all young children share, in their conception of the earth, two presuppositions, namely that the ground where they live is flat and that physical objects need some kind of support (Vosniadou, 1994). However, there is no compelling evidence that these presuppositions are coded in the genes. Beliefs in ground flatness and up-down gravity may be derived from early experience. Infants living from birth in an interplanetary station where space would be round and imponderable would probably develop different beliefs about their habitat. Thus, universality does not imply genetic determination. More crucial to the present question, lack of universality does not imply a cultural origin, either. It is well-known that blue eyes are not universal; however, they are genetically coded. The cognitive profile associated with Down syndrome is, fortunately, not universal, but it has a genetic origin. And it is possible, although it will never be demonstrated, that the exceedingly high abilities of Leonardo da Vinci, which have not been shared by many other humans, were due in part to his "biological" capacities.

This term "biological" is often used in the context of the debate on genetically vs. environmentally determined capacities. Of course, this term must cover capacities that do not emerge spontaneously from the genes. The genes probably specify nothing but the overall structure of the mind. The infant's mind develops mainly under the action of epigenetic processes, that is, interactions of genetic and environmental information. The establishment of the phonetic boundaries that are typical of the native language provides an example of the combined influence of a biological predisposition to categorical perception and a cultural setting. Indeed, the categorical perceiving mode is part of the innate structure of the mind, and the different boundary in the voiced/unvoiced distinction for stops in English and French is a cultural phenomenon. Most learning is thus both biological and cultural.

However, a characteristic of epigenetic regulations is that they occur within some critical period that corresponds, according to Newport, Bavelier, and Neville (chapter 27), to a "peak period of plasticity, occurring at some maturationally defined time in development, followed by reduced plasticity later in life." For this reason, whatever the amount of reductionism involved, we propose to take the critical period contingency as a criterion for distinguishing biological from cultural learning. Cultural learning is open-ended learning. An example of cultural learning is the acquisition of phoneme awareness: it can occur at any age (Morais et al., 1988). Cultural learning may be cumulative as, for example, the learning of new words throughout life. It may also include the learning of procedures and involve a change from controlled to automatic operations as, for example, in writing.

An important question concerns how much of the adult mind is due to biological learning and how much is due to cultural learning. Developmental psychologists tend to embrace one of two opposite lines of thought about this question. According to the "biopsychological" approach, the structures of the mind are a product of biological evolution and epigenesis (Pinker, 1998). According to the "sociocultural" approach, the mind is a product of sociocultural history (Vygotsky, 1978). Probably, neither does full justice to both biological and cultural contributions.

Cultural learning is erroneously believed to involve only general processes (cf. Cossu, 1999). This idea stems from a regrettable confusion between biological and modular capacities. An important lesson from the study of literacy is that one has to distinguish between biological modules, like the system for processing speech, and cultural modules. Reading, in particular, presents many important modular characteristics. It involves fast, mandatory processes. Moreover, it is supported by a precise network of brain areas (e.g., see Allison et al., 1994; Cohen et al., 2000; Peterson, et al., 1990). One knows less well, unfortunately, what are the functions of these areas in the illiterate adult; and one totally ignores whether ex-illiterates use the same brain system for literacy as adults who became literate in childhood. Interestingly, the learning of abnormal auditory-visual associations in the adult barn owl is magnified by previous juvenile learning (Knudsen, 1998). If priming effects were also found for the acquisition of literacy, this would appear to be more dependent on biological learning than is usually assumed.

It should be clear anyway that brain data, at least at a microanalytic level, do not provide an argument for biology and against cultural learning. Cummins and Cummins (1999) present as uncontroversial the following logical argument: "... the mind is what the brain does; the brain was shaped by evolution; the mind was shaped by evolution." It would be no less logical (and materialist) to rephrase it using "learning," or even "cultural learning," in place of "evolution." The two arguments are not mutually exclusive. Both evolution and culture shape the brain, even though culture can only occupy the small space of freedom allowed by evolutionary constraints. Indeed, neurogenesis and the establishment of new synapses occur throughout life, to a much smaller extent, indeed, in adulthood than in childhood. Cultural learning, including learning literacy, must modify brain structures to some extent. Besides, the consistency of these modifications across individuals is not surprising. Cultural learning of phoneme-grapheme representations may be expected to develop in brain areas close or connected to those dedicated to basic phonological representations and to abstract visual representations. Different areas will be activated in the learning of a distinct cultural module, namely musical reading (see Peretz, chapter 24, for a discussion of the biological vs. cultural nature of music), which involves other types of representations, so that dissociation can occur in case of brain damage.

Although the lack of universality of literacy does not imply that literacy mainly results from cultural learning, there is no evidence that the literate people are those who were born with the literacy gene or who benefited from the appropriate epigenetic regulations. Nevertheless, the idea that literacy results from cultural learning has been disputed. For instance, Cossu (1999) believes that "specific biological components are responsible for the acquisition of reading," in particular "a core component may have emerged in phylogeny which later became refined as a cross-modal device for the (automatic) connection between phonology and other perceptual domains" (p. 226). This conception deserves some critical remarks.

The development of a new functional system may take advantage of components of previously existing systems. These "exaptation" processes (Gould and Vrba, 1982) are at work in alphabetic literacy: the phoneme representations involved in speech perception become useful for the written representation of language, provided that conscious awareness and recoding (two further biologically based capacities) are available to operate on them. However, the idea that "the metaphonological parser may have emerged as an outcome of a general trend for intersensory integration" (Cossu, 1999, p. 232) is dismissed by the fact that conscious phoneme segmentation does not emerge spontaneously but is elicited by alphabetic teaching, so that it can appear at five years of age, at sixty, or never. Moreover, the notion of a "general trend for intersensory connection" is paradoxically reminiscent of the notion of a general-purpose device. General trends are not selected in evolution; what is selected is a specific organ or mechanism. While a specific cross-modal device involving phonology has probably been developed in the human species to take advantage of the correlation between heard and visual speech (lipreading), there is no evidence that a similar specific device was developed to match phonology to symbols, hand movements, or stones.

A great deal of evidence now indicates that most dyslexics suffer from highly specific phonological deficits, which result from some anomaly in their biologically determined system of speech perception (see discussion in, e.g., Galaburda, chapter 25; Morais, in press). Indeed, cultural learning may be hampered by developmental injury to the brain areas underlying the basic capacities to be recruited. It must be emphasized that to recognize the biological foundations of alphabetic literacy is not to say that this is a biological achievement. To depend on and to be an instance of are not the same notion. The literate mind is a product of cultural learning capitalizing on biological capacities.

What Illiterates Tell Us about the Literate Mind

We summarize below the main findings on the cognitive effects of literacy. The absence of a literacy effect can be demonstrated by comparing literate and illiterate people. By contrast, the presence of a selective effect of literacy requires the comparison of illiterates and ex-illiterates. Obviously, some of the mentioned effects, such as phoneme awareness, are specific to alphabetic literacy, whereas others, such as lexical growth, may concern all types of literacy.

Language
Cerebral specialization for language is innate (Bertoncini et al., 1989) and consequently is present in illiterate people (Castro and Morais, 1987).

The automatic extraction of phonetic information is unaffected by literacy. Similar patterns of results in literate and illiterate people were obtained for categorical perception (Castro, 1993), the McGurk effect (Morais, Castro, and Kolinsky, 1991), feature blendings (Morais et al., 1987), and migration errors (i.e., illusory conjunctions: Morais & Kolinsky, 1994; Kolinsky and Morais, 1996).

Specific literacy effects concern the use of a phonemic attentional recognition strategy under difficult listening conditions (Morais et al., 1987) and the susceptibility of phoneme integration to orthographic knowledge, as observed with the phonological fusion paradigm (Morais et al., 1991).

Both word and pseudoword repetition are poorer in illiterate than in literate people (Castro-Caldas et al., 1998), the effect being stronger for pseudowords. The illiterate, but not the literate people, failed to activate the anterior cingulate cortex and basal ganglia during pseudoword repetition. These regions have a role in language and attention. According to Frith (1998), "illiterates have not developed their capacity for phonological processing, and hence they rely more on lexical-semantic systems" (p. 1011). According to Castro-Caldas et al. (1998), "the absence of knowledge of orthography limits the ability of illiterate subjects to repeat pseudowords correctly" (p. 1060). These two explanations are not exclusive of each other. Yet, since illiterates are not poorer than ex-illiterates in repeating words or pseudowords (Morais and Mousty, 1992), it seems that automatic activation of orthographic representations requires a degree of literacy ability above that of ex-illiterates.

Among metaphonological abilities, phoneme awareness shows a huge effect of alphabetic literacy (Morais et al., 1979; Read et al., 1986). By contrast, awareness of syllables (Morais et al., 1986), rhyme (Morais et al., 1986), and phonological length (Kolinsky, Cary, and Morais, 1987) was observed in at least some of the illiterate people tested. Morais et al. (1989) found that, like literate people, illiterates detect syllabic targets better than nonsyllabic targets.

Lexical knowledge increases during the first years of schooling, mostly as a consequence of literacy. Reading activities would account for about half of the 3000 new words that a schoolchild acquires each year (Nagy, Anderson, and Herman, 1987). By fifth grade, the best 10 percent readers would read 200 more texts outside school than the 10 percent poorest readers (Anderson, Wilson, and Fielding, 1988).

Concerning the notion of word, the results obtained depended on task. In illiterates, the command to repeat, one word at a time, an orally presented sentence elicits a segmentation into main syntactic constituents (e.g., "the car / stands / in front of the door"; Cary, 1988). According to Cary and Verhaeghe (1991), illiterates only produced 4 percent word segmentations, whereas ex-illiterates produced 48 percent and literates 86 percent. However, illiterates scored 72 percent word responses in the repetition of the last "bit" of an interrupted sentence (Cary and Verhaeghe, 1991). Unfortunately, they were not compared with ex-illiterates. When tested subsequently in the segmentation task, the same illiterates still produced 80 percent subject-verb-complement segmentations. Illiterates may thus consciously access the word unit, but they are usually biased to process meaning.

Semantic Knowledge
Comparing semantic knowledge in illiterate people, collective-farm activists, and young people with one to two years of schooling, Luria (1976) claimed that illiterates, contrary to the other groups, have no taxonomic knowledge. However, Scribner and Cole (1981), studying Vai people in Liberia, found results that "discourage conclusions about a strong influence of literacy on categorization and abstraction" (p. 124). It also seems that, in free recall, illiterates can use category organization of the items when the categories are explicitly indicated to them or simply suggested by having them sort the items into piles (Cole et al., 1971; Scribner, 1974). This would not be possible if the illiterates had no preexistent semantic organization.

Moreover, the categorization capacity appears very early in development. Three-year-olds can understand the logic of class inclusion, for instance, that "car" and "bike" belong to the superordinate "vehicle" (Markman, 1984). Matching in terms of taxonomic category is almost perfect at four years of age. Presented with fish, car, and boat, the last two are put together, in spite of the fact that fish and boat are found in water (Rosch et al., 1976). In five-year-olds, categorical relationships led to more false recognitions of a probe in a sentence than the part-whole relationship (Mansfield, 1977). Testing six-year-olds, Radeau (1983) found an auditory semantic priming effect. About 80 percent of the pairs used were taxonomically organized (coordinate: "arm-leg"; or superordinate: fruit-apple). All these findings reflect the taxonomic organization of semantic memory in preliterate children.

The apparent discrepancy between these data and Luria's (1976) can be accounted for in two ways: either people who are not stimulated to think categorically lose this knowledge, or they keep taxonomic knowledge but develop a strong preference for practical schemes.

Recent work we carried out in Brazil, in collaboration with Scliar-Cabral, Monteiro, and Penido, supports the second interpretation. Both illiterates and ex-illiterates could make semantic as well as perceptual classifications and shift from one type of dimension to the other. In a classification test requiring the subjects to match a target with either a taxonomically related or an unrelated item, illiterates made as much taxonomic choices (93 percent and 82 percent for images and words, respectively) as poorly literate people did (these are adults who completed only four school grades in childhood). In another test requiring the subjects to group twelve drawings taken from four categories, the illiterates grouped the items correctly, and the majority of their justifications were of a taxonomic kind. Finally, in fluency tests, the illiterates produced, on average, 12.9 words per category and the poorly literate adults, 11.9. The ratio of subcategory repetitions (Bousfield, 1953), indicating taxonomic clustering, was similar in the two groups. Thus, illiterates display both categorical knowledge and a hierarchical organization of categories.

However, when presented with a choice between a taxonomic and a functional relationship, both illiterate and poorly literate participants chose less frequently than more educated people the taxonomic relationship. In some of the classification tests, they also offered functional justifications for most of their choices (e.g., arm and leg were matched "because if I don't have my arm, I can't scratch my leg"), consistent with Luria's (1976) observations. Categorical thinking is thus strongly stimulated by schooling. Unschooled people tend to develop stories or narrative imagining and, in this sense, even illiterates exhibit what Turner (1996) called a "literary mind."

Obviously, contents knowledge is influenced by both literacy (since much information is acquired through reading) and schooling.

Memory

Illiterates use phonological codes in short-term memory (Morais et al., 1986). However, the illiterates' and ex-illiterates' short-term memory span is far smaller than the span displayed by literate adults. The superior capacity demonstrated by the schooled people may be linked to experience and organizational processes obtained through schooling or through literacy activities or both. It is also possible that the ex-illiterates' low reading ability does not allow the automatic activation of orthographic representations in verbal memory tasks.

Comparing forward digit and Corsi's block spans, we found an interaction between schooling and material (work in progress in collaboration with Grimm-Cabral, Penido, and dos Passos): only unschooled subjects were better for the visuospatial than for the verbal span. Lack of automatic activation of orthographic representations provides a potential explanation of their verbal span inferiority. Besides, ex-illiterates did not exhibit the length effect displayed by literate participants in verbal span. Thus, there seems to be a schooling effect, but not a specific effect of literacy, on either verbal rehearsal or mapping of phonological representations to output plans (cf. recent data on the development of the word length effect in children in Henry et al., 2000).

In the running digit span task, involving two mechanisms, the phonological loop (necessary for sequential storage and rehearsal) and the central executive (concerned with information updating), illiterates and ex-illiterates obtained similar scores. Thus, the executive component of working memory would not be selectively affected by literacy.

Illiterates were compared to Qur'anic literates for recall of studied lists of words (Scribner and Cole, 1981). There was no literacy effect for recall in any order, but the literates, possibly influenced by the incremental method of learning the Qur'an, were better for ordered recall.

Finally, immediate sentence recall is as good in illiterate as in university people (Ardila, Rosselli, and Rosas, 1989).

Executive Functions

The ability to inhibit irrelevant information and to selectively attend to one dimension of the stimulus is affected by schooling. In visual speeded classification (cf. Garner, 1974), illiterate unschooled subjects showed some difficulty at selectively attending to the form dimension when color varied orthogonally (e.g., green or red squares vs. green or red circles) (Kolinsky, 1988). Given that most illiterates can identify digit symbols, we also used a digit Stroop test. Literate participants were significantly faster than illiterate and ex-illiterate ones, but the size of the interference effect observed in the incongruent condition (e.g., to respond "three" when presented with "222") and of the facilitation observed in the congruent condition (e.g., "22") did not vary between the groups.

Planning ability was evaluated with the Tower of London test (cf. Shallice, 1982). In this test, a start state, consisting of a set of three differently colored disks placed on pegs, has to be reconfigured into a goal state in the fewest moves possible. The main constraint is that participants can only move one disk at a time. Thus, a sequence of moves must be planned, monitored, and possibly revised. In some trials, an incorrect move suggested by a local resemblance with the goal state has to be inhibited. The results showed no significant difference between illiterate and ex-illiterate participants in either number of movements or time of execution. However, in number of movements, though not in time of execution, literate participants performed better than unschooled ones.

Reasoning

Goody (1968) considered literacy to be a precondition for syllogism. Luria (1976) reported that illiterates could not perceive the logical relation between the parts of the syllogism. According to Scribner and Cole (1981), logic problems demonstrated the strongest effects of schooling, but neither Vai nor Arabic literacy had an effect on the number of correct responses or on theoretical justifications. However, examining illiterates in Brazil, Tfouni (1988) noticed that some of them could understand and explain syllogisms after displaying the behavior described by Luria.

A clear case of schooling influence concerns "intelligence" tests, among which are the Raven Progressive Matrices. As indicated by Neisser

(1998), some Matrices items may require "a special form of visual analysis . . . : each entry must be dissected into the simple line segments of which it is composed before the process of abstraction can operate" (p. 10). As indicated below, this kind of visual analysis is strongly influenced by schooling so that the sources of the unschooled people's inferiority in intelligence tests may be multiple. More important, we did not obtain significant differences between illiterates and ex-illiterates in these tests. (Cary, 1988; Verhaeghe, 1999).

Visual Cognition
According to Luria (1976), in unschooled people "neither the processing of elementary visual information nor the analysis of visual objects conforms to the traditional laws of psychology" (p. 22).

The first part of this claim is clearly unmotivated. Examining low-level feature extraction in an indirect way (i.e., through the occurrence of illusory conjunctions), we observed a clear developmental effect (Kolinsky, 1989), but no difference at all between illiterate and literate adults (Kolinsky, Morais, and Verhaeghe, 1994).

The second part of Luria's claim deserves more extensive comments. When conscious analysis of the figure or selective attention to one of its parts is required by the task, differences between unschooled and schooled, literate, people are observed. This was the case for recognition of incomplete figures (Luria, 1976; Verhaeghe, 1999), identification of hierarchical figures (identified more frequently at the local than at the global level; Verhaeghe, 1999), and part verification, in which unschooled adults displayed poorer scores than second-graders (e.g., see Kolinsky et al., 1987). Recognition of superimposed figures seems far better succeeded: especially in a detection task on geometric material, the unschooled subjects' performance was almost perfect (Verhaeghe, 1999).

However, the reported differences between schooled and unschooled people on "visual cognition," that is, explicit visual analysis (Kolinsky and Morais, 1999), should not be interpreted as resulting from literacy, since illiterates and ex-illiterates were equally poor. There was one exception: ex-illiterates displayed better mirror-image discrimination skills than illiterates (Verhaeghe and Kolinsky, 1992). However, this literacy effect seems related to the fact that our writing system includes mirror-image letters like "b" vs. "d" (see Gibson, 1969, for a similar suggestion). Readers of a written system that does not incorporate mirror-image signs (the Tamil syllabary) are as poor as illiterates in discriminating mirror-images (Danziger and Pederson, 1999). More important, other activities (e.g., lace-making) drawing the observer's attention to the left-right orientation of the stimuli may promote mirror-image discrimination as well (Verhaeghe and Kolinsky, 1992). Thus, no genuine, specific effect of literacy is observed in visual cognition.

The Literate Mind and the Modular Organization of the Mental Capacities

Coltheart's (1999) view of the organization of the input cognitive systems includes separate modules for spoken language, written language, and image processing. Within each of these modules there are interlevel interactions, but the modules are encapsulated from each other.

Among other relevant findings, those reviewed here on specific literacy effects and on schooling effects cast doubt both on the autonomy of the three mentioned modules, especially between spoken and written language, and on absolute internal interactions. It seems more appropriate to consider, on the one hand, autonomous processes for spoken and for written language at the early perceptual level, and, on the other hand, interactions between spoken and written language at higher (postperceptual) levels of processing. The postperceptual processes include recognition strategies and intentional analyses. Spoken language postperceptual processes are affected by literacy. Nonlinguistic visual postperceptual processes may be influenced by schooling and other special training, but do not depend in a specific way on literacy.

Conclusion

The empirical evidence reviewed above, albeit fragmentary, should contribute to drawing the portrait of three distinct cognitive concepts, namely, of the literate mind, the educated mind, and the universal mind. The finding of significant differences in ex-illiterate people compared to illiterate people points to specific characteristics of the literate mind. The observation of significant differences in people who completed a high degree of schooling compared to unschooled people (who include both ex-illiterates and illiterates) demonstrates specific characteristics of the educated mind. Finally, similar data from all these populations argue for universal characteristics of the human mind.

It is only for metaphonological knowledge, especially phoneme awareness, as well as for some processes involved in speech recognition (and, obviously, for literacy abilities), that the illiterates differed clearly from the ex-illiterates. Thus, the specific characteristics of the literate mind seem to be restricted to only a few aspects of the language capacity. However, one should not forget that these aspects of language are of overwhelming importance. Moreover, literacy is a main avenue to education: according to present cultural standards, illiterate people can hardly be cultivated people. Most domains of expertise are out of reach of illiterates. We thus strongly invite cognitive psychologists to take the study of literacy effects into high consideration.

Schooling effects also deserve systematic investigation. As a matter of fact, dramatic differences were obtained in either performance or processing strategies comparing both illiterate and ex-illiterate people to "literate" ones, actually to people who have reached a high level of schooling. The superior abilities of the educated mind involved lexical knowledge, verbal repetition (the results mentioned may, however, be due to the insufficient literacy level of the ex-illiterate participants), categorical thinking, verbal memory (possibly organizational processes and rehearsal), resolution of logical problems, explicit analysis of visual features and dimensions, and some planning components of executive processes.

Finally, the present findings support the hypothesis of universality of a significant array of capacities. These include the early processes of speech and visual perception, as well as the organization of semantic information in terms of hierarchical taxonomic categories, and the inhibition of information that is incongruent with the one relevant to the task.

Given that our present aim was mainly to examine the literate mind, the set of studies mentioned here did not address the characteristics of either the educated or the universal mind in an exhaustive way. It is worth recalling here that Mehler and Dupoux (1994) offered a rather comprehensive account of the universal characteristics of the human mind. In particular, Mehler's own experimental work has shown how the study of neonate cognition is crucial to assessing these universal characteristics. Of course, evidence on the final, adult state is important as well. But we believe that the evidence reviewed here shows that future cognitive work on human adults should distinguish more carefully than has been done in the past between the literate, the educated, and the universal mind.

References

Allison, T., McCarthy, G., Nobre, A., Puce, A., and Belger, A. (1994). Human extrastriate visual cortex and the perception of faces, words, numbers, and colors. *Cerebral Cortex, 4,* 544–554.

Anderson, R. C., Wilson, P. T., and Fielding, L. G. (1988). Growth in reading and how children spend their time outside school. *Reading Research Quarterly, 23,* 285–303.

Ardila, A., Rosselli, M., and Rosas, P. (1989). Neuropsychological assessment in illiterates: Visuopatial and memory abilities. *Brain and Cognition, 11,* 147–166.

Bertoncini, J., Morais, J., Bijeljac-Babic, R., McAdams, S., Peretz, I. & Mehler, J. (1989). Dichotic perception and laterality in neonates. *Brain and Language, 37,* 591–605.

Bousfield, W. A. (1953). The occurrence of clustering in recall of randomly arranged associates. *Journal of General Psychology, 49,* 229–240.

Cary, L. (1988). *A analise explicita das unidades da fala nos adultos não alfabetizados.* Unpublished doctoral dissertation, University of Lisbon.

Cary, L, and Verhaeghe, A. (1991). Efeito da pratica da linguagem ou da alfabetização no conhecimento das fronteiras formais das unidades lexicais : Comparação de dois tipos de tarefas. *Actas das las Jornadas de Estudo dos Processos Cognitivos, 33–49.*

Castro, S. L. (1993). *Alfabetizaçao e percepçao da fala.* Oporto, Portugal: Instituto Nacional de Investigaçao Cientifica.

Castro, S. L., and Morais, J. (1987). Ear differences in illiterates. *Neuropsychologia, 25,* 409–417.

Castro-Caldas, A., Petersson, K. M., Reis, A., Stone-Elander, S., and Ingvar, M. (1998). The illiterate brain. Learning to read and write during childhood influences the functional organization of the adult brain. *Brain, 121,* 1053–1063.

Cohen, L., Dehaene, S., Naccache, L., Lehéricy, L. Dehaene-Lamberts, G., Hénaff, M.-A., and Michel, F. (2000). The visual word form area. Spatial and temporal characterization of an initial stage of reading in normal subjects and posterior split-brain patients. *Brain, 123,* 291–307.

Cole, M., Gay, J., Glick, J. A., and Sharp, D. W. (1971). *The cultural context of learning and thinking.* New York: Basic Books.

Coltheart, M. (1999). Modularity and cognition. *Trends in Cognitive Sciences*, 3, 115–120.

Cossu, G. (1999). Biological constraints on literacy acquisition. *Reading and Writing*, 11, 213–237.

Cummins, D. D., and Cummins, R. (1999). Biological preparedness and evolutionary explanation. *Cognition*, 73, B37–53.

Danziger, E., and Pederson, E. (1999). Through the looking-glass: Literacy, writing systems and mirror-image discrimination. *Written Language and Literacy*, 1, 153–164.

Dupoux, E., and Mehler, J. (1992). Unifying awareness and on-line studies of speech: A tentative framework. In J. Alegria, D. Holender, J. Junça de Morais, and M. Radeau (Eds.), *Analytic approaches to human cognition* (pp. 59–75). Amsterdam: Elsevier.

Frith, U. (1998). Literally changing the brain. *Brain*, 121, 1011–1012.

Garner, W. R. (1974). *The processing of information and structure*. Potomac, MD: Erlbaum.

Gibson, E. J. (1969). *Principles of perceptual learning and development*, New York: Appleton-Century-Crofts.

Goody, J. (1968). *Literacy in traditional societies*. Cambridge, UK: Cambridge University Press.

Gould, S. J., and Vrba, E. S. (1982). Exaptation—a missing term in the science of form. *Paleobiology*, 8, 4–15.

Henry, L. A., Turner, J. E., Smith, P. T., and Leather, C. (2000). Modality effects and the development of the word length effect in children. *Memory*, 8, 1–17.

Knudsen, E. I. (1998). Capacity for plasticity in the adult owl auditory system expended by juvenile experience. *Science*, 279, 1531–1533.

Kolinsky, R. (1988). *La séparabilité des propriétés dans la perception des formes*. Unpublished doctoral dissertation, Université Libre de Bruxelles.

Kolinsky, R. (1989) The development of separability in visual perception. *Cognition*, 33, 243–284.

Kolinsky, R., and Morais, J. (1996). Migrations in speech recognition. In *A guide to spoken word recognition paradigms. Language and Cognitive Processes*, 11, 611–619.

Kolinsky, R., and Morais, J. (1999). We all are Rembrandt experts—or how task dissociations in school learning effects support the discontinuity hypothesis. *Behavioral and Brain Sciences*, 22, 381–382.

Kolinsky, R. Cary, L., and Morais, J. (1987). Awareness of words as phonological entities: The role of literacy. *Applied Psycholinguistics*, 8, 223–232.

Kolinsky, R., Morais, J., Content, A., and Cary, L. (1987). Finding parts within figures: A developmental study. *Perception*, 16, 399–407.

Kolinsky, R., Morais, J., and Verhaeghe, A. (1994). Visual separability: A study on unschooled adults. *Perception*, 23, 471–486.

Luria, A. R. (1976). *Cognitive development. Its cultural and social foundations.* Cambridge, MA: Harvard University Press.

Mansfield, A. F. (1977). Semantic organization in the young child: Evidence for the development of semantic feature systems. *Journal of Experimental Child Psychology, 23,* 57–77.

Markman, E. M. (1984). The acquisition and hierarchical organization of categories by children. In M. Sophian (Ed.), *Origins of cognitive skills.* Hillsdale, NJ: Erlbaum.

Mehler, J., and Dupoux, E. (1994). *What infants know: The new cognitive science of early development.* Cambridge, MA: Blackwell.

Morais, J. (in press). Levels of phonological representation in skilled reading and in learning to read. *Reading and Writing.*

Morais, J., and Kolinsky, R. (1994). Perception and awareness in phonological processing: The case of the phoneme. *Cognition, 50,* 287–297.

Morais, J., and Mousty, P. (1992). The causes of phonemic awareness. In J. Alegria, D. Holender, J. Junça de Morais, and M. Radeau (Eds.), *Analytic approaches to human cognition* (pp. 193–211). Amsterdam: Elsevier.

Morais, J., Bertelson, P., Cary, L., and Alegria, J. (1986). Literacy training and speech analysis. *Cognition, 24,* 45–64.

Morais, J., Cary, L; Alegria, J., and Bertelson, P. (1979). Does awareness of speech as a sequence of phones arise spontaneously? *Cognition, 7,* 323–331.

Morais, J., Castro, S.- L., and Kolinsky, R. (1991). La reconnaissance des mots chez les adultes illettrés. In R. Kolinsky, J. Morais, and J. Segui (Eds.), *La reconnaissance des mots dans les différentes modalités sensorielles. Etudes de psycholinguistique cognitive* (pp. 59–80). Paris: Presses Universitaires de France.

Morais, J., Castro, S.-L., Scliar-Cabral, L., Kolinsky, R., and Content, A. (1987). The effects of literacy on the recognition of dichotic words. *Quarterly Journal of Experimental Psychology, 39A,* 451–465.

Morais, J., Content, A., Bertelson, P., Cary, L., and Kolinsky, R. (1988). Is there a critical period for the acquisition of segmental analysis ? *Cognitive Neuropsychology, 5,* 347–352.

Morais, J., Content, A., Cary, L., Mehler, J., and Segui, J. (1989). Syllabic segmentation and literacy. *Language and Cognitive Processes, 4,* 57–67.

Nagy, W. E., Anderson, R. C., and Herman, P.A. (1987). Learning word meanings from context during normal reading. *American Educational Research Journal, 24,* 237–270.

Neisser, U. (1998). Introduction: Rising test scores and what they mean. In U. Neisser (Ed.), *The rising curve. Long-term gains in IQ and related measures.* Washington, DC: American Psychological Association.

Petersen, S. E., Fox, P. T., Snyder, A. Z., and Raichle, M. E. (1990). Activation of extrastriate and frontal cortical areas by visual words and word-like stimuli. *Science, 249,* 1041–1043.

Pinker, S. (1998). *How the mind works*. London: Penguin Books.

Radeau, M. (1983). Semantic priming between spoken words in adults and children. *Canadian Journal of Psychology, 37,* 547–556.

Read, C., Zhang, Y., Nie, H., and Ding, B. (1986). The ability to manipulate speech sounds depends on knowing alphabetic writing. *Cognition, 24,* 31–44.

Rosch, E. Mervis, C. Gray, W. Johnson, D., and Boyes-Braem, P. (1976). Basic objects in natural categories. *Cognitive Psychology, 8,* 382–439.

Scribner, S. (1974). Developmental aspects of categorized recall in a West African society. *Cognitive Psychology, 6,* 475–494.

Scribner, S., and Cole, M. (1981). *The psychology of literacy*. Cambridge, MA : Harvard University Press.

Shallice, T. (1982). Specific impairments in planning. *Philosophical Transactions of the Royal Society of London. Series B: Biological Sciences, 298,* 199–209.

Tfouni, L. V. (1988). Adultos não alfabetizados: O avesso do avesso. Campinas, Brazil: Pontes.

Turner, M. (1996). *The literary mind*. Oxford: Oxford University Press.

Verhaeghe, A. (1999). *L'influence de la scolarisation et de l'alphabétisation sur les capacités de traitement visuel*. Unpublished doctoral dissertation, University of Lisbon.

Verhaeghe, A., and Kolinsky, R. (1992). Discriminação entre figuras orientadas em espelho em função do modo de apresentação em adultos escolarisados e adultos iletrados. In *Proceedings of the I jornadas de estudo dos processos cognitivos da sociedade portuguesa de psicologia* (pp. 51–67). Lisbon, Astoria.

Vosniadou, S. (1994). Universal and culturespecific properties of children's mental models of the earth. In L. A. Hirschfeld and S. A. Gelman (Eds.), *Mapping the mind. Domain specificity in cognition and culture* (pp. 412–430). Cambridge, MA: MIT Press.

Vygotsky, L. S. (1978). *Mind in society*. Cambridge, MA: Harvard University Press.

27

Critical Thinking about Critical Periods: Perspectives on a Critical Period for Language Acquisition

Elissa L. Newport, Daphne Bavelier, and Helen J. Neville

Over many years Jacques Mehler has provided us all with a wealth of surprising and complex results on both nature and nurture in language acquisition. He has shown that there are powerful and enduring effects of early (and even prenatal) experience on infant language perception, and also considerable prior knowledge that infants bring to the language acquisition task. He has shown strong age effects on second-language acquisition and its neural organization, and has also shown that proficiency predicts cerebral organization for the second language. In his honor, we focus here on one of the problems he has addressed—the notion of a critical period for language acquisition—and attempt to sort out the current state of the evidence.

In recent years there has been much discussion about whether there is a critical, or sensitive, period for language acquisition. Two issues are implicit in this discussion: First, what would constitute evidence for a critical period, particularly in humans, where the time scale for development is greater than that in the well-studied nonhuman cases, and where proficient behavioral outcomes might be achieved by more than one route? Second, what is the import of establishing, or failing to establish, such a critical period? What does this mean for our understanding of the computational and neural mechanisms underlying language acquisition?

In this chapter we address these issues explicitly, by briefly reviewing the available evidence on a critical period for human language acquisition, and then by asking whether the evidence meets the expected criteria for critical or sensitive periods seen in other well-studied domains in human and nonhuman development. We conclude by stating what we think the outcome of this issue means (and does not mean) for our understanding of language acquisition.

What Is a Critical or Sensitive Period?

Before beginning, we should state briefly what we (and others) mean by a critical or sensitive period. A critical or sensitive period for learning is shown when there is a relationship between the age (more technically, the developmental state of the organism) at which some crucial experience is presented to the organism and the amount of learning which results. In most domains with critical or sensitive periods, the privileged time for learning occurs during early development, but this is not necessarily the case (cf. bonding in sheep, which occurs immediately surrounding parturition). The important feature is that there is a peak period of plasticity, occurring at some maturationally defined time in development, followed by reduced plasticity later in life. In contrast, in many domains and systems, there may be plasticity uniformly throughout life (open-ended learning), or plasticity may increase with age as experience or higher-level cognitive skills increase.

As discussed in greater detail below, the mechanisms underlying critical periods are quite diverse in the systems in which they have been studied. Even without access to the underlying mechanisms, however, one can define and identify the relevant phenomena in behavioral terms. In this discussion we will not attempt to distinguish between a critical period and a sensitive period (sometimes distinguished by the abruptness of offset or the degree of plasticity remaining outside of the period); as we will also discuss in detail, most critical periods show more gradual offsets and more complex interactions between maturation and experiential factors than the original concept of a critical period included.

Overview of the Available Evidence Concerning a Critical or Sensitive Period for Language Acquisition

A number of lines of research, both behavioral and neural, suggest that there is a critical or sensitive period for language acquisition. First, many studies show a close relationship between the age of exposure to a language and the ultimate proficiency achieved in that language (see, e.g., Newport, 1990; Emmorey and Corina, 1990; Mayberry and Fischer, 1989; Johnson and Newport, 1989, 1991; Krashen, Long, and Scarcella, 1982; Long, 1990; Oyama, 1976; Pallier, Bosch, and Sebastián-Gallés, 1997; Patkowski, 1980; and others). Peak proficiency in the language, in terms of control over the sound system as well as the grammatical structure, is displayed by those whose exposure to that language begins in infancy or very early childhood. With increasing ages of exposure there is a decline in average proficiency, beginning as early as ages four to six and continuing until proficiency plateaus for adult learners (Johnson and Newport, 1989, 1991; Newport, 1990). Learners exposed to the language in adulthood show, on average, a lowered level of performance in many aspects of the language, though some individuals may approach the proficiency of early learners (Birdsong, 1992; Coppieters, 1987; White and Genesee, 1996).

These effects have been shown for both first and second languages, and for measures of proficiency including degree of accent, production and comprehension of morphology and syntax, grammaticality judgments for morphology and syntax, and syntactic processing speed and accuracy. For example, Johnson and Newport (1989, 1991) have shown that Chinese or Korean immigrants who move to the United States and become exposed to English as a second language show strong effects of their age of exposure to the language on their ability to judge its grammatical structure many years later, even when the number of years of exposure is matched. These effects are not due merely to interference of the first language on the learner's ability to acquire the second language: deaf adults, acquiring American Sign Language (ASL) as their primary language, show effects of age of exposure on their grammatical skills in ASL as much as 50 years later, even though they may not control any other language with great proficiency (Emmorey, 1991; Emmorey and Corina, 1990; Mayberry and Fischer, 1989; Mayberry and Eichen, 1991; Newport, 1990). These studies are also in accord with the famous case studies of individual feral or abused children, isolated from exposure to their first language until after puberty (Curtiss, 1977), where more extreme deficits in phonology, morphology, and syntax occur.

However, age of exposure does not affect all aspects of language learning equally. As reviewed in more detail below, the acquisition of vocabulary and semantic processing occur relatively normally in late learners. Critical period effects thus appear to focus on the formal properties of language and not the processing of meaning. Even within the formal properties of language, though, various aspects of the language may be more or less dependent on age of language exposure. For example, late learners acquire the basic word order of a language relatively well, but more complex aspects of grammar show strong effects of late acquisition (Johnson and Newport, 1989; Newport, 1990). Very recent studies have reported that late learners may pick up information about lexical stress with impunity but show deficits in acquiring the phonetic information important in native-like pronunciation (Sanders, Yamada, and Neville, 1999). Further research is needed to characterize the structures which do and do not show strong effects of age of learning.

Age of exposure has also been shown to affect the way language is represented in the brain. Positron emission tomography (PET), functional magnetic resonance imaging (fMRI), and event-related potential (ERP) studies all indicate strong left hemisphere activation for the native language in bilinguals (Dehaene et al., 1997; Perani et al., 1996; Yetkin et al., 1996). However, when second languages are learned late (after seven years), the regions and patterns of activation are partially or completely nonoverlapping with those for the native language. Neural organization for late-learned languages tends to be less lateralized and displays a high degree of variability from individual to individual (Dehaene et al., 1997; Kim et al., 1997; Perani et al., 1996; Weber-Fox and Neville, 1996; Yetkin et al., 1996). In contrast, the few studies that have observed early bilinguals or highly proficient late bilinguals report congruent results for native and second languages (Kim et al., 1997; Perani et al., 1998).

Some results indicate that there may be considerable specificity in these effects. For example, age of acquisition appears to have more pronounced effects on grammatical processing and its representation in the brain than on semantic processing (Weber-Fox and Neville, 1996). In Chinese-English bilinguals, delays of as long as sixteen years in exposure to English have very little effect on the organization of the brain systems active in lexical/semantic processing: when responding to the appropriateness of open-class content words in English sentences, all groups of learners show evoked potential components similarly distributed over the posterior regions of both hemispheres. In contrast, when judging the grammaticality of English syntactic constructions or the placement of closed-class function words in sentences, only early learners show the characteristic anterior left hemisphere ERP components; learners with delays of even four years show significantly more bilateral activation (Weber-Fox and Neville, 1996). Similar effects appear for signed languages (Neville et al., 1997).

Comparable results also appear in ERP studies of English sentence processing by congenitally deaf individuals who have learned English late and as a second language (ASL was their first language) (Neville, Mills, and Lawson, 1992). ERP responses in deaf subjects to English nouns and to semantically anomalous sentences in written English are like those of hearing native speakers of English. In contrast, for grammatical information in English (e.g., function words), deaf subjects do not display the specialization of the anterior regions of the left hemisphere characteristic of native speakers. These data suggest that the neural systems that mediate the processing of grammatical information are more modifiable and vulnerable in response to altered language experience than are those associated with lexical/semantic processing.

Taken together, these results provide fairly strong evidence for a critical or sensitive period in acquiring the phonological and grammatical patterns of the language and in organizing the neural mechanisms for handling these structures in a proficient way. Nonetheless, the question of whether there is a critical period for language acquisition continues to be controversial. In the sections below we address some of the theoretical issues concerning what a critical or sensitive period for language acquisition might look like.

Empirical and Theoretical Questions Concerning a Critical or Sensitive Period for Language Acquisition

In response to the findings cited above, several questions have been raised about whether these age effects represent the outcome of a critical or sensitive period, or rather whether they might arise from other variables correlated with age but not with maturation. One set of questions concerns whether the behavioral function has the correct shape for a critical or sensitive period. Does the decline in sensitivity extend over too many years to be called a critical period? Must a critical period involve an abrupt decline and a total loss of plasticity at the end?

A second set of questions concerns the range of proficiency achieved by late learners. In particular, does a critical period require that no late learner achieve native proficiency? What is the appropriate interpretation of finding some late-learning individuals with native or near-native proficiency? What is the significance of finding correlations between neural organization for language and the proficiency rather than age of the learner?

Third, how does one distinguish a critical period from an interference effect? Late learners of a second language have used their primary language for more years than early learners, and therefore may have a more "entrenched" proficiency in that language. Does this indicate that their difficulty acquiring a second language is due to interference rather than to a critical period?

Finally, if there are differing ages at which learning declines for different aspects of language (such as phonology vs. syntax, or different aspects of grammar), does this mean that there is no true critical period, or that there are multiple critical periods?

In the section below we address these questions by considering what the evidence and interpretations are for domains other than language, for example in the development of auditory localization or visual acuity, where our understanding of critical periods is believed to be more solid (or at least where there is somewhat less controversy in the field about whether the notion of a critical period is sensible). We then compare the evidence for language to that for these other domains, attempting to determine whether the data or outcomes for language are in fact different or the same as these more well-studied and familiar cases.

What Is the Status of These Issues in Our Understanding of Critical Periods in Other Domains?

Within the language acquisition literature, many researchers have expressed the expectation that, if there were a critical or sensitive period for learning, it would have to have a number of strong characteristics.

For example, investigators have expected that a critical or sensitive period for language learning should have an abrupt end, at a well-defined and consistent age; and this age should be related to the onset of physical puberty (e.g., at about age 12 or 13) (Bialystok and Hakuta, 1994). It should permit no individual to achieve native or near-native proficiency after the critical period ends (Bialystok and Hakuta, 1994; Birdsong, 1992, 1999; White and Genesee, 1996). It should show no variation across different first language–second language combinations (Birdsong and Molis, 2000). It should limit the learning of a new second language, but there should not be accompanying loss or decrement of an early acquired first language (otherwise it would be better interpreted as an interference rather than a critical period effect) (Jia, 1998).

However, many of these strong or absolute characteristics are not true of critical or sensitive periods in other domains. Many critical or sensitive periods show more complex properties than have been expected or demanded in the case of language, and ongoing work in developmental neurobiology and psychobiology continually reveals that plasticity and learning may show quite complex and interacting effects of maturation, experience, stimulus salience or preferences, and the like. As discussed below, critical or sensitive periods in most (if not all) behavioral domains involve gradual declines in learning, with some (reduced, but not absent) ability to learn by mature organisms, and with more learning achieved during the waning portion of the critical period if the organism is presented with extremely salient or strongly preferred stimuli, or with learning problems similar to those experienced early in life. If such complex phenomena are expected and routinely found within critical periods in other domains, they should also be expected for language learning.

The Shape and Length of a Critical Period

Offset Time: Effects of Maturation vs. Experience   As recent research has revealed, the time at which a critical period ends is influenced by both maturational and experiential factors, and therefore cannot be pegged to a precise age. The most widely studied effect of this type is that isolation will extend a critical period: animals reared in isolation from the stimuli crucial for learning during the sensitive period will retain the ability to learn, post-isolation, at ages during which there is little or no plasticity in normally reared individuals. This effect occurs for a wide range of domains showing critical periods, including the establishment of neuromuscular junctions (whose critical period is extended if neural activity is suppressed: Oppenheim and Haverkamp, 1986), the development of ocular dominance columns in visual cortex (Hubel and Wiesel, 1970), imprinting in ducks (Hess, 1973), and song learning in many avian species (Nordeen and Nordeen, 1997). Conversely, the critical period may be shortened if the animal is exposed to strongly preferred stimuli, such as imprinting to a live duck rather than a decoy (Hess, 1973), or auditory localization calibrated with the ears normally open rather than with one ear plugged (Knudsen, 1988). While experience may alter the timing of the critical period to some degree, however, eventually plasticity will decline even if the animal is never exposed to preferred stimuli; this suggests that there is a maturational component underlying the change of plasticity, even though experience can modulate its timing.

Length of Critical Periods It is also clear that the absolute length of critical periods varies, and therefore that one cannot evaluate a critical period by whether it is very short and rapidly concluded, or gradual and extended over several months or years. In part, such variations would seem to depend on the rate at which the organism develops: slowly developing organisms, like humans, would be expected to show critical periods lasting orders of magnitude longer than those of frogs or cats, with both relative to the general rate of maturation of the organism. There may also be a relationship between how long, or when in development, a particular system takes to develop normally, and how long its critical period will be: systems which develop very early may have short periods of plasticity, whereas late-developing systems may have long critical periods. For example, in cat retina, the Y cells (which are more abundant in the periphery) develop later and more slowly than the X cells (clustered in the fovea), and are altered by visual experience over a longer period of time (Sherman, 1985).

Sharpness of the Decline in Plasticity A related point is that critical periods vary in how abruptly they end. Some critical periods (e.g., that involved with the establishment of ocular dominance columns in cat visual cortex; Hubel and Wiesel, 1970) end very suddenly; differences of hours or days in the time of exposure to relevant input can lead to profound differences in the behavioral or physiological outcome. But the occurrence of such sudden and apparently complete termination of plasticity in some systems should not mislead us into expecting that every system with a sensitive period will likewise show such a dramatic change in receptivity to experience. Many well-studied critical or sensitive periods show a gradual decline in plasticity as the critical period closes, with learning during this period of decline being only partially successful or responsive only to the strongest stimuli. For example, auditory localization in the barn owl displays full adjustment to a monaural ear plug if the plug is inserted early; with later and later ages of plug insertion (during adolescence), there is a gradually reduced degree of error correction achieved; and adult barn owls show only a limited ability to correct for plugs (Knudsen, 1988). At the same time, the amount of correction shown at various ages is greater for removing an earplug than for inserting an earplug, suggesting that plasticity during the intermediate stages of development is also differentially responsive to experiences for which the system is best tuned (for the barn owl, apparently the "normal" settings, where both ears are open or balanced in the auditory input they receive) versus experiences for which the system is not. Similar gradual declines in learning, involving intermediate degrees of plasticity and differential sensitivity to strong or weak stimuli, are also found in imprinting (Hess, 1973) and in certain aspects of avian song learning (Eales, 1987; K. Nordeen, 2000, personal communication).

Plasticity Outside of the Critical Period

Degree of Plasticity after the Critical Period Ends   It is also the case that some systems may continue to show plasticity, though reduced, after the critical period is over and the organism has reached an asymptotic adult state. For example, while ocular dominance columns do not appear to retune themselves to changes in binocular experience during adulthood (Hubel and Wiesel, 1970), auditory localization in the barn owl does show some (limited) recalibration in the adult: while young barn owls can correct errors introduced by earplug insertion up to as much as twenty degrees of arc, adult barn owls can correct such errors up to five or ten degrees (Knudsen, 1988). Similarly, songbirds may acquire some new syllables after the sensitive period for sensory acquisition is over and depend critically on auditory feedback throughout life to maintain adult song patterns (Nordeen and Nordeen, 1992), though (in critical period learners) most of the acquisition of song structures occurs during the critical period.

Individual Success in Learning beyond the Critical Period Even though there may be some plasticity in many systems after their critical or sensitive periods close, it is argued by some in the language acquisition literature that it would not be compatible with the concept of a critical period if any individuals (or any substantial number of individuals) could achieve native proficiency from exposure after the critical period ends (Birdsong, 1992; Birdsong and Molis, 2000; White and Genesee, 1996). Virtually every study on this topic has shown that the number of individuals from an unselected population who are first immersed in the language during adulthood and yet who approach native proficiency is relatively small (Birdsong, 1992; Birdsong and Molis, 2000; Coppieters, 1987; Johnson and Newport, 1989, 1991; Newport, 1990; White and Genesee, 1996). Nonetheless, some investigators have questioned whether a critical or sensitive period would allow anyone (Joseph Conrad is a favorite example) to do so.

There are a number of reasons why such high proficiency might be achieved outside of a critical period by some individuals. First, while the precise mechanisms underlying a critical period for language learning are unknown, it is clear that maturational changes in this domain are gradual and probabilistic, with adults capable of learning many things about a new language. One should not be surprised, then, if, in addition, there is a normal distribution of variation among individuals in their capacity to learn, even though the mean outcome is substantially lower for learning in adulthood as compared with childhood. As with any other variable, individuals vary in the timing and extent of various maturational processes (e.g., the rate at which they undergo physical growth or experience menarche), and therefore should be expected to show variation in the timing and extent of whatever cellular and/or cognitive variables produce the decline in language learning during development. Individual variability appears in all animal studies, particularly during the waning portion of the critical period or after; see, for example, variation among zebra finches in the effects of heterospecific tutors (Eales, 1987) or of deafening on song maintenance. Second, in adult humans (though perhaps not in ducks or cats or birds, who lack our general brain power), there is always the possibility of learning by mechanisms other than the ones originally suited for the task, and other factors (such as a good "ear" for languages, strong formal analytic skills, or extensive conscious learning and practice) might be responsible for the skills of outstanding adult learners.

Finally, adult language learners in all of these studies are always facing a *second*-language learning task during adulthood, not a first.[1] Even though the first and second languages may be quite different from one another, the fact that all languages share many properties (and that the language systems have been engaged and used for language learning and analysis during childhood) is likely to mean that in many ways adult language learning is not an entirely new learning task; and re-learning or transfer during adulthood of skills experienced during early life might be expected to show some resilience and success. A somewhat related phenomenon has been demonstrated in auditory localization: barn owls that wear visual prisms (shifting accurate visual and auditory localization) for a brief period during early development show an unexpected ability to adjust to the same prisms applied during adulthood, after the usual critical period for recalibration is over. Moreover, early experience with one set of prisms permits some recalibration to other prisms in adulthood, if the two sets of prisms share the same direction and similar magnitude of shift (Knudsen, 1998).

Maturation vs. Interference

Whenever one finds age effects on the acquisition of proficiency in a domain, there are always a number of interpretations of these effects, and a number of underlying mechanisms which might be responsible. For example, there might be maturational changes in the plasticity of the system, so that learning is reduced or no longer possible after the critical period, no matter what the early experiences of the organism. There might also be effects of early experience which change the capacity of later experiences to affect the system further. In the language acquisition literature, such accounts (known as maturational vs. interference effects) have been considered competing alternatives. But in the developmental psychobiology literature, where it has become increasingly clear that early experience always interacts with maturation and produces biological consequences, both types of accounts are considered critical or sensitive period effects.

Bateson (1979) discusses these two types of explanations for the critical period for imprinting in birds. One of his explanations is that, with maturation, the imprinting mechanism no longer functions. A second explanation, however, is that early experiences specify (and thus narrow) the range of objects the bird considers familiar, and correspondingly lead novel objects first experienced late in life to elicit fear and flight (rather than following). On this account, birds isolated in their cages during the critical period are not deprived of early experience, but will learn from whatever they are exposed to. They become imprinted on the pattern of their cage bars and will fear other novel objects, and, as a result, they will not be capable of undergoing the crucial experience (following) required for imprinting to other objects. In short, no matter what the early experiences of the organism might be, learning from these experiences will occur and will have priority over learning late in life.

One important consequence of this view is that maturational change vs. interactions between early and later learning are extremely difficult, if not impossible, to distinguish, and may in fact be different descriptions of the same process. Researchers studying second-language acquisition have tried to argue against a critical or sensitive period for language acquisition by suggesting that age effects arise from interference of the first language on the second (Flege, 1999; McCandliss et al., 1998). On first consideration it would appear that this argument does not apply as well to the age effects seen in late first-language acquisition (Emmorey and Corina, 1990; Mayberry and Eichen, 1991; Mayberry and Fischer, 1989; Newport, 1990) as to those seen in late second-language acquisition: why should the late acquisition of a first language be limited if there has been no prior learning of another language? However, if one hypothesizes that early linguistic input (or its absence) inexorably narrows the language systems to what has been experienced, late first- or second-language acquisition should both suffer, and maturational and interference accounts merge. Both accounts share an assumption that early experience has primary foundational effects and thus are both critical period accounts.

Multiple Critical Periods vs. Stimulus Differences in Producing Plasticity
A final issue concerns whether there are multiple critical periods for distinct components of language acquisition, and more generally how to interpret finding different age functions for different types of linguistic properties.

Interaction between Quality of Stimuli and Degree of Plasticity   An important and interesting property of critical or sensitive periods is that, as plasticity declines and the critical period comes to a close, strong stimuli may still lead to learning, whereas weak stimuli do not. This was described briefly above, to illustrate the gradual nature of the decline in plasticity at the end of many critical periods. But more generally there is an interaction throughout the critical period between the age or maturational state of the learner and the strength or salience of the stimuli from which the animal might learn. Strong stimuli for which the system is best tuned produce strong learning during the peak of the critical period, continue to produce learning even when the critical period would otherwise be over, and may close the critical period to subsequent learning from weaker stimuli. In contrast, weak stimuli—those from which the system is capable of learning, but which are not preferred—may produce learning during the peak of the critical period but no learning at all during the waning portions of the period.

One example, already described, is that the barn owl can re-calibrate auditory-visual localization when an earplug is inserted in one ear (or when prisms are introduced on the eyes) up to about sixty days of age, but it can re-calibrate back to normal settings, when the earplug (or prism) is removed, up to 200 days (Knudsen, 1988). An even more striking example, discovered only recently, is that these ages limiting plasticity in the barn owl are those derived from procedures allowing the animal to recover in a small individual cage. In contrast, when the owl is allowed to recover in a group aviary, where it can fly and interact freely with other owls, successful re-calibration after the addition of prisms continues up to 200 days, while successful re-calibration after the removal of prisms can be done throughout life (Brainerd and Knudsen, 1998). In other words, the experience required to produce normal localization abilities changes with age, with much flexibility on environmental requirements for learning in early development but more stringent 'enriched' conditions required for learning in later life. Similarly, sparrows can learn from the playback of recorded sparrow song up to about fifty days of age, but can continue to learn later than this from a live tutor (Baptista and Petrinovich, 1986; Marler, 1970).

An important consequence of this interaction between the quality of stimuli and the degree of plasticity they produce is that the function relating age of exposure to degree of learning will look quite different when measured for strong, preferred stimuli than for weak, less preferred stimuli, and may give the impression of two quite different timetables for the critical period.

Discriminating Multiple Critical Periods from Complex Interactions between Plasticity, Stimuli, and Task In the language acquisition literature, a number of investigators have found different ages of decline in plasticity for syntax vs. phonology or for other aspects of language (Flege, Yeni-Komshian, and Liu, 1999) and have suggested that there are multiple critical periods for different aspects of language (Flege, Yeni-Komshian, and Liu, 1999; Hurford, 1991; Long, 1990; Scovel, 1988; Singleton, 1989). Even more striking contrasts have been found between the formal (phonological or grammatical) aspects of language vs. those that deal with meaning (semantic or lexical), where the former show strong changes in acquisition over age, while the latter appear to show little or no effect of age of learning (Johnson and Newport, 1989; Weber-Fox and Neville, 1996).

There are well-attested multiple critical periods, arising from separate neural mechanisms, outside of language. For example, in the visual system, the development of acuity, orientation, stereopsis, and photopic vs. scotopic vision show different critical periods for sensitivity to visual experience, and these different developmental timetables correspond to distinct psychophysical and neural subsystems (Harwerth et al., 1986). But in some areas of vision (e.g., for Vernier acuity vs. grating acuity) there are controversies about whether differences in age effects arise from separate subsystems, or rather result from the differing difficulty of tasks used to measure the function (Skoczenski and Norcia, 1999).

For language, it is not always clear when differing age effects are the result of distinct subsystems, each with its own critical period (or no critical period), and when they are the result of differing degrees of complexity in the linguistic structure tested (or the task used for testing) and therefore different levels of performance in a single critical period function. The contrast in developmental plasticity between formal (phonological and grammatical) vs. semantic aspects of language appears to be widespread and consistent with other types of evidence suggesting separately developing subsystems (Goldin-Meadow, 1978; Newport, 1981; Newport, Gleitman, and Gleitman, 1977). But it is less clear whether there are different critical periods for phonology and syntax. Does phonology appear to show an earlier decline in plasticity, as compared with syntax, because it truly has a different critical period? Or does it show these differences because the aspects of phonology we have tested are more difficult than those we have tested for syntax, or the measurements more refined? Future research will need to consider how to distinguish a contrast across subsystems, displaying different developmental timetables and types of plasticity, from effects of stimulus strength and complexity.

Summary

In sum, the properties of plasticity in language acquisition are similar to those in other well-studied systems believed to display critical or sensitive periods. First, there is a strong relationship between the age at which learners are exposed to a language and the proficiency they attain in its phonological and grammatical structures: early exposure results in peak proficiency, with a gradual decline in proficiency as age of exposure increases. As in other systems, the mean level of proficiency declines with age of exposure, but this effect combines with increasing individual variation as age increases, and with different degrees of proficiency achieved for different aspects of the language, depending on the type or complexity of the construction (in other systems, this is characterized as stimulus strength or preference) and the similarity to other experiences of early life.

Different aspects of language display somewhat different age functions; further research is needed to reveal whether these are separate critical periods or the result of different measures of plasticity at intermediate ages.

Finally, some (few) individuals may achieve native or near-native proficiency even though their exposure to the language does not occur until adulthood. While other systems do show individual variation during the end of the critical period and some remaining plasticity after the critical period closes, it is not common to observe full learning by individuals exposed to the experience only late in life. However, language learning in humans is also unlike imprinting, song learning, or vision in two important ways. First, humans bring many high-level cognitive abilities to the task of learning a language, and might be capable of using systems to acquire a second language other than the ones they use for primary language acquisition. Second, for all but the few case studies of feral children, human language learning late in life is the second (or third or fourth) learning experience in the domain, whereas critical periods in other domains are typically studied by observing learning after extreme deprivation of experience in the domain during early life. The level of proficiency achieved in adult language learning might thus be best compared to studies of adult barn owls whose prisms or earplugs are switched at various ages, rather than to those first exposed to sound during adulthood.

In short, we believe that the acquisition of formal systems in language does show a critical or sensitive period, like that of other well-studied systems.

What Does It Mean to Demonstrate a Critical or Sensitive Period from the Point of View of Overall Issues in Language Acquisition?

One of the reasons discussions of critical periods in language acquisition have been extremely heated is that the demonstration of a critical period for language is interpreted by many researchers to signal or support a particular view of the language acquisition mechanism. As a final note, it may be important to clarify what we think such a demonstration actually means about acquisition, what it leaves open, and what its significance might be for more general issues in language acquisition.

A convincing demonstration of a critical or sensitive period for language acquisition would necessarily entail that some maturational factors are crucial to the acquisition process, and that not every hard-working language learner can achieve native proficiency. It does not, however, support any one particular view of the acquisition mechanism (e.g., the one held by Chomsky, 1965, 1981, 1995, often associated with the concept of a critical period in the language acquisition literature). As already noted, there are often several different accounts of the possible mechanisms underlying a critical or sensitive period in any domain (cf. Bateson, 1979, for discussion of various mechanisms underlying the critical period for imprinting; and Newport, 1990, for discussion of the Less-is-More hypothesis for language acquisition), all of which might be compatible with observing a systematic relationship between age and plasticity. In addition, demonstrating that there is a critical or sensitive period for language acquisition does not show that experiential variables, such as length of experience, similarity between first and second language, or motivation, have no effect on acquisition. Careful studies of age effects have usually tried to match for or eliminate such variables, in order to see whether an age effect remains; but maturational effects in acquisition certainly coexist with effects of experiential variables in real learners. Moreover, as we have discussed above, critical period effects in most domains show strong interactions between age or maturational state and experiential variables like stimulus strength or length of isolation.

The main reason to be interested in whether there is a critical or sensitive period for language acquisition, in our opinion, is because it tells us something about the type of learning involved. It is, of course, a truism that all learning involves both nature and nurture, both biological and experiential factors. But within this obvious generalization, there are at least two broadly different types of systems. Some systems are extremely open-ended, with relatively little specified or favored in advance about how the system will be organized. Such systems typically can be formed or re-formed by experience at virtually any time in life, with little or no relationship between their developmental status and their ability to be molded by new experiences. Other systems, in contrast, are more narrowly predisposed and developmentally tuned. For these systems, certain stimuli are favored in learning (i.e., certain stimuli may naturally be more salient to learners or require fewer trials to achieve strong learning), and certain states may be the natural "settings" even without stimulus exposure. Most relevant to the present discussion, such systems are often ones with sensitive periods, during which they are open to fairly extensive modification by the environment, but with strong constraints on the developmental moments during which this plasticity is available.

The available evidence suggests that different parts of language may diverge in this dichotomy. It appears that certain basic aspects of the semantic and lexical parts of language may be plastic in the first sense. In contrast (and here there is much more evidence), the formal parts of language appear to be plastic in the second sense, with both favored stimuli and developmental limits on plasticity. To repeat a point already made above, this does not tell us precisely what type of biological mechanisms underlie the acquisition process, or whether (or where) they are entirely specific to language learning or derived from more general serial order or pattern-learning mechanisms. But it does direct us toward some general classes of mechanisms. Most important, it puts the formal aspects of language learning in the large and diverse category of systems—like imprinting in ducks, auditory localization in the barn owl, and song learning in birds—whose neural and behavioral plasticity is beginning to be understood.

Acknowledgments

We are grateful to Dick Aslin, Kathy Nordeen, Ernie Nordeen, and Emmanuel Dupoux for stimulating discussion and helpful comments. During the writing of this chapter we were supported in part by NIH grant DC00167 to E.L.N., a Charles A. Dana Foundation grant to D.B., and NIH grant DC00481 to H.J.N.

Note

1. Compared with birds reared in isolation from song or ducks reared in isolation from moving objects, even delayed first-language acquisition (the late acquisition of sign language by deaf adults who have no other full language) is relatively more similar to experiences of early life, which typically include exposure to nonlinguistic gesture and often the regular use of family home sign systems.

References

Baptista, L. F., and Petrinovich, L. (1986). Song development in the white-crowned sparrow: Social factors and sex differences. *Animal Behavior, 34,* 1359–1371.

Bateson, P. (1979). How do sensitive periods arise and what are they for? *Animal Behavior, 27,* 47–48.

Bialystok, E., and Hakuta K. (1994). *In other words: The science and psychology of second language acquisition.* New York: Basic Books.

Birdsong, D. (1992). Ultimate attainment in second language acquisition. *Language, 68,* 706–755.

Birdsong, D. (Ed.) (1999). *Second language acquisition and the critical period hypothesis.* Mahwah, NJ: Erlbaum.

Birdsong, D., and Molis, M. (2000). *On the evidence for maturational constraints in second language acquisition.* Unpublished manuscript, University of Texas, Austin.

Brainerd, M. S., and Knudsen, E. I. (1998). Sensitive periods for visual calibration of the auditory space map in the barn owl optic tectum. *Journal of Neuroscience, 18,* 3929–3942.

Chomsky, N. (1965). *Aspects of the theory of syntax.* Cambridge: MIT Press.

Chomsky, N. (1981). *Lectures on government and binding.* Dordrecht: Foris Publications.

Chomsky, N. (1995). *The minimalist program.* Cambridge: MIT Press.

Coppieters, R. (1987). Competence differences between native and near-native speakers. *Language, 63,* 544–573.

Curtiss, S. (1977). *Genie: A psycholinguistic study of a modern-day "wild child."* New York: Academic Press.

Dehaene, S., Dupoux, E., Mehler, J., Cohen, L., Perani, D., van de Moortele, P.-F., Lehérici, S., and Le Bihan, D. (1997). Anatomical variability in the cortical representation of first and second languages. *Neuroreport, 17,* 3809–3815.

Eales, L. A. (1987). Do zebra finch males that have been raised by another species still tend to select a conspecific song tutor? *Animal Behavior, 35,* 1347–1355.

Emmorey, K. (1991). Repetition priming with aspect and agreement morphology in American Sign Language. *Journal of Psycholinguistic Research, 20,* 365–388.

Emmorey, K., and Corina, D. (1990). Lexical recognition in sign language: Effects of phonetic structure and morphology. *Perceptual and Motor Skills 71,* 1227–1252.

Flege, J. E. (1999). Age of learning and second-language speech. In D. Birdsong (Ed.), *Second language acquisition and the critical period hypothesis.* Mahwah, NJ: Erlbaum.

Flege, J. E., Yeni-Komshian, G. H., and Liu, S. (1999). Age constraints on second language acquisition. *Journal of Memory and Language, 41,* 78–104.

Goldin-Meadow, S. (1978). A study in human capacities. *Science, 200,* 649–651.

Harwerth, R., Smith, E., Duncan, G., Crawford, M., and von Noorden, G. (1986). Multiple sensitive periods in the development of the primate visual system. *Science, 232,* 235–238.

Hess, E. H. (1973). *Imprinting: Early experience and the developmental psychobiology of attachment.* New York: Van Nostrand Reinhold.

Hubel, D. H., and Wiesel, T. N. (1970). The period of susceptibility to the physiological effects of unilateral eye closure in kittens. *Journal of Physiology, 206,* 419–436.

Hurford, J. R. (1991). The evolution of the critical period for language acquisition. *Cognition, 40,* 159–201.

Jia, G. X. (1998). *Beyond brain maturation: The critical period hypothesis in second language acquisition revisited.* Unpublished doctoral dissertation, New York University, New York.

Johnson, J. S., and Newport, E. L. (1989). Critical period effects in second language learning: The influence of maturational state on the acquisition of English as a second language. *Cognitive Psychology, 21,* 60–99.

Johnson, J., and Newport, E. L. (1991). Critical period effects on universal properties of language: The status of subjacency in the acquisition of a second language. *Cognition 39,* 215–258.

Kim, K. H. S., Relkin, N. R., Lee, K.-M., and Hirsch, J. (1997). Distinct cortical areas associated with native and second languages. *Nature, 388,* 171–174.

Knudsen, E. I. (1988). Sensitive and critical periods in the development of sound localization. In S. S. Easter, K. F. Barald, and B. M. Carlson (Eds.), *From message to mind: Directions in developmental neurobiology.* Sunderland MA: Sinauer Associates.

Knudsen, E. I. (1998). Capacity for plasticity in the adult owl auditory system expanded by juvenile experience. *Science, 279,* 1531–1533.

Krashen, S. D., Long, M. H., and Scarcella, R. C. (1982). Age, rate, and eventual attainment in second language acquisition. In S. Krashen, R. C. Scarcella, and M. Long (Eds.), *Child-adult differences in second language acquisition* (pp. 161–172). Rowley, MA: Newbury House.

Lenneberg, E. H. (1967). *Biological foundations of language.* New York: Wiley.

Long, M. (1990). Maturational constraints on language development. *Studies in Second Language Acquisition, 12,* 251–285.

Marler, P. (1970). A comparative approach to vocal learning: Song development in white-crowned sparrows. *Journal of Comparative Physiological Psychology, 71,* 1–25.

Mayberry, R., and Eichen, E. (1991). The long-lasting advantage of learning sign language in childhood: Another look at the critical period for language acquisition. *Journal of Memory and Language, 30,* 486–512.

Mayberry, R., and Fischer, S. D. (1989). Looking through phonological shape to lexical meaning: The bottleneck of non-native sign language processing. *Memory and Cognition, 17,* 740–754.

McCandliss, B. D., Fiez, J. A., Conway, M., Protopapas, A., and McClelland, J. L. (1998). Eliciting adult plasticity: Both adaptive and non-adaptive training improved Japanese adults' identification of English /r/ and /l/. *Society for Neuroscience Astracts, 24,* 1898.

Neville, H. J., Coffey, S. A., Lawson, D. S., Fischer, A., Emmorey, K., and Bellugi, U. (1997). Neural systems mediating American Sign Language: Effects of sensory experience and age of acquisition. *Brain and Language, 57,* 285–308.

Neville, H. J., Mills, D., and Lawson, D. (1992). Fractionating language: Different neural subsystems with different sensitive periods. *Cerebral Cortex, 2,* 244–258.

Newport, E. L. (1981). Constraints on structure: Evidence from American Sign Language and language learning. In W. A. Collins (Ed.), *Minnesota symposium on child psychology.* Hillsdale NJ: Erlbaum.

Newport, E. L. (1990). Maturational constraints on language learning. *Cognitive Science, 14,* 11–28.

Newport, E. L., Gleitman, H., and Gleitman, L. R. (1984). Mother, I'd rather do it myself: Some effects and non-effects of maternal speech style. In C. E. Snow and C. A. Ferguson (Eds.), *Talking to children: Language input and acquisition* (pp. 109–356). Cambridge, UK: Cambridge University Press.

Nordeen, K. W., and Nordeen, E. (1997). Anatomical and synaptic substrates for avian song learning. *Journal of Neurobiology, 33,* 532–548.

Oppenheim, R. W., and Haverkamp, L. (1986). Early development of behavior and the nervous system: An embryological perspective. In E. M. Blass (Ed.), *Handbook of behavioral neurobiology.* Vol. 8. New York: Plenum Press.

Oyama, S. (1976). A sensitive period for the acquisition of a nonnative phonological system. *Journal of Psycholinguistic Research, 5,* 261–283.

Pallier, C., Bosch, L., and Sebastián-Gallés, N. (1997). A limit on behavioral plasticity in speech perception. *Cognition, 64,* B9–B17.

Patkowski, M. (1980). The sensitive period for the acquisition of syntax in a second language. *Language Learning, 30,* 449–472.

Perani, D., Dehaene, S., Grassi, F., Cohen, L., Cappa, S. F., Dupoux, E., Fazio, F., and Mehler, J. (1996). Brain processing of native and foreign languages. *Neuroreport, 7,* 2439–2444.

Perani, D., Paulesu, E., Galles, N. S., Dupoux, E., Dehaene, S., Bettinardi, V., Cappa, S. F., Fazio, F., and Mehler, J. (1998). The bilingual brain: Proficiency and age of acquisition of the second language. *Brain, 121,* 1841–1852.

Sanders, L., Yamada, Y., and Neville, H. J. (1999). Speech segmentation by native and non-native speakers: An ERP study. *Society for Neuroscience Abstracts, 25,* 358.

Scovel, T. (1988). *A time to speak.* New York: Newbury House.

Sherman, S. M. (1985). Development of retinal projections to the cat's lateral geniculate nucleus. *Trends in Neuroscience, 86,* 350–355.

Singleton, D. (1989). *Language acquisition: The age factor.* Clevedon, England: Multilingual Matters.

Skoczenski, A. M., and Norcia, A. M. (1999). Development of VEP Vernier acuity and grating acuity in human infants. *Investigative Ophthalmology and Visual Science, 40,* 2411–2417.

Weber-Fox, C., and Neville, H. J. (1996). Maturational constraints on functional specializations for language processing: ERP and behavioral evidence in bilingual speakers. *Journal of Cognitive Neuroscience, 8,* 231–256.

White, L., and Genesee, F. (1996). How native is near-native? The issue of ultimate attainment in adult second language acquisition. *Second Language Research, 12,* 238–265.

Yetkin, O., Zerrin, Y. F., Haughton, V. M., and Cox, R. W. (1996). Use of functional MR to map language in multilingual volunteers. *American Journal of Neuroradiology, 17,* 473–477.

28

Cognition and Neuroscience: Where Were We?

John C. Marshall

Jacques Mehler is almost as renowned as a polyglot traveler than as a cognitive neuroscientist (the self-designation we one-time psychologists are now expected to adopt). Whenever foreign colleagues visit or local colleagues return from far-flung places, they almost invariably bring news and greetings from Jacques. It seems appropriate, then, on the occasion of this festschrift, to range a little more widely in time, space, and culture than the mere thirty years of Jacques's Parisian editorship of *Cognition*.

I hence begin 400 km southeast of Kiev, whence comes the earliest known evidence of intra vitam trepanation. The Mesolithic skulls found there in the Varilyevka II cemetery site are between eight and nine thousand years old. One of the skulls shows complete closure of the left frontal trepanation hole, indicating that this individual, a male who was at least fifty years old at the time of death, had survived for some considerable period after his neurosurgery (Lillie, 1998). Moving to a Neolithic burial site at Ensisheim (in what is now Alsace), we find another man of approximately fifty, this time with two trepanations, one frontal and one parietal. Again, there is definite evidence of significant bone regrowth. These are fully healed trepanations from a very drastic operation carried out over seven thousand years ago (Alt et al., 1997). More recently still, four thousand years ago, the major European centers were in the south of France, and in the Nuraghic culture of Sardinia, where again a reasonable track record of low patient mortality associated with trepanning was obtained.

It would thus appear that some of our moderately distant ancestors had both strong constitutions and good access to skilled neurosurgeons. But sadly, there is no written record of Mesolithic neuropsychologists describing the patients' impaired cognitive status before surgery and following their postoperative progress. Even the great literate civilizations of Mesopotamia, Egypt, and South America (where the Incas were great masters of trepanation) have left little or no trace of how they conceptualized the association between head injury and consequent behavioral disorder. Concerning the intervening variables that early physicians may have postulated to mediate the relationship, we remain in the dark. The Assyrians recorded impairment of consciousness "when a man's brain holds fire, and his eyes are dim" (Thompson, 1908). And for these "diseases of the head" they prescribed a combination of pharmacology and prayer (albeit presumably without the benefit of double-blind clinical trials). But the relevant cuneiform tablets say nothing further about either the disease or the treatment: the physician grinds the medicine and intones the incantation *tout court*. Any resemblance to current psychiatric practice is, of course, entirely coincidental.

Nonetheless, what limited evidence there is clearly suggests that the intimate relations among brain, behavior, and cognition have long been recognized, even if the connections were seen within a predominantly magicoritualistic framework. An Egyptian surgical papyrus from three and a half thousand years ago informs us that "the breath of an outside god or death" has entered the patient's brain and that he henceforth became "silent in sadness" (Breasted, 1930). And one accordingly presumes that the neurosurgeons must have trepanned in order to provide a larger exit door that would encourage the god's departure (but cf. Sullivan, 1996a). The sophisticated technology of drilling is thus associated with a distinctly premodern view of the brain that does not even associate different gods with attacks upon different cognitive functions or brain structures.

And then five or six centuries before the common era, the Ionian Enlightenment and its intellectual descendents create our scientific world (Marshall, 1977). In particular, early Greek theories of cognition and brain function drive the progress of Western neuroscience for the two millennia up to and including the present. The specifics may have been superseded, but (with one or two crucial exceptions) the overall framework remains the one created between the time of Thales (639–544 BCE) and of Heron (first century CE). During this period, the lure of the gods as explanatory agents gave way to humbler mechanical principles that promised a deeper understanding of how the world worked.

In this scenario, European science began at Miletus on the western coast of Asia Minor, "a cross roads of the Near East, where colonists of the Ionian branch of the Greeks had settled and intermarried with the older Asiatic population" (Farrington, 1947). "The central illumination" of the Milesian engineers, Thales, Anaximenes (fl. 544 BCE), and Anaximander (611–547 BCE), "was the notion that the whole universe works in the same way as the little bits of it that are under man's control" (Farrington, 1947). "The vast phenomena of nature," Farrington continues, "so awe-inspiring in their regularity or their capriciousness, in their beneficence or their destructiveness, had been the domain of myth. Now they were seen to be not essentially different from the familiar processes engaged in by the cook, the farmer, the potter, and the smith" (p. 3).

Once our own praxis and technology had been deployed to interpret the "natural" world, an even bolder conjecture could be put forward. Perhaps the human psyche itself was explicable by physical concepts? Thales himself seems to have taken a tentative step along that path with the hypothesis that the ability of the "soul" to move the "body" toward or away from other objects was akin to the action of a magnet: a clever idea although it is not entirely clear in which direction Thales intended his simile to point. As Aristotle (384–322 BCE) remarks (in *De Anima*), one could take Thales to mean either that the "magnetic" soul is the cause of bodily action, or that, by virtue of their capacity to move iron, magnets have souls. Aristotle's jeu d'esprit is oddly reminiscent of somewhat more recent arguments between the proponents of weak vs. strong artificial intelligence.

More important, the Greek physicians took up the Milesian turn. Here is Hippocrates (460?–377 BCE) expressing (in *Regimen*) his total commitment to the central dogma of the Ionian Enlightenment: "Men do not understand how to observe the invisible by means of the visible. Their techniques resemble the physiological processes of man, but men do not know this . . . Though men understand the technical processes, they fail to understand the natural processes imitated by the techniques." It was this emphasis upon "natural processes" that led the physicians to "naturalistic" (and in many cases explicitly physical or mechanistic) explanations of sensation, perception, and (eventually) cognition. Here, for example, is Alcmaeon of Croton on the perception of sound: "Hearing is by means of the ears, he says, because within them is an empty space, and this space resounds. A kind of noise is produced by the cavity, and the internal air re-echoes this sound" (as summarized by Theophrastus, in *De Sensibus*). But the terminus of all the senses (and ultimate interpreter thereof) is the brain. Again Theophrastus is clear that Alcmaeon taught that "All the senses are connected in some way with the brain; consequently they are incapable of action if the brain is disturbed or shifts its position, for this organ stops up the passages through which the senses act" (*De Sensibus*).

And once sensation becomes conceptualized as a function of the sense organs *and* the brain, it is a (relatively) short step for Hippocrates to argue (in *The Sacred Disease*) that the brain is responsible for the ganze megillah of cognitive, affective, and conative processes: "It ought to be generally known that the source of our pleasure, merriment, laughter, and amusement, as of our grief, pain, anxiety, and tears, is none other than the brain. It is specially the organ which enables us to think, see, and hear, and to distinguish the ugly and the beautiful, the bad and the good, pleasant and unpleasant . . . It is the brain too which is the seat of madness and delirium, of the fears and frights which assail us, often by night, but sometimes even by day; it is there where lies the cause of insomnia and sleep-walking, of thoughts that will not come, forgotten duties, and eccentricities" (Gross, 1998, p. 13).

One upshot, then, of the revolution provoked by the Milesians (the first mechanization of the world-picture) was that the superstructure of neuroscientific inquiry was built by novel interactions between engineers, cognitive scientists, and physicians. The more detailed consequences of this interaction included first, that psychological theory came to be dominated by metaphors drawn from the high (and sometimes the low) technology of the day; second, that physicians deployed these mechanical metaphors in order to understand the functions of the brain; and third, that robotic engineers attempted explicitly to model the cognitive and motive powers of the psyche.

A pertinent example of interaction between low(ish) technology and brain anatomy/physiology concerns the study of recognition memory. Both Plato (427?–347? BCE) in the *Theaetetus* and later Aristotle in *De memoria et reminiscentia* conjectured that sensory images impress themselves on the mind in a fashion akin to a stylus inscribing figures on a wax writing tablet or a signet ring leaving an impression on sealing wax. "Whatever is imprinted", Plato argues, "this we remember and know, as long as its image remains; but when it is effaced, or can no longer be imprinted, we forget and do not know it." Is Plato serious? That depends. He does not, of course, literally believe that the soul contains wax: His phrasing shows no sign of straightforward category error: "Suppose, then, I beg, *for the sake of argument* (my italics), that we have in our souls a waxen tablet . . ." But Plato is equally clear that he regards this way of looking at how fleeting sensations can leave a (more or less) permanent record is a useful *model* of the psychological functions of interest. Specifically, the model deploys template matching over the wax impressions to capture how recognition memory might work. When we attempt to remember "things that we have seen or heard, or have ourselves thought of," we put the previously impressed waxen tablet "under our perceptions and thoughts." If the new stimulus matches the prior impression, the stimulus is recognized. Furthermore, false positives can arise when an approximately matching stimulus is taken as a perfect fit.

Plato makes this notion even more homely by suggesting that this error is like not noticing that you have put your left and right sandals on the wrong feet. In such ways, then, the model domesticates a set of psychological phenomena that had previously been resistant to explanation. The nature of the wax tablet model helps to further naturalize a range of individual differences in the accuracy of recognition memory. If the tablet is too small, the templates produced by sensory images will start to overlap with consequent degradation of recognition. The consistency of the wax also has consequences. If the wax is too hard, one will find it difficult to learn. Many trials will be required to make a good impression on the wax, but once the impression is stamped in, it will persist over time with relatively little loss of definition. Per contra, if the wax is too soft, one will learn to recognize objects quickly (one trial will leave a good impression), but over time forgetting will be more rapid as the image in the more pliable wax becomes progressively more indistinct.

The emphasis that Plato places on the dichotomy (or parameter) of soft and hard found echoes in the humoral theory of bodily health as balance and disease as imbalance between the constituent elements of matter. The extension into matters of mind and brain enabled the Greek physicians to argue (in Aristotle's summary) that the short memory of young children was due to the soft, moist nature of their brains, while the inability of the elderly to learn new tricks was due to the rigidity and hardness of their brains (see Marshall and Fryer, 1978). The truth (or otherwise) of these conjectures is not, for present purposes, the point: There would be another two or three millennia to start getting some of the details right. The crucial issue is rather that here we have an explanatory framework that, at very least, suggests that the mind/brain can be studied by the normal processes of empirical science (however little the Greeks themselves may have dirtied their hands with the hard slog of observational and experimental biology). The humoral theory had other long-term consequences: How could one develop ideas about neurotransmitters (norepinephrine, dopamine, serotonin, etc.) and their imbalance in neuropsychiatric disorders unless the ground had been long since prepared by the hippocratic balance of blood, phlegm, yellow bile, and black bile?

A further example, this time of interaction between Greek robotics and neuroscience, illustrates again Farrington's point about the transition from myth to technology in Greek thought. Just as the Hebrews told the story of how the Lord God animated the golem Adam by breathing air into him, so the equivalent Greek narrative told how Prometheus animated with fire the first clay man and woman that he had made. But no naturalistic understanding of such motive forces was possible until the great robotic engineers at AIT (Alexandria Institute of Technology) showed how complex automata could be powered by fire, air, and water. Ktesibios (c. 300 BCE), the son of a barber, developed the piston and cylinder (i.e., a pump) in order to power pneumatically an adjustable mirror in his father's shop. But he went on to devise a wide range of pneumatic and hydraulic automata, including singing birds, and animals that drink and move. In later developments, Philon (c. 220 BCE) constructed yet more elaborate hydraulic automata, including clepsydra (water clocks), while Heron managed to devise entire automated scenes: Hercules shooting a snake, or clubbing a dragon; an automated theatre that depicted the Trojan War (in five acts!).

In this environment, it is hardly surprising that physicians should become enamored of the notion that the brain controlled human actions by similar means. We accordingly find that Herophilus (c. 300 BCE) and Erasistratus (c. 260 BCE) propagate the notion that animal spirits (*pneuma psychikon*) should flow along the sensory nerves to give rise to sensation and along the motor nerves to cause action. Even the (to us somewhat surprising) conjecture that while the veins contain blood, the arteries contain air (*pneuma*) becomes comprehensible in such an intellectual climate (Majno, 1975).

Again, I emphasize, it is not the *accuracy* of Greek neuroscience that concerns me here, but rather the "leading ideas" thereof. It is surely the structure of those ideas that enabled later scholars to approach more closely to the actual mechanisms involved. It is surely the entirely unobvious idea of nervous *transmission* (unobvious, i.e., before the Greeks) that enabled Galvani, Du Bois Reymond, and Helmholtz to explore the action potential as an electrical phenomenon, and later still for Hodgkin and Huxley to propose the sodium-potassium *pump* account of nervous transmission (see Glynn, 1999, for a somewhat different interpretation).

These ideas of transmission and serial processing (i.e., successive mental operations take place in successive physical locales) were then taken up into the brain itself. Once Herophilus and Erasistratus had discovered the cerebral ventricles (a series of large, "water"-filled cavities, placed centrally in the brain), what better loci could there possibly be for the physical instantiation of Aristotle's psychology: the first ventricle creates percepts from sensory impressions; the second ventricle performs further cognitive operations on those percepts, and the final representation is taken away to the faculty of memory in the third ventricle. For us, (visual) sensation gives rise to perceptual representations in striate and extrastriate cortex, whence further cognitive transformations are effected in either the dorsal (parietal) or ventral (temporal) processing stream before final storage in hippocampal regions. The material substrate has changed over two millennia, but not the overall structure of the model.

The anatomy of the ventricular system with its central (midline) locus in the brain did not lend itself to notions of *hemispheric* specialization. Nonetheless, the Greeks were acutely aware that small differences in size or temperature could result in (roughly) bilaterally symmetrical organs having somewhat distinct functions. On such grounds, Anaxagoras (500?–428 BCE) argued that male children resulted from the seed of the right testicle, whereas females developed from the seed of the left testicle (Sullivan, 1996b), Although the hippocratic school failed to transfer this insight into the cerebral hemispheres, other physicians did. Somewhere between Diocles of Carystus (in the fourth century BCE) and Soranus of Ephesus (in the second century BCE) the notion of functional hemispheric asymmetry was put forward: "Accordingly, there are two brains in the head: one which gives understanding, and another one which provides sense perception. That is to say, the one which is lying on the right side is the one which perceives; with the left one, however, we understand" (Lokhorst, 1982). Sadly, the somewhat corrupt medieval manuscript in which the doctrine has come down to us does not discuss the evidence that led to this account of lateral specialization The similarity to the conjectures of John Hughlings Jackson (in the late nineteenth century) is nonetheless striking.

What, then, was *not* prefigured in Greek neuroscience? I can think of only three leading ideas that arose later. The first, and most obvious, is Franz Joseph Gall's radically new account in the late eighteenth (and early nineteenth) century of how mental faculties should be construed. Gall's conjecture that mental organs "are individuated solely in terms of the specific *content domains* with which they respectively deal" (Marshall, 1984) is only faintly prefigured in Thomist psychology. There is then the question of how Gallist mental organs are instantiated physically. As is well-known, Gall himself started a long tradition in neuropsychology whereby psychological functions had a punctate locus in the cerebral cortex (as indeed pre-Gallist faculties of mind had a punctate localization in the ventricles).

The second leading idea I consider novel is thus Charlton Bastian's late-nineteenth century claim that per contra the brain may operate by parallel distributed processing: "The fundamental question of the existence, or not, of real 'localizations' of function (after some fashion) in the brain must be kept altogether apart from another secondary question, which, though usually not so much attended to, is no less real and worthy of our separate attention. It is this: Whether, in the event of 'localization' being a reality, the several mental operations or faculties are dependent

(a) upon separate areas of brain-substance, or (b) whether the 'localization' is one characterized by mere distinctness of cells and fibres which, however, so far as position is concerned, may be interblended with others having different functions. Have we, in fact, to do with *topographically separate areas of brain-tissue* or merely with *distinct cell and fibre mechanisms existing in a more or less diffuse and mutually interblended manner?*" (Bastian, 1880). And finally, the leading idea of the twentieth century was clearly that the mind/brain makes infinite use of finite means. Prefigured by von Humboldt, this notion of recursive function theory had its most profound consequences in the initial formulations of transformational generative grammar (Chomsky, 1955/1975).

I shall leave Jacques (and the other contributors to this book) to tell me where we're going next in neuroscience. I cannot, however, resist the temptation to wonder if the path from Greek neuroscience to the present is really the best approach to the human mind (Marshall, 2000). For what if Aristotle was right after all? The human brain evolved to cool the blood just as insect and bird wings evolved not for flight but to cool the body.

References

Alt, K. W., Jeunesse, C., Buitrago-Téllez, C. H., Wächter, R., Boës, E., and Pichler, S. I. (1997). Evidence for stone age cranial surgery. *Nature, 387,* 360.

Bastian, H. C. (1880). *The brain as an organ of mind.* London: Kegan Paul.

Breasted, J. H. (1930). *The Edwin Smith surgical papyrus.* Chicago: University of Chicago Press.

Chomsky, N. (1955/1975). The logical structure of linguistic theory. New York: Plenum Press.

Farrington, B. (1947). *Head and hand in Ancient Greece.* London: Watts.

Glynn, I. (1999). Two millennia of animal spirits. *Nature, 402,* 353.

Gross, C. C. (1998). *Brain, vision, memory: Tales in the history of neuroscience.* Cambridge, MA: MIT Press.

Lillie, M. C. (1998). Cranial surgery dated back to Mesolithic. *Nature, 391,* 354.

Lokhorst, G.-J. (1982). An ancient Greek theory of hemispheric specialization. *Clio Medica, 17,* 33–38.

Majno, G. (1975). The healing hand: Man and wound in the ancient world. Cambridge, MA: Harvard University Press.

Marshall J. C. (1977). Minds, machines and metaphors. *Social Studies of Science, 7,* 475–488.

Marshall, J. C. (1984). Multiple perspectives on modularity. *Cognition, 17,* 209–242.

Marshall, J. C. (2000). Planum of the apes. *Brain and Language, 71,* 145–148.

Marshall, J. C., and Fryer, D. (1978). Speak, memory. An introduction to some historic studies of remembering and forgetting. In M. M. Gruneberg and P. Morris (Eds.), *Aspects of memory.* London: Methuen.

Sullivan, R. (1996a). The identity and work of the ancient Egyptian surgeon. *Journal of the Royal Society of Medicine, 89,* 467–473.

Sullivan, R. (1996b). Thales to Galen: A brief journey through rational medical philosophy in ancient Greece. Part I. Pre-Hippocratic medicine. *Proceedings of the Royal College of Physicians of Edinburgh, 26,* 135–142.

Thompson, R. C. (1908). Assyrian prescriptions for diseases of the head. *American Journal of Semitic Languages, 24,* 323–353.

Appendix
Short Biography of Jacques Mehler

Born, August 17, 1936, Barcelona, Spain. Divorced. Two children.

Education

Instituto Libre de Segunda Ensenanza, Buenos Aires, Argentina, 1948–51
University of Buenos Aires, Argentina, 1952–58
Oxford University, 1958–59
University College London, 1959–61
Harvard University, 1961–64

Degrees

Quimico, University of Buenos Aires, 1957
Licenciatura en Ciencias Quimicas, University of Buenos Aires, 1958
B. Sc., London University, 1961
Ph. D. in Psychology, Harvard University, 1964

Professional Associations

European Society of Cognitive Psychology
Association de Psychologie Scientifique de Langue Française
Cercle de Psycholinguistique de Paris
Association Européenne de Psycholinguistique
Acting Fellow Rodin Remediation Academy
Member Academia Europaea—Subject Group Chairperson
Psychonomic Society
Latin American Society of Neuropsychology
Vice President of the Club de Neuro Audio Acoustique
Member of the Psychonomic Society, Inc.
Member of Scientific Committee of the Fyssen Foundation

Member of the Advisory Council of the International Association for the Study of Attention and Performance

Awards and Distinctions

French Academy of Sciences Prize (Prix Fanny Emden), 1988
Ipsen Foundation Prize (Neuronal Plasticity), 1995
Docteur Honoris Causa, Université Libre de Bruxelles, 1997

Professional Positions Held

Present

Directeur de Recherche C.N.R.S., since 1980
Director of the Laboratoire de Sciences Cognitives et Psycholinguistique, C.N.R.S., France, 1986–98
Professore Chiara Fama, Scuola Internazionale Superiore di Studi Avanzati, Trieste, since 1999
Directeur d'Etudes l'Ecole des Hautes Etudes en Sciences Sociales, Paris, since 1982
Editor in Chief, *Cognition, International Journal of Cognitive Science*, since 1972
Professor, Doctoral Program in Cognitive Sciences at the University of Paris VI, EHESS and the Ecole Polytechnique, since 1989

Past

Research fellow in Cognitive Studies, Harvard University, 1964–65
Research Associate, Department of Psychology, Massachusetts Institute of Technology, 1965–67
Visiting Fellow, Centre International d'Epistémologie Génétique, University of Geneva, Switzerland, 1966 (from March to September)
Visiting Professor, Université Libre de Bruxelles, Brussels, Belgium, 1973–75 and 1977–78
Visiting Professor, Ph.D. Program in Psychology, The Graduate School and University Center of the City University of New York, 1974 (from March to June)
Professor, Centre Universitaire de Vincennes, Paris VIII, 1969–76
Visiting Researcher, Program in Cognitive Science, University of California, San Diego, 1980 (from June to September)
Head, Module Court en Sciences Cognitives, Magistère de Biologie, Ecole Normale Supèrieure, Paris, 1986, 1987
Professor, D.E.A. de Neurosciences, Université Pierre et Marie Curie, Paris, 1987–89
Research Affiliate, Department of Psychology, Massachusetts Institute of Technology, 1989–90 (from July to June)
Distinguished Lecturer, Inauguration of the Cognitive Science Program at the University of Texas, Dallas, October 1990

Visiting Professor, Rutgers Center for Cognitive Science, 1991, 1992, 1993 (September to December)
Visiting Professor, Universites of Madrid, Barcelona, and Tenerife, April and May 1992
Visiting Professor, Summer School on Language and Cognitive Science, Italy, June 1992
Visiting Professor, Psychology Department, University of Pennsylvania, September to December 1992
Visiting Professor, CUNY Graduate School , September to December 1993
Visiting Professor, Universitat de Barcelona, July and October 1995
Visiting Professor, Universitat de Barcelona, May and June 1996
Visiting Professor, Universitat de Barcelona, June 1997
Visiting Professor, Rutgers Center for Cognitive Science, 1998 (September to December)
Visiting Professor, University of Tenerife, December 1999

Other
Member of the Comité National du C.N.R.S., Section XXVI Psychophysiologie et Psychologie, 1976–81
Cofounder and Member of the Board of the Association Européenne de Psycholinguistique, since 1978
Member of the President's Committee on the Future of Scientific Research in France, 1980
Member of the Editorial Committee of "Construire l'Avenir Livre Blanc sur la Recherche," presented to the President of France, 1980
Member of the Scientific Council of the Max Planck Institut fûr Psycholinguistik from 1982 to 1993
Research Affiliate, MIT Center for Cognitive Science, 1983 to present
Professor, European Summer School in Psycholinguistics, 1985
Member of the Conseil Scientifique de Biologie, Ecole Normale Supérieure, Paris, since 1985
Member of the Comité National du C.N.R.S., Section XXX Psychologie et Physiologie, 1987–92
Acting Fellow, Academia Rodin Remediato, since 1987
Member of the Scientific Advisory Committee on Cognitive Science, SISSA, Trieste, since 1988
Member of the Scientific Advisory Committee on the Neurosciences, Cognition, and Computation, 1988–89
Member of the Advisory Committee, French Ministère de la Recherche et de la Technologie, "Sciences de la Cognition," 1989–93
Member of the Scientific Advisory Committee of the Fyssen Foundation, 1989–97
Principal Researcher, HFSP Consortium "Processing Consequences of Contrasting Language Phonologies," 1990–93
Member of the Scientific Advisory Committee of the Paris chapter of Cognisciences, since 1990

Coordinator of the European Communities Human Capital and Mobility project "Language as a Cognitive Capacity, Perception, and Acquisition," 1992–94

Member of the Human Frontiers Science Program Review Committee for Fellowships and Workshops in Brain Functions, 1991–94

Principal Researcher, HFSP consortium "Processing Consequences of Contrasting Language Phonologies," 1995–98

Member of the Panel on Development of Language and Cognition, Sackler Foundation, 1999–

Member of the Advisory Board on Brain, Mind, and Behavior, McDonnell Pew Foundation for Cognitive Neuroscience, 1999–

Bibliography of Jacques Mehler

Articles

Mehler, J. (1963). Some effects of grammatical transformations on the recall of English sentences. *Journal of Verbal Learning and Verbal Behavior, 2,* 346–351.

Mehler, J., and Miller, G. A. (1964). Retroactive interference in the recall of simple sentences. *British Journal of Psychology, 55,* 295–301.

Mehler, J., Bever, T. G., and Carey, P. (1967). What we look at when we read. *Perception and Psychophysics, 2,* 213–218.

Mehler, J., and Carey, P. (1967). The role of surface and base structure in the perception of sentences, *Journal of Verbal Learning and Verbal Behavior, 6,* 335–338.

Mehler, J., and Bever, T. G. (1967). Cognitive capacity of very young children. *Science, 158,* 141–142.

Mehler, J., and Carey, P. (1968). The interaction of veracity and syntax in the processing of sentences. *Perception and Psychophysics, 3,* 109–111.

Mehler, J. (1968). The genesis of language. *Psychology Today, 1,* 18.

Mehler, J., and Bever, T. G. (1968). The study of competence in cognitive psychology. *International Journal of Psychology, 3,* 273–280.

Bever, T. G., Mehler, J., and Epstein, J. (1968). What children do in spite of what they know. *Science, 162,* 979–981.

Mehler, J., and Bever, T. G. (1968). Quantification, conservation and nativism. *Science, 162,* 979–981.

Mehler, J. (1968). La grammaire générative a-t-elle une réalité psychologique? *Psychologie Française, 13,* 137–156.

Mehler, J. (Ed.) (1969). Psycholinguistique et grammaire générative, *Langages,* 16.

Mehler, J. (1969). Introduction, *Langages, 16,* 3–15.

Mehler, J., and Carey, P. (1969). De la structure profonde et de la structure de surface dans la perception des phrases. *Langages, 16,* 106–110.

Bardies, B. de, and Mehler, J. (1969). Psycholinguistique, messages et codage verbal I: L'Acquisition du langage. *L'Année Psychologique, 69,* 562–598.

Carey, P., Mehler, J., and Bever, T. G. (1970). Judging the veracity of ambiguous sentences. *Journal of Verbal Learning and Verbal Behavior, 9,* 243–254.

Mehler, J., and Muchnik, M. (1971). Puede una pseudofilosofía aclarar el concepto de pseudociencia? *Ciencia Nueva, 12,* 47–49.

Mehler, J., and Bardies, B. de (1971). Psycholinguistique, messages et codage verbal: Deuxième partie: Etudes sur le rappel de phrases. *L'Année Psychologique,* 547–581.

Mehler, J., and Bever, T. G. (1972). Editorial. *Cognition, 1,* 9–10.

Mehler, J., and Bever, T. G. (1973). Editorial. *Cognition, 2,* 7–11.

Mehler, J., and Bever, T. G. (1975). Reason and unreason. *Cognition, 3,* 83–92.

Mehler, J., Barrière, M., and Jassik-Gerschenfeld, D. (1976). Reconnaissance de la voix maternelle par le nourrisson. *La Recherche, 7,* 786–788.

Mehler, J., Bever, T. G., and Franck, S. (1976). Editorial. *Cognition, 4,* 7–12.

Mehler, J. (1976). Psychologie et psychanalyse: Quelques remarques. *Revue Française de Psychanalyse, 4,* 605–622.

Mehler, J., Pittet, M., and Segui, J. (1978). Strategies for sentence perception. *Journal of Psycholinguistic Research, 7,* 3–16.

Mehler, J., Carey, P., and Segui, J. (1978). Tails of words: Monitoring ambiguity. *Journal of Verbal Learning and Verbal Behavior, 17,* 29–35.

Mehler, J. (1978). La perception des stimuli acoustiques verbaux et non verbaux chez les jeunes enfants. *La Recherche, 88,* 324–330.

Mehler, J., Bertoncini, J., Barrière, M., and Jassik-Gerschenfeld, D. (1978). Infant perception of mother's voice. *Perception, 7,* 491–497.

Dommergues, J. Y., Frauenfelder, U., Mehler, J., and Segui, J. (1978/79). L'Intégration perceptive des phrases. *Bulletin de Psychologie, 32,* 893–902.

Mehler, J. (1979). Quand la science du comportement devient psychologie de la connaissance. *La Recherche, 10,* 100.

Frauenfelder, U., Segui, J., and Mehler, J. (1980). Monitoring around the relative clause. *Journal of Verbal Learning and Verbal Behavior, 19,* 328–337.

Mehler, J., and Bertoncini, J. (1981). Quelques remarques à propos de la perception chez le nouveau-né. *Revue du Praticien, 31,* 387–396.

Mehler, J., and Bertoncini, J. (1981). Syllables as units in infant perception, *Infant Behavior and Development, 4,* 271–284.

Mehler, J., Dommergues, J. Y., Frauenfelder, U., and Segui, J. (1981). The syllable's role in speech segmentation. *Journal of Verbal Learning and Verbal Behavior, 20,* 298–305.

Mehler, J. (1981). The role of syllables in speech processing: Infant and adult data. *Philosophical Transactions of the Royal Society of London. Series B: Biological Sciences, 295,* 333–352.

Mehler, J., and Franck, S. (1981). Editorial, *Tenth anniversary volume. Cognition, 10,* 1–5.

Segui, J., Frauenfelder, U., and Mehler, J. (1981). Phoneme monitoring, syllable monitoring and lexical access. *British Journal of Psychology, 72,* 471–477.

Segui, J., Mehler, J., Frauenfelder, U., and Morton, J. (1982). The word frequency effect on lexical access. *Neuropsychologia, 20,* 615–627.

Frauenfelder, U., Dommergues, J. Y., Mehler, J., and Segui, J. (1982). L'Intégration perceptive des phrases: Aspects syntaxiques et sémantiques. *Bulletin de Psychologie, 356,* 893–902.

Cutler, A., Mehler, J., Norris, D., and Segui, J. (1983). A language-specific comprehension strategy. *Nature, 304,* 159–160.

Mehler, J., Morton, J., and Jusczyk, P. (1984). On reducing language to biology. *Cognitive Neuropsychology, 1,* 83–116.

Mehler, J., and Bertoncini, J. (1984). La recherche sur L'etat initial: Quelques réflexions. *Neuropsychiatrie de L'Enfance et de L'Adolescence, 32,* 497–510.

Lecours, A. R., Mehler, J., and Parente, M. A. (1985). Au pied de la lettre. *Union Médicale du Canada, 114,* 1020–1026.

Cutler, A., Mehler, J., Norris, D., and Segui, J. (1986). The syllable's differing role in the segmentation of French and English. *Journal of Memory and Language, 25,* 385–400.

Mehler, J. (1986). Review of Philip Lieberman's *The Biology and Evolution of Language. Journal of the Acoustical Society of America, 80,* 1558–1560.

Segui, J., and Mehler, J. (1986). Psicologia cognitiva y percepcion del lenguaje: Contribucion al estudio experimental del habla. *Revista Argentina de Linguistica, 2,* 317–342.

Segui, U., Frauenfelder, U., Laine, C., and Mehler, J. (1987). The word frequency effect for open-and-closed class items. *Cognitive Neuropsychology, 4,* 33–44.

Mehler, J., Lambertz, G., Jusczyk, P. W., and Amiel-Tison, C. (1987). Discrimination de la langue maternelle par le nouveau-né. *Comptes Rendus de L'Académie des Sciences. Serie III, Science de la Vie, 303,* 637–640.

Lecours, A. R., Mehler, J., and Parente, M. A. (1987). Illiteracy and brain damage—1. Aphasia testing in culturally contrasted populations. *Neuropsychologia, 25,* 231–245.

Cutler, A., Mehler, J., Norris, D., and Segui, J. (1987). Phoneme identification and the lexicon. *Cognitive Psychology, 19,* 141–177.

Bertoncini, J., Babic-Bijeljac, R., Blumstein, S., and Mehler, J. (1987). Discrimination by neonates of very short CVs. *Journal of the Acoustical Society of America, 82,* 31–37.

Lecours, A. R., Mehler, J., and Parente, M. A. (1987). Illiteracy and brain damage—2. Manifestations of unilateral neglect in testing "auditory comprehension" with iconographic materials. *Brain and Cognition, 6,* 243–266.

Mehler, J., and Dupoux, E. (1987). De la psychologie à la science cognitive. *Le Débat, 47,* 65–87.

Mehler, J., and Bertoncini, J. (1988). Development—A question of properties, not change? *Revue Internationale de Sciences Sociales, UNESCO* [also in French, Spanish, and Arabic], *115,* 121–135.

Lecours, A. R., Mehler, J., and Parente, M. A. (1988). Illiteracy and brain damage—3. A contribution to the study of speech and language disorders in illiterates with unilateral brain damage (initial testing). *Neuropsychologia, 26,* 575–589.

Bertoncini, J., Bijeljac-Babic, R., Jusczyk, P. W., Kennedy, L., & Mehler, J. (1988). An investigation of young infants' perceptual representations of speech sounds. *Journal of Experimental Psychology: General, 117,* 21–33.

Nespoulous, J.-L., Dordain, M., Perron, C., Ska, B., Bub, D., Caplan, D., Mehler, J., and Roch Lecours, A. (1988). Agrammatism in sentence production without comprehension deficits: Reduced availability of syntactic structures and/or of grammatical morphemes? A case study. *Brain and Language, 33,* 273–295.

Mehler, J., Jusczyk, P., Lambertz, G., Halsted, N., Bertoncini, J., and Amiel-Tison, C. (1988). A precursor of language acquisition in young infants. *Cognition, 29,* 143–178.

Pinker, S., and Mehler, J. (Eds.) (1988). Connectionism and symbol systems. *Cognition, 28,* 1–2.

Bertoncini, J., Morais, J., Bijeljac-Babic, R., McAdams, S., Peretz, I., and Mehler, J. (1989). Dichotic perception and laterality in neonates. *Brain and Language, 37,* 591–605.

Cutler, A., Mehler, J., Norris, D., and Segui, J. (1989). Limits on bilingualism. *Nature, 340,* 229–230.

Morais, J., Content, A., Cary, L., Mehler, J., and Segui, J. (1989). Syllabic segmentation and literacy. *Language and Cognitive Processes, 4,* 57–67.

Mehler, J., and Cutler, A. (1990). Psycholinguistic implications of phonological diversity among languages. *Golem, 1,* 119–134.

Dehaene, S., Dupoux, E., & Mehler, J. (1990). Is numerical comparison digital? Analogical and symbolic effects in two-digit number comparison. *Journal of Experimental Psychology: Human Perception and Performance, 16,* 626–641.

Jusczyk, P. W., Bertoncini, J., Bijeljac-Babic, R., Kennedy, L. J., and Mehler, J. (1990). The role of attention in speech perception by young infants. *Cognitive Development, 5,* 265–286.

Dupoux, E., and Mehler, J. (1990). Monitoring the lexicon with normal and compressed speech: Frequency effects and the prelexical code. *Journal of Memory and Language, 29,* 316–335.

Christophe, A., Pallier, C., Bertoncini, J., and Mehler, J. (1991). A la recherche d'une unité: Segmentation et traitement de la parole. *L'Année Psychologique, 91,* 59–86.

Segui, J., Sebastián-Gallés, N., and Mehler, J. (1991) Estructura fonologica y percepcion del habla. *Revista Argentina de Linguistica, 7,* 89–107.

Sebastian, N., Dupoux, E., Segui, J., and Mehler, J. (1992). Contrasting syllabic effects in Catalan and Spanish: The role of stress. *Journal of Memory and Language, 31,* 18–32.

Dehaene, S., and Mehler, J. (1992). Cross-linguistic regularities in the frequency of number words. *Cognition, 43,* 1–29.

Cutler, A., Mehler, J., Norris, D., and Segui, J. (1992). The monolingual nature of speech segmentation by bilinguals. *Cognitive Psychology, 24,* 381–410.

Mehler, J., Sebastian, N., Altmann, G., Dupoux, E., Christophe, A., and Pallier, C. (1993). Understanding compressed sentences: The role of rhythm and meaning. In P. Tallal, A. M. Galaburda, R. Llinas, and C. Von Euler (Eds.), *Temporal Information Processing in the Nervous System: Special Reference to Dyslexia and Dysphasia. Annals of the New York Academy of Sciences, 682,* 272–282.

Cutler, A., and Mehler, J. (1993). The periodicity bias. *Journal of Phonetics, 21,* 103–108.

Pallier, C., Sebastián y Gallés, N., Felguera, T., Christophe, A., and Mehler, J. (1993). Attentional allocation within the syllabic structure of spoken words. *Journal of Memory and Language, 32,* 373–389.

Bijeljac-Babic, R., Bertoncini, J., and Mehler, J. (1993). How do four-day-old infants categorize multisyllabic utterances? *Developmental Psychology, 29,* 711–723.

Otake, T., Hatano, G., Cutler, A., and Mehler, J. (1993). Mora or syllable? Speech segmentation in Japanese. *Journal of Memory and Language, 32,* 258–278.

Mazoyer, B. M., Dehaene, S., Tzourio, N., Frak, V., Murayama, N., Cohen, L., Levrier, O., Salamon, G., Syrota, A., and Mehler, J. (1993). The cortical representation of speech. *Journal of Cognitive Neuroscience, 32,* 373–389.

Mehler, J., Dupoux, E., Pallier, C., and Dehaene-Lambertz, G. (1994). Cross-linguistic approaches to speech processing? *Current Opinions in Neurobiology, 4,* 171–176.

Christophe, A., Dupoux, E., Bertoncini, J., and Mehler, J. (1994). Do infants perceive word boundaries? An empirical approach to the bootstrapping problem for lexical acquisition. *Journal of the Acoustical Society of America, 95,* 1570–1580.

Mehler, J. (1994). Editorial. *Cognition, 50,* 1–6.

Mehler, J. (1994). A few words about cognition. *Cognitive Studies, 1,* 8–9.

Mehler, J., and Christophe, A. (1994). Language in the Infant's Mind. *Philosophical Transactions of the Royal Society of London. Series B: Biological Sciences, 346,* 13–20.

Mehler, J., Bertoncini, J., Dupoux, E., and Pallier, C. (1994). The role of suprasegmentals in speech perception and acquisition. *Dokkyo International Review, 7,* 338–343.

Bertoncini, J., Floccia, C., Nazzi, T., and Mehler, J. (1995). Morae and syllables: Rythmical basis of speech representations in neonates. *Language and Speech, 38,* 311–329.

Cohen, L., and Mehler, J. (1996). Click monitoring revisted: An on-line study of sentence comprehension. *Memory and Cognition, 24,* 94–102.

Deheane, S., Tzourio, N., Frak, V., Raynaud, L., Cohen, L., Mehler, J., and Mazoyer, B. (1996). Cerebral activations during number multiplication and comparison: A PET study. *Neuropsychologia, 34,* 1097–1106.

Mehler, J., Pallier, C., and Christophe, A. (1996). Psychologie cognitive et acquisition des langues. *Médecine Sciences, 12,* 94–99.

Perani, D., Dehaene, S., Grassi, F., Cohen, L., Cappa, S. F., Dupoux, E., Fazio, F., and Mehler, J. (1996). Brain processing of native and foreign languages. *NeuroReport, 7,* 2439–2444.

Fazio, F., Perani, D., Dehaene, S., Grassi, F., Cohen, L., Cappa, S. F., Dupoux, E., and Mehler (1997). Brain processing of native and foreign languages. *Human Brain Mapping Supplement,* 583.

Dupoux, E., Pallier, C., Sebastian, N., and Mehler, J. (1997). A destressing "deafness" in French? *Journal of Memory and Language, 36,* 406–421.

Pallier, C., Christophe, A., and Mehler, J. (1997). Language-specific listening. *Trends in Cognitive Science, 1,* 129–132.

Dehaene, S., Dupoux, E., Mehler, J., Cohen, L., Perani, D., van de Moortele, P. F., Leherici, S., and Le Bihan, D. (1997). Anatomical variability in the representation of first and second languages. *Neuroreport, 17,* 3809–3815.

Koechlin, E., Dehaene, S., and Mehler, J. (1997). Numerical transformations in five month old human infants. *Mathematical Cognition, 2,* 89–104.

van Ooyen, B., Bertoncini, J., Sansavini, A., and Mehler, J. (1997). Do weak syllables count for newborns? *Journal of the Acoustical Society of America, 102,* 3735–3741.

Nazzi, T., Bertoncini, J., and Mehler, J. (1998). Language discrimination by newborns. Towards an understanding of the role of rhythm. *Journal of Experimental Psychology: Human Perception and Performance, 24,* 1–11.

Paulesu, E., and Mehler, J. M. (1998). Right on in sign language. *Nature—News and Views, 392,* 233–234.

Pallier, C., Sebastian, N., Dupoux, E., Christophe, A., and Mehler, J. (1998). Perceptual adjustment to time-compressed speech: A cross-linguistic study. *Memory and Cognition, 26,* 844–851.

Bachoud-Lévi, A-C., Cohen L., Dupoux, E., and Mehler. (1998). Where is the length effect? A cross linguistic study. *Journal of Memory and Language, 39,* 331–346.

Perani, D., Paulesu, E., Sebastián-Gallés, N., Dupoux, E., Dehaene, S., Bettinardi, V., Cappa, S. F., Fazio, F., and Mehler, J. (1998). The bilingual brain: Proficiency and age of acquisition of the second language. *Brain, 121,* 1841–1852.

Ramus, F., and Mehler (1999). Language identification with suprasegmental cues : A study based on speech resynthesis. *Journal of the Acoustical Society of America, 105,* 512–521.

Dupoux, E., Kakehi, K., Hirose, Y., Pallier, C., and Mehler, J. (1999). Epenthetic vowels in Japanese: A perceptual illusion? *Journal of Experimental Psychology: Human Perception and Performance, 25,* 1568–1578.

Ramus, F., Nespor, M., and Mehler, J. (1999). Correlates of linguistic rhythm in the speech signal. *Cognition, 73,* 265–292.

Peperkamp, S., and Mehler, J. (1999). Spoken and signed language: A unique underlying system? *Language and Speech, 42,* 333–346.

Ramus, F., Hauser, M. D., Miller, C., Morris, D., and Mehler, J. (2000). Language discrimination by human newborns and by cotton-top tamarin monkeys. *Science, 288,* 340–351.

LeClec', G., Dehaene, S., Cohen, L., Mehler, J., Dupoux, E., Poline, J. B., Léhericy, S., van de Moortele, P. F., and LeBihan, D. (2000). Distinct cortical areas for names of numbers and body parts independent of language and input modality. *NeuroImage, 12,* 381–391.

Christophe, A., Mehler, J., and Sebastián-Gallés, N. (2001). Perception of prosodic boundary correlates by newborn infants. *Infancy, 2,* 385–394.

Books and Book Chapters

Mehler, J. (1966). Some effects of grammatical transformations on the recall of English sentences. In D. H. Kausler (Ed.), *Reading in verbal learning, contemporary theory and research.* New York: Wiley.

Mehler, J. (1968). The structure of writing and reading strategies. In *Proceedings of the Thirteenth Annual Convention of the International Reading Association,* Boston.

Carey, P., Mehler, J., and Bever, T. G. (1970). When do we compute all the interpretations of an ambiguous sentence? In G. Flores d'Arcais and W. Levelt (Eds.), *Advances in psycholinguistics* (pp. 61–75). Amsterdam: North Holland.

Mehler, J. (1971). Sentence completion tasks. In R. Huxley and E. Ingram (Eds.), *Studies in language acquisition: Methods and models* (pp. 137–146). New York: Academic Press.

Mehler, J. (1971). Studies in language and thought development. In R. Huxley and E. Ingram (Eds.), *Studies in language acquisition: Methods and models* (pp. 201–229). New York: Academic Press.

Mehler, J. (1971). Le développement des heuristiques perceptives chez le très jeune enfant. In H. Hécaen (Ed.), *Neuropsychologie de la perception visuelle* (pp. 154–167). Paris: Masson.

Mehler, J. (1973). Some remarks on psycholinguistic studies. In M. Gross, M. Halle, and M. Schutzenberger (Eds.), *Formal analysis of natural languages* (pp. 296–326). The Hague: Mouton.

Mehler, J. (1973). Language and perception : A few observations. In *Actes des 4èmes journées d'etudes de la communication parlée*. Brussels: Galf.

Mehler, J., and Noizet, G. (1974). Introduction. In J. Mehler and G. Noizet (Eds.), *Pour une psychologie linguistique* (pp. 7–22). The Hague: Mouton.

Mehler, J., and Noizet, G. (Eds.) (1974). *Pour une psychologie linguistique*. The Hague: Mouton.

Mehler, J., and Bever, T. G. (1974). La mémorisation des phrases peut se faire selon leur structure syntaxique de base. In J. Mehler and G. Noizet (Eds.), *Pour une psychologie linguistique* (pp. 421–450). The Hague: Mouton.

Mehler, J. (1974). Connaître par désapprentissage. In M. Piattelli-Palmarini and E. Morin (Eds.), *L'Unité de L'homme* (pp. 287–299). Paris: Le Seuil.

Mehler, J. (1974). A propos du développement cognitif. In M. Piattelli-Palmarini and E. Morin (Eds.), *L'Unité de L'homme* (pp. 300–301). Paris: Le Seuil.

Mehler, J., Barriere, M., Ruwet, N., and Segui, J. (1974). Comparing comparatives. In *Problèmes actuels en psycholinguistique* (pp. 431–447). Paris, C.N.R.S. (Actes des Colloques Internationaux).

Mehler, J. (1978). Infants' perception of speech and other acoustic stimuli. In J. Morton and J. Marshall (Eds.), *Psycholinguistics series II* (pp. 69–105). London: Elek Scientific Books.

Mehler, J. (1979). Psychologie et Psycholinguistique. In M. Piattelli-Palmarini, (Ed.), *Théories du langage, théories de l'apprentissage* (pp. 483–496). Paris: Le Seuil.

Mehler, J., and Bertoncini, J. (1980). Predisposizioni linguistiche nel lattante. In G. Braga, V. Braitenberg, C. Cipolli, E. Coseriu, S. Crespi-Reghizzi, J. Mehler, and R. Titone (Eds.), *L'accostamento interdisciplinare allo studio del linguaggio* (pp. 139–150). Milan: Fraco Angelli.

Eimas, P., and Mehler, J. (1980). The structuring of language by developmental processes. In U. Bellugi and M. Studdert-Kennedy, (Eds.), *Sign language and spoken language: Biological constraints on linguistic form*. Berlin: F. L. Verlag Chimie.

Mehler, J. (1980). Introduction et conclusion. In M. Piattelli-Palmarini (Ed.), *On language and learning: A debate between Chomsky and Piaget*. Cambridge, MA: Harvard University Press.

Mehler, J., Segui, J., and Frauenfelder, U. (1981). The role of the syllable in language acquisition and perception. In T. Myers, J. Laver, and J. Anderson (Eds.), *The cognitive representation of speech* (pp. 295–305). Amsterdam: North Holland.

Mehler, J. (1981). Language dispositions in the infant: Studies in cerebral asymmetries. In *Toronto Semiotic Circle Prepublication Series*. Vols. 1 and 2) (pp. 25–48).

Mehler, J. (1982). Studies in the development of cognitive processes. In S. Strauss (Ed.), *U-shaped behavioral growth* (pp. 271–293). New York: Academic Press.

Mehler, J., Garrett, M., and Walker, E. (Eds.) (1982). *Perspectives on mental representation.* Hillsdale, NJ: Erlbaum.

Mehler, J., Garrett, M., and Walker, E. (1982). Introduction. In J. Mehler, M. Garrett, and E. Walker (Eds.), *Perspectives on mental representation,* Hillsdale, NJ: Erlbaum.

Mehler, J. (1982). Unlearning: Dips and drops—A theory of cognitive development. In T. G. Bever (Ed.), *Regressions in development: Basic phenomena and theoretical alternatives* (pp. 133–152). Hillsdale, NJ: Erlbaum.

Segui, J., Dommergues, J. Y., Frauenfelder, U., and Mehler, J. (1982). The perceptual integration of sentences: Syntactic and semantic aspects. In J. F. Le Ny and W. Kintsch (Eds.), *Language and comprehension* (pp. 63–72). Amsterdam: North Holland.

Mehler, J. (1983). La connaissance avant l'apprentissage. In S. de Schonen (Ed.), *Le développement dans la première année* (pp. 129–155). Paris: Presses Universitaires de France.

Mehler, J., & Fox, R. (Eds.) (1984). *Neonate cognition: Beyond the buzzing, blooming confusion.* Hillsdale, NJ: Erlbaum.

Mehler, J. (1984). Introduction: Some reflections on initial state research. In J. Mehler and R. Fox (Eds.), *Neonate cognition: Beyond the buzzing, blooming confusion* (pp. 1–5). Hillsdale, NJ: Erlbaum.

Mehler, J. (1984). Language related dispositions in early infancy. In J. Mehler and R. Fox (Eds.), *Neonate cognition: Beyond the buzzing, blooming confusion* (pp. 7–26). Hillsdale, NJ: Erlbaum.

Mehler, J. (1984). Observations psycholinguistiques: Acquisition et perception du langage. In *Actes du Congrès International D'Orthophonie, Octobre 1983* (pp. 115–138). Bordeaux: l'Onadrio.

Lecours, A. R., Mehler, J., Parente, M. A., and Vadeboncoeur, A. (1984). Alphabétisation et cerveau. In *Pour Comprendre 1984/Understanding 1984* (pp. 176–189). Paris: Commission Canadienne de L'Unesco.

Mehler, J., and Segui, J. (1987). English and French speech processing: Some psycholinguistic investigations. In M. E. H. Schouten (Ed.), *The psychophysics of speech perception* (pp. 405–418). Dordrecht, Netherlands: Martinus Nijhoff.

Mehler, J., Juszczyk, P. W., Dehaene-Lambertz, G., Bertoncini, J., and Amiel-Tison, C. (1988). Un premier stade de l'acquisition du langage chez le nouveau-né de quatre jours. In *Les Cahiers du CTNRHI.* Vol. 41 (pp. 25–29).

Mehler, J. (1988). Language use and linguistics diversity. In W. Hirst (Ed.), *The making of cognitive science* (pp. 153–166). Cambridge, UK: Cambridge University Press.

Mehler, J., and Pinker, S. (Eds.) (1988). *Connectionism and symbol systems.* Cambridge, MA: MIT Press.

Mehler, J. (1990). Language at the initial state. In A. Galaburda (Ed.), *From reading to neurons* (pp. 189–216). Cambridge, MA: MIT Press.

Mehler, J., and Dupoux, E. (1990). *Naître/humain*. Paris: Odile Jacob. [Translated into Chinese, Japanese, Greek, Italian, Portuguese, and Spanish.]

Mehler, J., Dupoux, E., and Segui, J. (1990). Constraining models of lexical access : The onset of word recognition. In G. Altmann (Ed.), *Cognitive models of speech processing—Psycholinguistic and computational perspectives* (pp. 236–262). Cambridge, MA: MIT Press.

Segui, J., Dupoux, E., and Mehler, J. (1990). The syllable: Its role in speech segmentation and lexical access. In G. Altmann (Ed.), *Cognitive models of speech processing—Psycholinguistic and computational perspectives* (pp. 263–280). Cambridge, MA: MIT Press.

Mehler, J., and Christophe, A. (1992). Speech processing and segmentation in romance languages. In Y. Tohkura, Y. Saigisaka, and E. Vatikiotis-Bateson (Eds.), *Speech perception, production and linguistic structure* (pp. 221–238). Tokyo: Ohm.

Dupoux, E., and Mehler, J. (1992). Unifying awareness and on-line structure of speech: A tentative framework. In J. Alegria, D. Holender, J. Junca de Morais, and M. Radeau (Eds.), *Analytic approaches to human cognition* (pp. 59–75). Amsterdam: Elsevier.

Christophe, A., Dupoux, E., and Mehler, J. (1993). How do infants extract words from the speech stream? A discussion of the bootstrapping problem for lexical acquisition. In *Proceedings of the Child Language Research Forum* (pp. 209–224). Stanford, CA: Cognitive Science Linguistic Institute.

Mehler, J., and Dupoux, E. (1994). *What infants know*. Cambridge, MA: Blackwell.

Pallier, C., and Mehler, J. (1994). Language acquisition: Psychobiological data. In *Fourth Refresher Course of the ESNR* (pp. 23–26). Udine, Italy: Edizioni Del Centauro.

Nazzi, T., and Mehler, J. (1994). Premiers pas vers l'acquisition d'une langue. In M. Azoulay and R. Mezin (Eds.), *Journées de Techniques Avancées en Gynécologie Obstetrique et Périnatologie* (pp. 485–497). Mayenne, France: Imprimerie de la Manutention.

Mehler, J. (1994). Panoramica sulle scienze cognitive. In P. Budinich, S. Fantoni, and S. Prattico (Eds.), *Panoramica Sulle Scienze Cognitive* (pp. 5–19). Naples: Cuen.

Mehler, J., Bertoncini, J., and Christophe, A. (1994). Environmental effects on the biological determinants of language. In C. Amiel-Tison and A. Stewart (Eds.), *The newborn infant: One brain for life* (pp. 81–92). Paris: Editions de l'INSERM.

Mehler, J., Bertoncini, J., and Christophe, A. (1994). Environnement et déterminants biologiques du langage. In C. Amiel-Tison and A. Stewart (Eds.), *L'Enfant nouveau-né: Un cerveau pour la vie* (pp. 95–105). Paris: Editions de l'INSERM.

Mehler, J., and Christophe, A. (1995). Maturation and learning of language in the first year of life. In M. S. Gazzaniga (Ed.), *The cognitive neurosciences* (pp. 943–953). Cambridge, MA: MIT Press.

Mehler, J., and Franck. S. (Eds.) (1995). *Cognition on cognition.* Cambridge, MA: MIT Press.

Mehler, J., and Franck, S. (1995). Building COGNITION. In J. Mehler and S. Franck (Eds.), *Cognition on cognition* (pp. 7–11). Cambridge, MA: MIT Press.

Mehler, J., Bertoncini, J., Dupoux, E., and Pallier, C. (1996). The role of suprasegmentals in speech perception and acquisition. In T. Otake and A. Cutler (Eds.), *Phonological structure and language processing—Cross-linguistic studies, Speech Research 12*, Berlin: Mouton De Gruyter.

Mehler, J., Dehaene-Lambertz, G., Dupoux, E., and Nazzi, T. (1996). Coping with linguistic diversity: The infant's viewpoint. In J. Morgan and K. Demuth (Eds.), *Signal to syntax* (pp. 101–116). Hillsdale, NJ: Erlbaum.

Mehler, J. (1996). The renaissance of psycholinguistics. In W. Levelt (Ed.), *Bressanone revisited* (pp. 1–6). Nijmegen, Netherlands: Max Planck Publications.

Mehler, J., and Ramus, F. (1997). La psychologie cognitive peut-elle contribuer à l'étude du raisonnement moral? In J. P. Changeux (Ed.), *Universalisme éthique, diversité culturelle et éducation* (pp. 119–136). Paris: Odile Jacob.

Mehler, J., Pallier, C., and Christophe, A. (1998). Language and cognition. In M. Sabourin, F. M. I. Craik, and M. Robert (Eds.), *Cognitive and Biological Aspects.* Vol. 1. *Proceedings of the Twenty-sixth International Congress of Psychology* (pp. 381–398). Hove, UK: Psychology Press.

Mehler, J., Pallier, C., and Christophe, A. (1998). Langage et cognition. In Y. Christen, L. Collet, and M-T Droy-Lefaix (Eds.), *Rencontres IPSEN en ORL.* Vol. 2. (pp. 1–16). Paris: IPSEN.

Dupoux, E., and Mehler, J. (1999). Non-developmental studies of development: Examples from newborn research, bilingualism and brain imaging. In C. Rovee-Collier, L. P. Lipsitt, and Harlene Hayne (Eds.), *Advances in infancy research.* Vol. 12. (pp. 375–406). Stamford, CT: Ablex.

Mehler, J., and Christophe, A. (2000). Acquisition of language: Infant and adult data. In M. S. Gazzaniga (Ed.), *The cognitive neurosciences,* 2d ed. (pp. 897–908). Cambridge, MA: MIT Press.

Ramus, F., and Mehler, J. (in press). In praise of funtional psychology: Dismissing the skeptics. In S. M. Kosslyn, A. M. Galaburda, and Y. Christen (Eds.), *Languages of the brain.* Cambridge, MA: Harvard University Press.

Mehler, J., Christophe, A., and Ramus, F. (2000). What we know about the initial state for language. In A. Marantz, Y. Miyashite, and W. O'Neil (Eds.), *Image, language, brain: Papers from the first mind-articulation project symposium* (pp. 51–75). Cambridge, MA: MIT Press.

Dehaene-Lambertz, G., Mehler, J., and Pena, M. (in press). Cerebral bases of language acquisition. In A. F. Kalverboer and A. Gramsbergen (Eds.), *Handbook of brain and behaviour in human development.* Dordrecht, Netherlands: Kluwer.

Other

How some sentences are remembered. Unpublished PhD thesis, Harvard University, Center for Cognitive Studies, Cambridge, MA.

Mehler, J., and Bever, T. G. (1966). La structure de la psycholinguistique. In Compte rendus des conférences données au Symposium International d'Epistémologie Génétique.

Mehler, J. (1966). Psicologia gramatica Y lenguaje. In *Proceedings of the First Inter-American Congress of Language and Hypoacusia,* Buenos Aires, 209–226.

Mehler, J., and Bever, T. G. (1967). Logic of innocence [interview]. *Technical Review, 70.*

Mehler, J. (1973). Scienza dei valori o valore della scienza, Rome, Consiglio Nazionale Delle Richerche.

Mehler, J. (1981). Faire de la psychologie une science exacte [interview]. *Le Monde,* 15 March 1981.

Mehler, J. (1984). Les processus du langage. *Courrier Du C.N.R.S.,* April–June 1984, 55–56.

Mehler, J. (1984). Des questions troublantes. In Les aventures de la raison dans la pensée et la science contemporaines. *Le Monde,* August 1984, 26–27.

Mehler, J. (1987). Cares d'una psicolingüistica [Interview with Llisterri, J., and Serra, M.]. *Limits, 2,* 61–78.

Mehler, J. (1989). Les berceaux de babel [interview]. *Le Monde,* 13 September 1989.

Mehler, J. (1990). Sciences cognitives: Une discipline en quête d'intelligence [interview]. *La Recherche, 225,* October 1990.

Dupoux, E., and Mehler, J. (1992). La segmentation de la parole. *Courrier du CNRS, 79,* 10.

Christophe, A., and Mehler, J. (1996). Comment le langage, le calcul et le raisonnement viennent à l'homme. *Rendez-Vous Scientifiques 94/95—Nouvelles des Laboratoires des Sciences de la Vie,* Paris: CNRS, 101–103.

Afterword

Most of the chapters in this book were presented and discussed at the International Conference on Language, Brain and Cognitive Development, which took place at the Collège de France in Paris from May 3 to May 5, 2001. This conference was held in honor of Jacques Mehler, who was also its Honorary President.

The scientific advisory commitee of the conference was made up of Alain Berthoz, Jean-Pierre Changeux, Marc Jeannerod, Juan Segui, Dan Sperber, Albert Galaburda, Lila Gleitman, Steve Pinker, and Elizabeth Spelke. We thank all of them for their time, help, and encouragement.

This conference could not have taken place without the generous support of the Fondation de France, the Action Cognitique of the French Ministry of Research, INSERM, the Réseau de Sciences Cognitives d'Ile-de-France, the Ecole des Hautes Etudes en Sciences Sociales, NIH, MIT Press, Editions Odile Jacob and the Hugot, IPSEN, and EDF foundations. All these institutions very kindly contributed time, money, or both. We would also like to thank our industrial partners: Candel & Partners, IdVector, Serono, and EISAI/PFIZER.

Finally, we would like to point out that the conference benefitted from the invaluable and faultless technical support of Michel Dutat and Vireak Ul. Thanks to them all, there were no hitches in any of the presentations.

For those of us who were lucky enough to have attended the conference, it will remain in our minds as an event rich in intellectual excitement, reflection, and emotion. We would therefore (and once again!) like to express our thanks to Jacques for having given us the opportunity to take part in such a rare event.

The Conference Organizing Committee
Laurent Cohen
Stanislas Dehaene
Emmanuel Dupoux
Susana Franck

Note: Archives of the conference including program, participants, and some of the presentations can be found at the following web site: http://www.lscp.net/conference2001/.

Index

Abstraction, 470
Accumulator model, 285–289, 310, 311
Action meanings, cognitive model of grasping, 274, 275
Agrammatism, 168–169
Alarm call, 425
Amusics, congenital, 437–438
Analog magnitude representations of numbers, 308–312
Analysis-by-synthesis model, 151–154
Aristotle, 505–509
Arithmetic reasoning principles, 283–284
Articulatory mechanism and "articulatory buffer," 244
Association, as real, 148–149
Associationism, 148–149
Associationist model, Rummelhart-McClelland (RM), 158–160, 162–163, 173–174
Associative information, 150
Atran, Scott, 13, 14
Attention, 409–410
  objects competing for, 333–334
Auditory images, 253
Autism, 269–270, 437
Automaticity, 408

Bach-y-Rita, 117–118
Bantu languages, 190–191
Bastian, Charlton, 510–511

Behavior, why we cannot rely on, 263–265
Behaviorism, 148
  stimulus-response, 35n2
Behind/not-behind distinction, 343
Biases, 4
Bilingualism/multilingualism, 379–385. See also Critical period(s), for language acquisition
  becoming bilingual in school, 386–388
  future of a multilingual world, 388–390
  how languages are represented in brain of bilingual people, 389
Biological and cognitive levels, relations between, 268–273
Biology. See also Neuroscience
  contribution to psychology, 397–398, 409, 411
Bodily sensations, representation of, 113–115, 117
Bootstrapping parameters of grammar, 136
Brain. See also Neuroscience
  causal influences from cognition to, 273–274
  sublevels of description in, 268
  synthesis between mind and, 409
Broca's area, 252, 253, 403

Cardinal numerons, 281, 288
Cardinal principle, 288

Catalan, 219, 381–388
Categorical perception, 419–429
Categories
  event, 344–349
  phonetic, 365
Categorization, 470–471
Causal modeling of developmental disorders, 263, 268–270
Chinese, 295
Chomsky, Noam, 12–14, 24–26, 418
Cognition
  history of the term, ix
  why we need, 263–265, 275
Cognition (journal), 19, 20, 32–35, 159
  antiestablishmentarianism and, 28–31
  history, 27–35, 275
  most cited articles in, 1971–2000, 35–38
  numbers of citations and papers, 32
Cognition on Cognition (Mehler and Franck), 26
Cognitive capacities that cannot be explained through learning, 10
Cognitive deficits, 18–19
Cognitive development, 8. *See also specific topics*
Cognitive devices, 51
Cognitive differences, and physical differences, 9–10
Cognitive domains, 10–11
Cognitive effect, 54
Cognitive journals, 23, 24. *See also Cognition*
  numbers of citations and papers, 32
Cognitive science, ix, 41–42
  classical
    demise of the old guard, and of, 19–20
  "nonclassical" and, 17–19
Competence, continuity *vs.* discontinuity and children's, 325–326
Competence-performance distinction, 326

Competition hypothesis, 333–334
Complement-head language, 136
Comprehension, 153
  as involving formation of two surface representations, 154
  phases/stages in, 151
Computational theory of mind, 53, 54
Conceptual development, continuity vs. discontinuity in, 303–304, 325, 335–336. *See also* Continuity
Connectionism, 148, 149
Connectionist pattern-associator memories, 163. *See also* Pattern-associator models
Connectionist theories of past tense, 158, 160, 164, 166, 167, 171. *See also* Past tense
Consciousness, Dennett's "multiple stage" model of, 154
"Conservation" and conservation stage, 8
Consonant clusters, 384
Consonant-vowel syllables, 189–190
Contact, type and amount of, 343
Contact/no-contact distinction, 342–343
Containers and containment events, 344–347, 349–355
Contextual similarity, 204–205
Continuity. *See also* Conceptual development
  ontogenetic, 281
  phylogenetic, 279–281
Continuity thesis/hypothesis, 306, 326
  Wynn's ease-of-learning argument against, 306–308
Continuity violations, infants' failures to detect, 351–357
Continuity vs. discontinuity in physical world, 341, 344
Correspondence principle, one-one, 288, 305

Counting, 281. *See also* Number(s)
  classification and set-theoretic approach, 282
  domain-specific approach, 283–285
  language dependency approach, 282–283
  non-numerical mechanism and, 290–291
  nonverbal, 285
    in adults, 289–290
    in animals, 285, 287–289
    discontinuity between verbal counting and, 290–294
    Gelman and Gallistel's continuity hypothesis, 306–308, 326
Counting principles
  and arithmetic operations and reasoning, 283–284
  how-to, 288
Count lists, 295
Count words, 297
Critical period(s), 465–466, 482, 486–487
  individual success in learning beyond, 490–491
  for language acquisition, 481, 486–487, 496–498
    empirical and theoretical questions concerning, 485–486
    evidence concerning, 482–485
    maturation vs. interference, 491–493
  length, 488
  maturational *vs.* experiential factors and, 487–488
  multiple
    discriminating between complex interactions between plasticity, stimuli, and task and, 494–495
    vs. stimulus differences in producing plasticity, 493
  nature of, 482
  plasticity after ending of, 489–490
  sharpness of decline of plasticity, 488–489

Deafness
  phonological, 196–197
  tone, 437
Derivational theory of complexity (DTC), 42–43
Determiner selection
  grammatical gender and, 211–219
  as special, 220–224
Determiner systems, 220
Development, how to study, 259–262
Developmental disorders, 263. *See also specific disorders*
  of biological origin, causal model of, 263, 268–270
Discrimination, 292, 304
  language, 368–369, 381, 382
  mirror-image, 474–475
  numerosity, 310 (*see also* Number representation)
Discrimination deficits, sound, 454
Dissociations, neuropsychological, 168–169, 436–437
Distracter words, 221, 247
Domain specificity, 47–48
"Dorsal pathway," 252
Dupoux, Emmanuel, 110
Dutch, 368, 369, 381
Dyslexia, 267–268, 447–449
  canonical phonological deficit theory of, 265–267
  phonological, 447–448
  rodent models of, 449, 451–458
  surface, 448
Dyslexic brain
  cortical malformations, 450
  anatomical and functional features, 451–457
  sexual dimorphism, 451, 453
  symmetrical cortical areas, 454–455

Ectopias, 449–455, 457
Effort, mental, 54
Encapsulation, 51–52
Epileptic patients, 438
Erasistratus, 509

Error-driven, learning as, 358n6
Errors
  malapropism, 228
  overregularization, 166–168
  speech, 210–211
  word-form-based, 229, 233–237
  word-substitution, 233, 234
  in word substitution, semantic, 233–235
Event categories, infants' formation of, 344–349
Event-specific expectations, 344
Evolution, 399
Evolutionary perspective, 49–50
Executive functions, 473
Expression raiser (EXPRAIS), 269–271
Extensional reasoning, 95
Externalism, 104–105, 110, 111–118

Fodor, Jerry, 4
  modularity and, 48–55
Form, word. *See* Word form
Fractions, 297
French, 219, 366–368
Functionalism, 103, 105, 107, 264

Gall, Franz, Joseph, 510
Gender congruency effects, 212–215, 218, 219, 223
Gender feature (competition), 213
Gender relatedness, 212
Gender selection, grammatical, 211–219
Generalization(s)
  conditioned, 157
  reliance on similarity, 162–163
  stimuli, 158
Genetic modularity, 49
German, 381
Government and binding (GB), theory of, 3
Grammar
  generative, 128–131
  initialization, 135–137

Grammatical components
  critical period for acquiring, 483–485
  interfaces between, 131–135
Grammatical gender, 211–219
Grammatical organization of native language, 372

Habituation and dishabituation, 318, 425–427
Head-complement language, 136
Head-complement parameters, 136–138
Height information, infants' perception of, 344–346
Herophilus, 509
Heuristics, 4
Hidden objects. *See* Object visibility; Occluded objects
Hippocrates, 505–506
Homophones, 230–231
Horizontal similarity, 204
"Hyperonym problem," 244
Hyperonyms, 244

Illusion(s), 99
  cognitive, 6–8
  perceptual, 51
Imagery, mental
  intrinsic and extrinsic constraints, 63–70
  representations underlying, as prepositional, 66
Imagery debate, 59, 61–69
  current state of, 61, 77–79
  historical background, 60–61
  neuropsychological evidence and "new stage" of, 70–72
Images, as depictive vs. descriptive, 78
Inflection, 175
Initial state, 364–366
Innatism, 4
Integers. *See* Number(s), natural
Internalism, 104–111
International Classification of Diseases (ICD), 264

Intonational phrases, 131–134, 136–139
Intrusion/intruding words, 229. *See also* Target-intrusion asymmetries

Japanese, 198, 427
Journal of Verbal Learning and Verbal Behavior (*JVLVB*), 23, 24

Kuhl, Patricia, 421–422

Language. *See also specific topics*
  biology and, 399, 400
  as cognitive domain, 10–11
  learning a second (*see* Critical period(s), for language acquisition)
  literacy, illiteracy, and, 468–470
  study of, 124–126, 139
Language acquisition. *See also* Plasticity
  critical or sensitive period for, 481–482, 486–487
    empirical and theoretical questions concerning, 485–486
    evidence concerning, 482–485
    nature of, 482
  what illiterates tell us about, 468–475
Language processing, 123–124
Languages
  distinguishing between unfamiliar, 368–369
  late selection, 217
  syllable- vs. stress-timed, 380, 381
Language switching, 389. *See also* Bilingualism/multilingualism
Language users, 241, 243
Lashley, Karl, 405, 406
Learning
  cognitive capacities that cannot be explained through, 10
  as error-driven, 358n6
Learning curve, U-shaped, 8–9
Lemmas/lemma structures, 216, 229–230, 244–246
  defined, 244

Lévi-Strauss, Claude, 443
Lexical access, frequency account for, 231
Lexical competition, 185
Lexical concepts, 244–246
Lexical frames, 246
Lexical frequency, 227–229, 231
Lexical frequency dilemmas, 232–238
Lexical frequency effects, 238
  locus of, 230
Lexical insertion, timing of, 235
Lexical networks, 245, 246
Lexical processing levels, 229
Lexical retrieval
  issues in, 228–232
  retrieval vs. insertion processes, 236
Lexical segmentation. *See* Segmentation
Linguistic data, different, 128–130
Linguistic diversity, 371–372
Linguistics, 127–128, 139. *See also specific topics*
Linguistic structures, complex structures underlying, 5
Literacy, 463, 475–477
  as biological vs. cultural function, 464–468
  nature of, 464
Literary mind, 471
Literate mind, 475–476
  and modular organization of mental capacities, 475
  what illiterates tell us about, 468–475
Localization, 405–406, 438–439
Logogens and logogen model, 272
Looking, preferential, 326–327, 331
  as reflecting purely perceptual representations, 328–329
Looking vs. reaching tasks, 330–331. *See also* Preferential looking

Malapropism errors, 228
Massachusetts Institute of Technology (MIT), 12, 17, 34

Massive modularity. *See* Modularity, massive
Maximum consistency principle, 223–224
Meaning, conditioning theory of, 157
Medial geniculate nucleus (MGN), 452–453
Mehler, Jacques (JM). *See also specific topics*
  bias against developmental studies, 364, 365
  bibliography, 516–527
  career and professional contributions, ix–x, 23–26, 123, 127, 159
  dissertation, 146–147
  encounters with, 4–5
  getting to know, 6
  on imagery, 59
  and infant speech perception research, 363–365
  and multilingualism, 379–382, 386, 389
  Piaget's ban of, 13–14
  Piaget's exchange with, 303–304
  on proper way to study development, 259–262
  short biography, 513–516
Memory
  literacy, illiteracy, and, 472
Mental imagery. *See* Imagery
Mental models, theory of, 86–87, 99. *See also* Reasoning
  evidence for, 98–99
Microgyrus in dyslexics, 450, 452–455, 457
Miller, George, 157–159
Mind, subdivision into specific domains, 10–11
Minimalism, 3
Modularity (mental), 10–11, 18–19
  evolutionary perspective on, 49–50
  Fodor on, 48–55
  levels at which it can be envisaged, 48–49
  massive, 48, 52, 55

Modular organization of mental capacities, 475
Modules, mental, 50–51
Monod, Jacques, 11–12
Motor images, 253
Multilingualism. *See* Bilingualism/multilingualism
Music
  biological foundations, 435–436, 442–443
  domain specificity, 437
  functions, 441–442
Musical abilities, 439–440
Musical savants, 437
Musical scales, 440
Musicogenic epilepsy, 438
Music-specific neural networks, 436–438
  content, 439–441
  localization, 438–439

Naming latencies, 222
  picture, 216–218
Native language, effects of, 198
Neonate cognition, 9–10
Neural structures, dedicated, 436
Neuroimaging, 248–249, 404, 412–413
Neurolinguistics, 127, 128
Neurons, mirror, 271, 273
Neurophysiological data, 397–398. *See also specific topics*
Neuroscience, cognitive, 397–400, 403–405. *See also* Brain; *specific topics*
  future goals, 411–413
  historical perspective on, 503–511
Noam, Jacques, 13
Noun phrase (NP) production, 209, 210, 216, 222
  determiner selection and, 217, 219–221, 223
  gender congruency and, 212–215, 220–221, 223 (*see also* Gender selection)
  with picture-word naming task, 212–215

Noun phrases (NPs)
  Italian, 216–217
  phonologically consistent and phonologically inconsistent, 221–222
  singular vs. plural, 213–215
  time course, 211
Number conservation task, 325
Number representation, 304
  prelinguistic representational systems, 308
    analog magnitude representations of, 308–312
    parallel individuation of small sets, 312–320
    questions that remain regarding, 321
Number(s). *See also* Counting
  learning about verbal and other symbolic re-representations of, 294–298
  natural, 305–306
  nature of, 281–282
  use of the term, in claims of prelinguistic representation thereof, 320–321
Number sense faculty, 279, 298n1
Numerical cognition, 280–281
Numerons, 306
  cardinal, 281, 288

Object concept development, 336–337
  task-dependence in, 326–327
Object-file representations, 313–320
Object images, 254
Object permanence tasks, 334
Object representations
  as competitive, 333–334
  as important component of "consciousness," 154–155
  precision, 330–333
    showing qualitative developmental continuity but increasing as infants grow, 334–335
Objects competing for attention, 333–334

Object tracking, 313–314
Object visibility, 330–333
Occluded objects, 329–334. *See also* Object visibility; Preferential looking
Occlusion events, 343–347, 349–351, 353–356
Offset time (critical periods), 487–488
One-one (correspondence) principle, 288, 305
Ontogenetic continuity, 281
Ordering principle, 288
Orgasm, 113–115, 117
Orientation times to languages, 383, 384
Orthography, English, 267
Osgood, Charles, 35n2
Overregularization errors, childhood, 166–168

Pain, 117–118, 273
Parallel distributed processing (PDP), 158, 168. *See also* Connectionism
Parameters (grammar), 136
Past tense, connectionist and symbolic theories of, 172–176
Past tense debate, 158–160
Past tense forms
  comparing models of generation of, 161–162
  cross-linguistic comparisons, 169–172
  for novel-sounding verbs, 163
  recognition of, 161
Pattern-associator models, 160, 161, 163, 170, 175
Perceptual Assimilation Model (PAM), 197, 198, 203–204
Perceptual constancy, 195. *See also* Object permanence tasks
Perceptual loop, 243–244, 249
Phenomenal character, 112–113, 117
Phenomenality, 106
Phenomenism, 103–104, 106, 114, 115

Phonemes and phonemic system of language, 129, 130, 421
Phonemic gating experiments, 200
Phonetic categories and sequences, 365
Phonetic encoding, 252–254
Phonological awareness, 455
Phonological bootstrapping hypothesis, 135
Phonological codes, 248, 251–252
Phonological components of language, critical period for acquiring, 483–485
Phonological constituents, 131, 132
Phonological deafness, 196–197
Phonological dyslexia, 447–448
Phonologically controlled processes, 232
Phonological phrases, 132–133, 136–138, 217
Phonological processing deficit, 267
Phonological representation, 161
Phonological system, native
 influence on speech perception, 196–197
Phonology, 19, 139
 prosodic, 131
Phonotactically illegal sequences/clusters, 198–201, 204
Phonotactic assimilation, 198–199, 204–205
 experimental studies, 199–204
Phonotactic constraints, 365
Phosphene-experiences, 115–117
Phrases. *See also* Noun phrases
 intonational, 131–134, 136–139
 phonological, 132–133, 136–138, 217
Phylogenetic continuity, 279–281
Phylogenetic continuity hypothesis, reasons for challenging a strong, 281
Phylogeny of primate groups, 423, 424

Physical events, infants' representations of, 352
 assumptions about, 352–353
Physical representations of infants
 how they are enriched, 354–357
 impoverished, 353–354
Physical world, infants learning about, 342–351
 identifying variables, 348–351
 availability of data on relevant conditions, 349
 exposure to relevant outcomes, 348–349
 infants' identification of initial concepts and variables, 342
Piaget, Jean, 8, 325, 336
 ban of Mehler, 13–14
 exchange with Mehler, 303–304
Piagetian search tasks, reflecting developing capacities for means-ends coordination, 329–330
Picture-theories of mental imagery, 64–66, 70, 77, 80n6. *See also* Visual cortex, primary, pictures in
Picture-word interference task/paradigm, 211–212, 246, 247
Picture-word naming paradigm, 216–218
Pitch (music), 440, 442
Planning ability, 473
Plasticity, 408, 412, 488, 495–496
 after critical period ends, 489–490
 interaction between quality and degree of, 493–494
 sharpness of decline of, in critical periods, 488–489
Plato, 506–508
Plural form of words, 169–170
Possible word constraint (PWC), 186–191
Post-traumatic stress disorder (PTSD), 273
 model for, 272
Preferential looking, 326–327, 331
 as reflecting purely perceptual representations, 328–329

Prepercepts, 144
Priming, 357, 408–409, 471
Probability(ies), 94
  conditional, 95–97
Prominence (grammar), 138–139
Propositional reasoning, 87–91
Prosodic features of language, 365–369
Prototype effect, 421, 422
Psycholinguistics, 127–129, 139
  comprehension vs. production perspectives, 242
  history, 241
"Psycholinguists, The" (Miller), 157–158

Qualia, 103, 104, 118n1
  defined, 118n1
Quantitative mechanism, 279

Reaching vs. looking tasks, 330–331
Reading, 410
Real-time analysis, 406–407
Reasoning, 85–86, 473–474
  modal, 93–95
  probabilistic, 94–97
  qualitative vs. quantitative, 349
  with quantifiers, 91–92
  sentential, 87–91
Reasoning principles, arithmetic, 283–284
Recapitulation, 154
  of object representation, 154
Regularization, systematic, 163–166
Regular vs. irregular words, 169–172
Relevance, 54
Representational content of thought, 103, 105
Representationalism, 103–104
  externalist, 111–118
  internalist, 104–111
Representations (mental), 43–45, 62, 144
  formation of dual meaning, 153
Rhythmic properties of languages, 380, 381

Royaumont Center, 12–13
Rummelhart-McClelland (RM) associationist model, 158–160, 162–163, 173–174

Saliency, 54
SARAH model, 367
Schooling, and literacy, 472–474, 476
Scuola Internazionale Superiore di Studi Avanzati (SISSA), 14, 15, 17
Segmentation cues, 186, 187
Segmentation (speech), 185–187, 191, 367, 369–370. See also Syllable(s)
Self-monitoring, 244, 249–251
Semantically controlled processes, 232, 237
Semantic errors in word substitution, 233–235
Semantic knowledge, 470–472
Semantics, 169
  lexical, 3–4
Sensitive periods. See Critical period(s)
Sentences, comprehension and clarity of, 144–146
Sentential connectives, models for, 89
Sentential reasoning, 87–91
Sesotho, 190–191
Set-size signature, 313–315, 320
Shortlist model, 184–187
Similarity, horizontal, vertical, and contextual, 204–205
Solidity of objects, 341
Solidity violations, infants' failures to detect, 351–357
Spanish, 219, 381–388
Speech
  categorical perception and, 419–422
  "internal," 244
Speech error distributions, 247
Speech errors, 210–211
Speech-is-special debate, historical background of, 418–419
"Speech-is-special" problem, 417

Speech perception
  infant, 363–366
    what we have learned about, 365–370
    what we still need to know about, 370–373
  in a model of production, 243–248
  native phonological system and, 196–197
  neurophysiological observations, 248–254
  phonotactic constraints in, 195–199, 203–206
    experimental studies, 199–204
Speech processing. *See also* Speech perception
  across languages, differential, 16, 368–372
Speech production, neurophysiological observations regarding, 248–254
Speech research, new approaches in, 422–429
Stable state, 364–366
Stimuli generalization, theory of, 158
Stimulus-response behaviorism, 35n2
Strauss, Sydney, 6–7
Subitizing, 280, 292
Successor principle, 296, 297
Supervenience, 112–113
Support events, 342–343
Syllabary, 248
Syllabification, 15–16, 250–254
Syllable(s), 15–16, 367
  how it is a possible word, 184–187
  how it plays a language-universal role, 189–191
  role in speech processing, 181
  as universal prelexical unit, 182–183
  where it plays a language-specific role, 187–189
  why it is not the unit of perception, 181–184, 189
Syllable- vs. stress-timed languages, 380, 381

Symbolic systems, 149
Symbolic theories of past tense. *See* Past tense
Syntactically controlled processes, 232
Syntactic(al) parameters, 136, 137
Syntactic theory, 3, 139
Syntax, 131, 133, 134, 138, 145
  bootstrapping, 135
  derivational nature, 150–153
  minimalist theories, 151
  as real, 146–147

Target-intrusion asymmetries, 232–235
Target-intrusion pairs and frequencies, 229, 233, 234, 236, 237
Task-dependent performance, 325–326
Temporal optimization principle, 222–223
TIGRE model, 368
Tip-of-tongue (TOT) states, 232
"Token-token identity," 274
Tone-deafness, 437
Tracking, object, 313–314
Transparency information, infants' perception of, 346–347
Trieste Encounters in Cognitive Science (TECS), 14–15
Trieste style, 14–17
Truth, principle of, 87, 98

Universal human mind, 463
U-shaped curve, 8–9

Verbs. *See also* Past tense
  regular vs. irregular, 169, 170
  systematic regularization, 163–164
Vertical similarity, 204–205
Visibility hypothesis. *See* Object visibility
Visual cognition, 474–475
Visual cortex, 71, 80n5
  primary, 71
  pictures in, 72–77

Visual sensations. *See* Phosphene-experiences
Visual system, 70–71

WEAVER lexical network, 245, 246, 253
Weber-fraction signature, 309, 312, 313, 319
Wernicke's area, 251, 253, 254
Word and morpheme exchanges, 232, 235–236
Word boundaries. *See* Syllable(s)
Word form, 229–230
Word-form-based errors, 229, 233–237
Word-form-based insertion processes, 236
Word-form errors in word substitution, 233, 235, 237
Word form level, connections at, 247–248
Words. *See also specific topics*
 distinction between rules and, 173
 headless, 165
 open- and closed-class, 209–211, 223, 231
 spaces between, 145
Words-and-rules theory, 160–162, 168, 172
Word segmentation. *See* Segmentation
Word-substitution errors, 233, 234
Word substitution process, post-retrieval mechanism in, 235